INORGANIC SYNTHESES

Volume 33

●●●●●●●

Editor-in-Chief
DIMITRI COUCOUVANIS
University of Michigan

●●●●●●●●●●●●●●●●●●●●●●●●●●●●●●●●●●●●●●

INORGANIC SYNTHESES

Volume 33

A Wiley-Interscience Publication
JOHN WILEY & SONS, INC.

Library of Congress Catalog Number: 39–23015

ISBN 0-471-20825-6

Printed in the United States of America.

10 9 8 7 6 5 4 3 2 1

PREFACE

Advances in Inorganic Chemistry over the last two decades reflect the new directions our discipline is following. The present volume of Inorganic Syntheses attempts to place some emphasis in these directions.

Many of the new areas of interest, are characterized by complexity in design and molecules containing specific functional groups. These characteristics, inspired by either metalloenzymology or the need for materials with specific properties and function, appear to be essential features of synthetic "supramolecules" or clusters.

Supramolecular syntheses have not been emphasized in an Inorganic Syntheses volume in the past, however, as interest in this area increases, the need becomes apparent for synthetic methodology, and real examples of supramolecular assemblies. There are various reasons why supramolecular methodology and specific syntheses of supramolecules were not emphasized previously in an Inorganic Syntheses volume. Included among these are: the inherent difficulty in the synthesis of multi-unit assemblies, the possible lack of general interest for specific supramolecules and lack of a clear demonstration of utility and function for many of these molecules.

In the first chapter of this volume are included procedures and methodology for the synthesis of molecules that may serve as magnetic building blocks, light gathering units and ditopic activators or receptors. We expect this chapter of Volume 33 to inspire similar syntheses of new materials using similar methodologies.

In the synthesis of new coordination compounds, whether classic or supramolecular, reagents that serve as convenient sources of metal ions or metal containing groups are of the utmost importance. In Chapters 2 and 4 are included a number of molecules useful as reagents or building blocks.

Low temperature solid state synthesis is an approach that shows great promise for the synthesis of materials with unusual interesting properties. Solid state synthesis has been the subject of Volume 30 of Inorganic Syntheses in the past. Continuing interes in solid state materials and particularly those obtained by relatively low temperature procedures has prompted us to include another series of such procedures and syntheses in Chapter 3 of this volume.

I would like to thank those who submitted the syntheses for Volume 33 and also the checkers who often dedicated excessive time and great efforts in checking complicated procedures. To the few contributors that will not see their syntheses in this volume, I apologize for not being able to find a person willing or able to check their syntheses.

In the prescreening of manuscripts, the members of the Inorganic Syntheses board were very helpful in evaluating the contributions, making suggestions and proposing possible reviewers.

Finally, the extensive, detailed, editorial screening of nearly every one of the manuscripts, by Duward Shriver, is appreciated and gratefully acknowledged.

DIMITRI COUCOUVANIS
Ann Arbor, Michigan

NOTICE TO CONTRIBUTORS
AND CHECKERS

The *Inorganic Syntheses* series is published to provide all users of inorganic substances with detailed and foolproof procedures for the preparation of important and timely compounds. Thus the series is the concern of the entire scientific community. The Editorial Board hopes that all chemists will share in the responsibility of producing *Inorganic Syntheses* by offering their advice and assistance in both the formulation and the laboratory evaluation of outstanding syntheses. Help of this kind will be invaluable in achieving excellence and pertinence to current scientific interests.

There is no rigid definition of what constitutes a suitable synthesis. The major criterion by which syntheses are judged is the potential value to the scientific community. An ideal synthesis is one that presents a new or revised experimental procedure applicable to a variety of related compounds, at least one of which is critically important in current research. However, syntheses of individual compounds that are of interest or importance are also acceptable. Syntheses of compounds that are readily available commercially at reasonable prices are not acceptable. Corrections and improvements of syntheses already appearing in *Inorganic Syntheses* are suitable for inclusion.

The Editorial Board lists the following criteria of content for submitted manuscripts. Style should conform with that of previous volumes of *Inorganic Syntheses*. The introductory section should incude a concise and critical summary of the available procedures for synthesis of the product in question. It should also include an estimate of the time required for the synthesis, an indication of the importance and utility of the product, and an admonition if any potential hazards are associated with the procedure. The Procedure section should present detailed and unambiguous laboratory directions and be written so that it anticipates possible mistakes and misunderstandings on the part of the person who attempts to duplicate the procedure. Any unusual equipment or procedure should be clearly described. Line drawings should be included when they can be helpful. All safety measures should be stated clearly. Sources of unusual starting materials must be given, and, if possible, minimal standards of purity of reagents and solvents should be stated. The scale shold be reasonable for normal laboratory operation, and any problems involved in scaling the procedure either up or down should be discussed. The criteria for judging the purity of the final product should be delineated clearly. The Properties section should supply and discuss those physical and chemical characteristics that are relevant to judging the purity of the product and to permitting its handling and use in an intelligent manner. Under References, all pertinent literature citations should be listed in order. A style sheet is available from the Secretary of the Editorial Board.

The Editorial Board determines whether sumitted syntheses meet the general specifications outlined above. Every procedure will be checked in an independent laboratory, and publication is contingent on satisfactory duplication of the syntheses.

Each manuscript should be submitted in duplicate to the Secretary of the Editorial Board Professor Stanton Ching, Department of Chemistry, Connecticut College, New London, CT 06320. The manuscript should be typewritten in English. Nomenclature should be consistent and should follow the recommendations presented in *Nomenclature of Inorganic Chemistry*, 2nd ed., Butterworths & Co, London, 1970 and in *Pure and Applied Chemistry*, Volume 28, No. 1 (1971). Abbreviations should conform to those used in publications of the American Chemical Society, particularly *Inorganic Chemistry*.

Chemists willing to check syntheses should contact the editor of a future volume or make this information known to Professor Ching.

TOXIC SUBSTANCES AND LABORATORY HAZARDS

Chemicals and chemistry are by their very nature hazardous. Chemical reactivity implies that reagents have the ability to combine. This process can be sufficiently vigorous as to cause flame, an explosion, or, often less immediately obvious, a toxic reaction.

The obvious hazards in the syntheses reported in this volume are delineated, where appropriate, in the experimental procedure. It is impossible, however, to foresee every eventuality, such as a new biological effect of a common laboratory reagent. As a consequence, *all* chemicals used and *all* reactions described in this volume should be viewed as potentially hazardous. Care should be taken to avoid inhalation or other physical contact with all reagents and solvents used in this volume. In addition, particular attention should be paid to avoiding sparks, open flames, or other potential sources which could set fire to combustible vapors or gases.

A list of 400 toxic substances may be found in the *Federal Register*, Volume 40, No. 23072, May 28, 1975. An abbreviated list may be obtained from *Inorganic Syntheses*, Vol. 18, p. xv, 1978. A current assessment of the hazards associated with a particular chemical is available in the most recent edition of *Threshold Limit Values for Chemical Substances* and *Physical Agents in the Workroom Environment* published by the American Conference of Governmental Industrial Hygienists.

The drying of impure ethers can produce a violent explosion. Further information about this hazard may be found in *Inorganic Syntheses*, Volume 12, p. 317.

CONTENTS

Chapter Two USEFUL REAGENTS AND LIGANDS

Chapter Three   SOLID-STATE MATERIALS AND CLUSTERS

Chapter Four   COMPOUNDS OF GENERAL INTEREST

INORGANIC SYNTHESES

Volume 33

Chapter One

SYNTHESES OF SELECTED SUPRAMOLECULES

1. TOWARD MAGNETIC BUILDING BLOCKS: SYNTHESIS OF A PLANAR Co(III)–CATION RADICAL–COBALT(III) COMPLEX OF THE BINUCLEATING LIGAND 1,2,4,5-TETRAKIS(2-HYDROXY-2-METHYLPROPANAMIDO)BENZENE

Submitted by SCOTT W. GORDON-WYLIE,* WYNDHAM B. BLANTON,* BRIAN L. CLAUS,* COLIN P. HORWITZ,* and TERRENCE J. COLLINS*
Checked by COLETTE BOSKOVIC[†] and GEORGE CHRISTOU[†]

The design of supramolecular architectures has focused on the synthesis of small subunits that bind to each other in a predictable fashion to form extended solids. How such molecular building blocks may be constructed has been discussed in detail.[1–9] In the design of molecular magnetic structures, the key factors necessary to promote magnetic exchange couplings of a particular type between two or more spin carriers have also been stated eloquently.[10–15] Despite significant advances in the general research area, there remains a dearth of magnetic building blocks suitable for the construction of extended, well-ordered, magnetic solids. Herein is reported the synthesis of a potentially useful magnetic building block, the Co(III) complex[16] of the multinucleating ligand 1,2,4,5-tetrakis(2-hydroxy-2-methylpropanamido)benzene [$H_8(\kappa^4:\kappa^4$-t-HMPA-B)].

* Department of Chemistry, Carnegie Mellon University, Pittsburgh, PA 15213.
† Indiana University, Bloomington, Indiana 47405-4001.

1

Five features of the Co_2^{III} (κ^4:κ4-t-HMPA-B) system considered to be important for the preparation of a supramolecular magnetic solid are as follows:

1. Bimetallic complexes with planar four-coordinate ions in the primary diamido-N-dialkoxido sites are capable of achieving planarity across the primary metal, the ligand donor atoms, and the central aromatic ring. This planarity is useful for reducing the complexity of derivative network solids.

2. The ligand strongly coordinates two primary metal ions, each in a strong σ-donor environment, stabilizing the intermediate spin, $S = 1$, Co(III) state.

3. Metallocomplexes such as the bis-Co(III) dianion possess two bidentate coordination sites capable of binding to additional metal ions. Using a similar ligand, with only one primary site and one secondary site, the primary–secondary metal ion exchange interaction has been shown to be ferromagnetic, and it has been postulated that the sign of the exchange intereaction is a consequence of the ligand donor interactions at Co(III).[17]

4. When coordinated to primary ions such as Co(III), the aryl part of the ligand can be oxidized to form a stable ligand cation radical structure that fosters magnetic communication between the two primary ions via the intermediacy of an organic $S = \frac{1}{2}$ unit.[16]

5. Since the ligand is a rigid rectangular unit and the secondary coordination sites are arranged at opposite ends of the rectangular unit, one can calculate the different possible structures (i.e., linear chains, hexagonal sheets, three-dimensional helical structures) that can arise when the ligand in primary metal-coordinated form binds to different secondary ions.[16]

Syntheses of the ligand, the di-Co(III) complex, and the one-electron oxidized di-Co(III) complex are relatively straightforward. The overall yields are good [~60–70% for $H_8(\kappa^4$:κ^4-t-HMPA-B) and >50% for each metallated complex], and the compounds can be easily prepared on relatively large scales.

A. 1,2,4,5-TETRAKIS(2-ACETATE-2-METHYLPROPANAMIDO) BENZENE (*t*-AcMPA-B)

t-AcMPA-B

Procedure

■ **Caution.** *Toxic and highly flammable solvents and reagents are used throughout this procedure and those that follow. All reactions and manipulations described herein should be performed in a well-ventilated fume hood.*

An amount of 1,2,4,5-tetraaminobenzene tetrahydrochloride (15 g, 0.053 mol, Fluka), 300 mL of 1,2-dichloroethane, and a magnetic stir bar are placed in a 500-mL, two-necked, round-bottomed flask. The flask is then fitted with a reflux condenser, which is attached to an N_2 source, a mineral oil bubbler, and a pressure-equalizing addition funnel capped with a rubber septum. The system is purged with N_2 (20 min), and 1-chlorocarbonyl-1-methylethyl acetate (46 mL, 0.32 mol, Aldrich) is cannulated into the addition funnel using positive N_2 pressure and added slowly (20 min) to the stirred solution. After all the 1-chlorocarbonyl-1-methylethyl acetate is added, Et_3N (100 mL, 0.69 mol, Aldrich) is cannulated into the addition funnel using positive N_2 pressure and added dropwise to the stirred solution (1 h). After the Et_3N addition is complete, the funnel is removed and replaced with a greased ground-glass stopper. The system is heated to reflux with a heating mantle and stirred under N_2 for 5 days, during which time a mixture of $Et_3N \cdot HCl$ and the crude product precipitate from the solution.

After cooling the solution to room temperature, the solvent is removed under reduced pressure using a rotary evaporator. The remaining yellow solid is dissolved in CH_2Cl_2 (800–1000 mL) and the solution transferred to a separatory funnel without filtering. The solution is first washed with 1.2 M HCl (4×400 mL) and then 1 M Na_2CO_3 (3×400 mL). The CH_2Cl_2 solution is gravity-filtered through filter paper (Whatman 1) into a 1-L round-bottomed flask and the solvent is removed under reduced pressure using a rotary evaporator. The residue, a tan solid, 1,2,4,5-tetrakis-(2-acetate-2-methylpropanamido)benzene (*t*-AcMPA-B), is scraped from the sides of the flask and slurried with pentane. The *t*-AcMPA-B is separated from the pentane by suction filtration using a medium-porosity glass frit and is air-dried (31 g; checkers obtained 30 g). This material is sufficiently pure to be used for the next step, or it can be recrystallized from acetone as follows. Impure *t*-AcMPA-B (6 g) is added to a 150-mL Erlenmeyer flask containing 110 mL of acetone and a magnetic stirring bar. The slurry is brought to a boil while stirring on a stirrer/hotplate, during which time most of the solid dissolves. The hot solution is rapidly suction-filtered through a coarse-porosity glass frit and the filtrate is placed in a freezer (approximately −20°C for 1–2 h). The white crystalline solid that precipitates is suction-filtered using a medium-porosity glass-fritted funnel, washed with diethyl ether, and air-dried to yield pure *t*-AcMPA-B (5.61 g, after recrystallization, yield 93.5%; checkers obtained 4.45 g, 74.2%).

Anal. Calcd. for $C_{30}H_{42}N_4O_{12}$ [MW (molecular weight) 650.68]: C, 55.38; H, 6.51; N, 8.61. Found: C, 55.48; H, 6.49; N, 8.57.

Properties

The pure material is a white solid. ^1H NMR (CD_3CN) δ ppm) = 8.45 (s, 4H, N*H*), 7.77 (s, 2H, C_6H_2), 2.08 [s, 12H, C(O)C*H*$_3$], 1.61 (s, 24H, CC*H*$_3$). $^{13}C\{^1H\}$ NMR (CD_3CN) δ pm = 173.4 (amide *C*O), 171.2 (acetyl *C*O), 129.3 (Ar *C*NHR), 121.8 (Ar *C*H), 81.5 (alkyl quaternary *C*) 25.2 (alkyl *C*H$_3$), 22.1 (acetyl *C*H$_3$). IR (Nujol) ν cm^{-1}) = 3275 (s, str, br, amide NH) 1740 (s, str, acetyl CO), 1667 (s, str, amide), 1610 (sh, w, aryl ring/amide). The checkers obtained ^1H NMR (CD_3CN) δ ppm) = 8.45 (s 3.7H), 7.77 (s 1.4H), 2.08 (s, 12.4H), 1.61 (s, 24.4H). $^{13}C\{^1H\}$ NMR (CD_3CN) δ ppm) = 173.2, 171.1, 128.9, 121.4, 81.2, 24.9, 22.0. IR (KBr disk) ν (cm^{-1}) = 3308, 1738, 1670, 1616.

B. 1,2,4,5-TETRAKIS(2-HYDROXY-2-METHYLPROPANAMIDO) BENZENE, [H$_8$(*t*-HMPA-B)]

t-AcMPA-B   H$_8$(*t*-HMPA-B)

Procedure

Solid *t*-AcMPA-B (31.0 g) is added to a 2-L, single-necked round-bottomed flask containing CH_3OH (1000 mL), NaOH (10.3 g, 0.25 mol, $\approx$4.4 equiv assuming pure *t*-AcMPA-B), and a magnetic stirring bar. The flask is fitted with a reflux condenser that is attached to an N_2 source and a mineral oil bubbler. The system is flushed with N_2 (10 min) and then the mixture is stirred and brought to reflux using a heating mantle under static N_2 (24 h). After cooling to room temperature, the CH_3OH is removed under reduced pressure using a rotary evaporator, yielding a tan solid, a mixture of 1,2,4,5-tetrakis(2-hydroxy-2-methylpropanamido) benzene [H$_8$(*t*-HMPA-B)] and sodium acetate. The solid is placed in a 600-mL beaker containing a 2 : 1 CH_3OH/H_2O mixture (the checkers used 300 mL of the solvent mixture) and a magnetic stir bar. The slurry is stirred and heated to boiling using a stirrer/hotplate. It is then cooled to 10–15°C in a refrigerator. Most of the H_8(*t*-HMPA-B) is not soluble in the CH_3OH/H_2O mixture, but sodium acetate is soluble. The white CH_3OH/H_2O slurry is suction-filtered while still at ~ 15°C using a medium-porosity glass-fritted funnel, and the white solid product is allowed to air-dry (17.8 g, 0.037 mol; checkers obtained 12.7 g). Overall yields from 1,2,4,5-tetraaminobenzene-tetrahydrochloride are typically 60–70%.

Anal. Calcd. for $C_{22}H_{34}N_4O_8$ (MW 482.53): C, 54.76; H, 7.10; N, 11.6%; Found: C, 54.64; H, 7.05; N, 11.55%.

Properties

H_8(*t*-HMPA-B) is a white powder that is very insoluble in most solvents except hot CH_3OH, likely attributable to the presence of an extensive hydrogen bonding network. It is slightly soluble in DMSO. ^1H NMR (DMSO-d^6) δ (ppm) = 9.5 (s, 4H, N*H*), 7.7 (s, 2H, C_6H_2), 5.8 (s, 4H, CO*H*), 1.2 (s, 24H, CCH_3). IR (Nujol) ν (cm^{-1}) = 3449, 3305, 3219 (OH alcohol, NH amide), 1656, 1630 (amide). The checkers obtained ^1H NMR (DMSO-d^6) δ (ppm) = 9.5 (s, 4.8H), 7.7 (s, 1.7H), 5.7 (s, 3.3H), 1.3 (s, 24H). IR (KBr disk) ν (cm^{-1}) = 3447, 3307, 3223, 1655, 1629.

C. BISTETRAPHENYLPHOSPHONIUM-DI-COBALT(III)-κ^4:κ^4- [1,2,4,5-TETRAKIS(2-OXY-2-METHYLPROPANAMIDO) BENZENE] {[PPh$_4$]$_2$[Co$_2^{III}$(κ^4:κ^4-*t*-HMPA-B)]}

H$_8$(*t*-HMPA-B)   [Co$^{III}_2$(κ^4:κ^4-*t*-HMPA-B)]$^{2-}$

Procedure

H_8(*t*-HMPA-B) (2.01 g, 4.11 mmol), 400 mL CH_3OH, anhydrous $CoCl_2$ (1.3 g, 10.0 mmol, Aldrich), NaOH [1.18 g, 29.5 mmol (although 8 equiv of NaOH are stoichiometrically correct, better yields were obtained using $\sim$7.2 equiv under aerobic conditions)], and a magnetic stirring bar are combined in a 500-mL Erlenmeyer flask. The resulting slurry is stirred at room temperature, *in air*, for 2 days. During this time, the H_8(*t*-HMPA-B) dissolves and a homogeneous green-brown solution of the sodium salt of the Co(III) complex forms. Tetraphenylphosphonium chloride, [PPh$_4$]Cl, (4.61 g, 12.3 mmol, Aldrich) is then added to the solution, which is stirred for an additional 2 h [3 equiv of (PPh$_4$)Cl are used here; $\leq$5.5 equiv of (PPh$_4$)Cl have been used, with an improved product yield of >70%, but multiple washings and recrystallizations, described below, are required to purify this higher-yield product]. The reaction mixture is then transferred to a 1-L round-bottomed flask and the CH_3OH is removed under reduced pressure using a rotary evaporator leaving a dark green-brown solid. CH_2Cl_2 (800 ml) and a magnetic stir bar are added to the dark green solid, the flask is capped, and the slurry is stirred (24 h). During this time, most of the solid dissolves, giving a dark green solution. The solution is gravity-filtered through filter paper (Whatman 1) and the CH_2Cl_2 is removed under reduced pressure using a rotary evaporator. The resulting dark green solid is triturated in dry diethyl ether (the checkers used 100 mL of diethyl ether) suction-filtered through a medium-porosity glass-fritted funnel, and air-dried to yield a mixture of (PPh$_4$)$_2$[Co(III)$_2$(κ^4:κ^4-*t*-HMPA-B)] and (PPh$_4$)Cl. The crude dark green powder is placed in a 100-mL beaker containing CH_3CN (50 mL) and a magnetic stir bar. The resulting slurry is stirred and heated to a boil on a stirrer/ hotplate for 10 min (the checkers reduced the volume to 10 mL, and then cooled to $-20°C$ in a freezer). On cooling, more dark green powder precipitates. The dark green precipitate is suction-filtered through a medium-porosity glass-fritted funnel and air-dried. The solid is washed one more time with CH_3CN to yield a dark green solid, 2.65 g, 2.08 mmol, yield = 50.7% (checkers obtained 3.44 g, 65.8%).

A second crop of material is obtained by combining the CH_3CN washes, reducing the solution volume to 30–40 mL, and then allowing diethyl ether vapor to diffuse into the CH_3CN solution. After $\sim$5–7 days, the dark green crystals are isolated from the mother liquor and air-dried to yield the second crop of the compound (0.2 g, 0.15 mmol, total yield = 54.2%). Typical total yields range from 50 to 60%. X-ray diffraction quality crystals are grown by allowing diethyl ether to vapor-diffuse into a concentrated solution of purified (PPh$_4$)$_2$[Co$_2^{III}$(κ^4:κ^4-*t*-HMPA-B)] in CH_3CN.

Anal. Calcd. for $C_{70}H_{66}N_4O_8Co_2P_2 \cdot 1.5H_2O$ (MW 1298): C, 64.77; H, 5.36; N, 4.32; P, 4.77. Found: C, 64.86; H, 5.35; N, 4.41; P, 4.55. The number of water molecules varies from zero to two depending on recrystallization and storage conditions as shown by elemental analysis and NMR data. The stoichiometry of the analytical sample was established by ^{1}H NMR. The crystal used for crystallographic analysis had two water molecules in the lattice for each $(PPh_4)_2[Co_2^{III}(\kappa^4{:}\kappa^4{\text{-}}t{\text{-HMPA-B}})]$.

Properties

Crystals of $(PPh_4)_2[Co_2^{III}(\kappa^4{:}\kappa^4{\text{-}}t{\text{-HMPA-B}})]$ are green-black rectangular parallelepipeds and are comparatively air-stable. They are mildly hygroscopic and thus are best stored in a desiccator over $CaSO_4$. ^{1}H NMR (CD$_3$CN) δ (ppm) = 9.8 (s, 24H, alkyl CH$_3$), 7.4, 7.22, 7.18, 7.155 (40H, aryl of phosphonium cation), 3.2 (H$_2$O), -0.8 (s, 2H, ArH). IR (Nujol) ν (cm^{-1}) = 1632 (amide). UV–vis (95% ethanol) λ_{max} (nm) as in Table I, Syn. 2 (on p. 34) (ε, M^{-1} cm^{-1}) = 238 (68,000), 376 (10,000), 468 (7300), 766 nm (2600). The cyclic voltammogram of the complex shows two well-resolved, reversible one-electron oxidation processes at 290 mV (ΔE_p = 60 mV) and 885 mV (ΔE_p = 82 mV) and two closely spaced, reversible or quasireversible[18] one-electron reductions at -1110 mV (ΔE_p = 60 mV) and -1240 mV (ΔE_p = 76 mV). [*Conditions*: Pt working electrode, Pt counterelectrode, saturated sodium chloride calomel reference electrode (SSCE) and 0.1 M (n-Bu$_4$N)(PF$_6$)/CH$_3$CN, puriss grade (n-Bu$_4$N)(PF$_6$) was purchased from Fluka; CH$_3$CN was freshly distilled from CaH$_2$.] The reversibility of the second oxidation couple is sensitive to solvent purity, becoming totally irreversible if the CH$_3$CN is wet. The checkers obtained ^{1}H NMR (CD$_3$CN) δ (ppm) = 10.0 (s, 23.4H), 7.5, 7.32, 7.30, 7.26 (40H), 2.8 (H$_2$O), -0.72 (s, 2.6H). IR (KBr disk) ν (cm^{-1}) = 1633. UV–vis (95% ethanol) λ_{max} (nm) (ε, M^{-1} cm^{-1}) = 228 (87,000), 262 (44,000), 374 (8900), 476 (7200), 780 (3600) with ε values based on MW = 1298. $E_{1/2}$ values were measured against a ferrocene/ferrocenium reference (+320 mV vs. SSCE[19]) mV (ΔE_p) = 482 (125), -109 (84), -1482 (78), -1599 (98).

D. TETRAPHENYLPHOSPHONIUM-DI-COBALT(III)-κ^4:κ^4-[1,2,4,5-TETRAKIS(2-OXY-2-METHYLPROPANAMIDO)BENZENE] {[PPh$_4$][Co$_2^{III}\kappa^4$:κ^4-t-HMPA-B)]}

$[Co^{III}_2(\kappa^4{:}\kappa^4{\text{-}}t{\text{-HMPA-B}})]^{2-}$ $[Co^{III}_2(\kappa^4{:}\kappa^4{\text{-}}t{\text{-HMPA-B}})]^{-}$

Procedure

The compounds $(PPh_4)_2[Co_2^{III}(\kappa^4:\kappa^4\text{-}t\text{-HMPA-B})]$ (243 mg, 0.19 mmol), CH_2Cl_2 (15 mL), and a magnetic stir bar are placed in a 50-mL, two-necked, round-bottomed flask. One neck of the flask is fitted with a gas inlet valve that is attached to a conventional Schlenk vacuum/N_2 line and the other with a rubber septum. The green solution is stirred and N_2 is bubbled through the solution ($\geq$10 min) to remove O_2. The rubber septum is removed, a sample of freshly prepared, solid ferrocenium hexafluorophosphate, $(Cp_2Fe)(PF_6)$* (63 mg, 0.19 mmol) is added to the stirred solution under a flow of N_2, and the rubber septum is replaced. [*$(Cp_2Fe)(PF_6)$ was prepared from ferrocene[20] by dissolution in conc. H_2SO_4 with aerial oxidation followed by dilution and precipitation with a concentrated aqueous solution of $(NH_4)(PF_6)$. Solid $(Cp_2Fe)(PF_6)$ is modestly air-stable, and can be stored for weeks to months under an inert atmosphere. Solutions of $(Cp_2Fe)(PF_6)$, however, are relatively unstable under aerobic conditions.] After addition of $(Cp_2Fe)(PF_6)$, the solution immediately turns deep purple. The mixture is stirred under static N_2 (1 h) and then exposed to air (30 min). A small amount of white solid precipitate in the dark purple solution is removed by suction filtration through a medium-porosity glass-fritted funnel in air and then the CH_2Cl_2 is removed in vacuo. The oxidized product appears to be somewhat water-sensitive, so it is best to filter and remove the CH_2Cl_2 using the Schlenk line rather than using equipment attached to a water aspirator. The remaining purple-black solid product is washed copiously with toluene (500–700 mL) to remove ferrocene and then diethyl ether ($\sim$200 mL). The purple powder is dissolved in a minimum amount of 1,2-dichloroethane and pentane vapor is diffused into the 1,2-dichloroethane solution. After a few days, the purple-black crystalline product is collected by suction filtration on a medium-porosity glass-fritted funnel, washed with pentane ($\sim$100 mL), and air-dried, 148 mg, 75% yield (checkers obtained 160 mg, 81%).

Anal. Calcd. for $C_{46}H_{46}N_4O_8Co_2P \cdot C_2H_4Cl_2$ (MW 1030): C, 55.94; H, 4.89; N, 5.44. Found: C; 55.68, H; 4.89, N; 5.42. Presence of 1,2-dichloroethane in the solid is confirmed by a 9.6% weight loss on thermogravimetric analysis.

Properties

In the solid state, the cation radical containing compound $(PPh_4)(Co_2^{III}(\kappa^4:\kappa^4\text{-}t\text{-HMPA-B})]$ is best stored in an inert-atmosphere glovebox under nitrogen or argon. Storage of the purple solid in a desiccator charged with $CaSO_4$ is also acceptable for short time periods (days–weeks). Solutions are most stable when prepared using halogenated solvents such as 1,2-dichloroethane. IR (KBr disk) ν (cm^{-1}) = 1687 (s, amide). UV–vis (1,2-dichloroethane) λ_{max} (nm) (ε, M^{-1} cm^{-1}) = 274 (37,400), 300 (21,000), 327 (20,700), 387 (10,000), 508 (8300), 548 (15,000), 657 (29,800), 816 (2000). EPR and SQUID (superconducting quantum interference device) magnetic susceptibility data reveal relatively strong antiferromagnetic coupling between the two rhombic $S = 1$ Co(III) ions and the $S = \frac{1}{2}$ ligand cation radical, $J \approx 75$ cm^{-1} (using $H = JS_i \cdot S_j$), yielding a ground spin state for the complex of $S_T = \frac{3}{2}$. The checkers obtained IR (KBr disk) ν (cm^{-1}) = 1696. UV–vis (1,2-dichloroethane) λ_{max} (nm) (ε, M^{-1} cm^{-1}) = 276 (27,500), 302 (15,500), 328 (14,800), 388 (7400), 506 (5400), 546 (10,500), 658 (21,900), 816 (1600) with ε values based on MW 1030.

References

1. J.-M. Lehn, *Supramolecular Chemistry*, VCH, Weinheim, 1995.
2. M. Meyer, A. A. M. Albrecht, D. C. O. Buchecker, and J. P. Sauvage, *J. Am. Chem. Soc.* **119**, 4599–4607 (1997).
3. P. R. Ashton, C. G. Claessens, W. Hayes, S. Menzer, J. F. Stoddart, A. J. P. White, and D. J. Williams, *Angew. Chem., Int. Ed. Engl.* **34**, 1862–1864 (1995).
4. P. R. Ashton, A. N. Collins, M. C. T. Fyfe, S. Menzer, J. F. Stoddart, and D. J. Williams, *Angew. Chem., Int. Ed. Engl.* **36**, 735–739 (1997).
5. P. R. Ashton, A. N. Collins, M. C. T. Fyfe, P. T. Glink, S. Menzer, J. F. Stoddart, and D. J. Williams, *Angew. Chem., Int. Ed. Engl.* **36**, 59–62 (1997).
6. A. Williams, *Chem. Eur. J.* **3**, 15–19 (1997).
7. W. T. S. Huck, R. Hulst, P. Timmerman, F. C. J. M. van Veggel, and D. N. Reinhoudt, *Angew. Chem., Int. Ed. Engl.* **36**, 1006–1008 (1997).
8. S. W. Keller, *Angew. Chem., Int. Ed. Engl.* **36**, 247–248 (1997).
9. B. Grossman, J. Heinze, E. Herdtweck, F. H. Köhler, H. Nöth, H. Schwenk, M. Spiegler, W. Wachter, and B. Weber, *Angew. Chem., Int. Ed. Engl.* **36**, 387–389 (1997).
10. O. Kahn, *Magnetism: A Supramolecular Function*, Kluwer Academic, Dordrecht, 1996, Vol. 484.
11. R. Willett, D. Gatteschi, and O. Kahn, *Magneto-Structural Correlation in Exchange Coupled Systems*, Kluwer Academic, Dordrecht, 1985, Vol. 140.
12. D. Gatteschi, O. Kahn, J. S. Miller, and F. Palacio, *Magnetic Molecular Materials*, Kluwer Academic, Dordrecht, 1991, Vol. 198.
13. J. S. Miller and A. J. Epstein, *Angew. Chem., Int. Ed. Engl.* **33**, 385–415 (1994).
14. S. Decurtins, H. W. Schmalle, P. Schneuwly, and H. R. Oswald, *Inorg. Chem.* **32**, 1888–1892 (1993).
15. S. Decurtins, H. W. Schmalle, P. Schneuwly, J. Ensling, and P. Gütlich, *J. Am. Chem. Soc.* **116**, 9521–9528 (1994).
16. T. J. Collins, S. W. Gordon-Wylie, E. L. Bominaar, C. P. Horwitz, and G. Yee, *Magnetism: A Supramolecular Function*, O. Kahn, ed., Kluwer Academic, Dordrecht, 1996.

17. S. W. Gordon-Wylie, E. L. Bominaar, T. J. Collins, J. M. Workman, B. L. Claus, R. E. Patterson, S. A. Williams, B. J. Conklin, G. T. Yee, and S. T. Weintraub, *Chem. Eur. J.* **1**, 528–537 (1995).
18. A. J. Bard and L. R. Faulkner, *Electrochemical Methods*, Wiley, New York, 1980.
19. N. G. Conelly and W. E. Geiger, *Chem. Rev.* **96**, 877 (1996).
20. E. S. Yang, M.-S. Chan, and A. C. Wahl, *J. Phys. Chem.* **79**, 2049–2051 (1975).

2. A LUMINESCENT DECANUCLEAR RUTHENIUM(II) POLYPYRIDINE COMPLEX: A CONVERGENT APPROACH TO A DENDRITIC STRUCTURE EMPLOYING THE "COMPLEXES AS METALS/COMPLEXES AS LIGANDS" SYNTHETIC STRATEGY

Submitted by SCOLASTICA SERRONI,[*,†] **SEBASTIANO CAMPAGNA,**[*] **FAUSTO PUNTORIERO,**[*] **ALBERTO JURIS,**[‡] **GIANFRANCO DENTI,**[†] **VINCENZO BALZANI,**[‡] **and MARGHERITA VENTURI**[‡]
Checked by DARON JANZEN and KENT R. MANN[§]

The "complexes as metals/complexes as ligands" strategy is a general method for the synthesis of polynuclear metal complexes.[1] Such a strategy can also be applied to the convergent synthesis of metal-based dendritic species. Herein the preparation of a dendritic decanuclear Ru(II) complex is reported. In a typical convergent approach to dendritic structures, preformed arms are mounted on a central core.[2] In the present syntheses, the complex [Ru(2,3-dpp)$_3$](PF$_6$)$_2$ (**1**; 2,3-dpp = 2,3-bis(2-pyridyl)pyrazine is used as the central core. The presence of three free chelating sites in **1** makes it a "complex ligand."

The arms that can be mounted on this core are complexes carrying labile ligands ("complex metals") that can be easily substituted by the free chelating sites of the core. Chlorides have been found to be effective labile ligands in these processes. The use of mononuclear arms such as M(L)$_2$Cl$_2$ [M = Ru(II),Os(II); L = 2,2'-bipyridine (bpy) or 2,2'-biquinoline] allows for the syntheses of various

[*] Dipartimento di Chimica Inorganica, Chimica Analitica e Chimica Fisica, Università di Messina, I-98166 Messina, Italy.
[†] Istituto di Chimica Agraria, Università di Pisa, I-56124 Pisa, Italy.
[‡] Dipartimento di Chimica "G. Ciamician," Università di Bologna, I-40126 Bologna, Italy.
[§] Department of Chemistry, University of Minnesota, 207 Pleasant Street SE, MN 55455-0431.

tetranuclear species:[3,4]

$$[\mathrm{Ru(2,3\text{-}dpp)_3}](PF_6)_2 + 3M(L)_2Cl_2 \rightarrow [\mathrm{Ru\{(\mu\text{-}2,3\text{-}dpp)M(L)_2\}_3}](PF_6)_8 \quad (1)$$

For the synthesis of dendritic complexes of higher nuclearity, polynuclear arms must be used in reaction (1).

The preparation of complex metals like $M(L)_2Cl_2$ is based on the reduction of the appropriate metal precursor (e.g., metal chloride) by means of the solvent, with coordination of the L ligand.[5] In order to inhibit the formation of $[M(L)_3]^{2+}$ species, the reaction must be carried out in the presence of an excess of chloride ions provided either by the metal precursor itself (e.g., in the case of K_2OsCl_6) or by added LiCl. On the basis of these considerations, the experimental conditions are set for the reaction between the metal precursor $RuCl_3 \cdot xH_2O$ and the mononuclear complex ligand $[\mathrm{Ru(bpy)_2(2,3\text{-}dpp)}](PF_6)_2$ (2) to obtain the trinuclear system 3. The latter contains two labile chloride ligands on the central metal. This reaction is represented in Eq. (2), where, for clarity, the bridging ligands are illustrated graphically.

By using the complex metal 3 [Eq. (2)] it is possible to introduce trinuclear subunits in suitable ligand substrates; for example, 3 can be used in the reaction that gives, in good yield, the title compound 4:

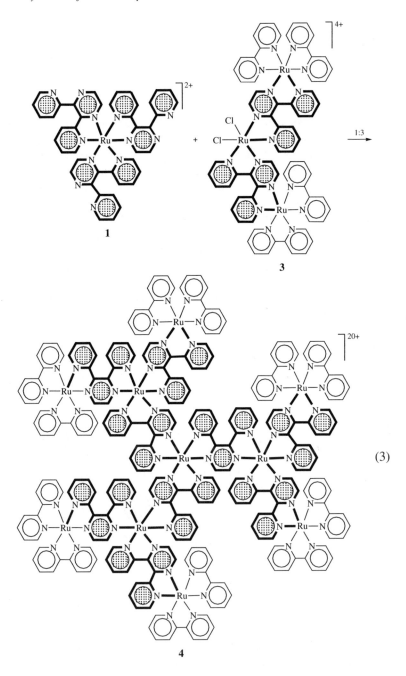

(3)

Materials and General Procedures

$RuCl_3 \cdot xH_2O$ is purchased from Aldrich Chemicals. 2,3-dpp,[6] $Ru(bpy)_2Cl_2 \cdot 2H_2O$,[5a] and $[Ru(2,3-dpp)_3](PF_6)_2$[3] are synthesized according to literature methods. It is also possible to purchase 2,3-dpp and $Ru(bpy)_2Cl_2 \cdot 2H_2O$ from Aldrich Chemicals, but sometimes further purification before use is necessary. Although 2,3-dpp can be purified by sublimation (200°C, under vacuum), $Ru(bpy)_2Cl_2 \cdot 2H_2O$ has to be washed with H_2O until the yellow-orange color, due to $Ru(bpy)_3^{2+}$, disappears from the washing liquor.

A. BISHEXAFLUOROPHOSPHATE-BISBIPYRIDYL-2,3-BIS (2-PYRIDYLPYRAZINE)RUTHENIUM(II) {[Ru(bpy)₂(2,3-dpp)](PF₆)₂}

$$Ru(bpy)_2Cl_2 \cdot 2H_2O + 2,3\text{-dpp} \rightarrow [Ru(bpy)_2(2,3\text{-dpp})](PF_6)_2 + 2Cl^- + 2H_2O$$

Procedure

■ **Caution.** *Because of the toxic nature of acetonitrile, it should be handled under a well-ventilated hood.*

A nitrogen-flushed apparatus is set up (Fig. 1) and used as follows. The 50-mL, two-necked, round-bottomed flask is charged with 67 mg (0.29 mmol) of 2,3-dpp, 5 mL of deaerated 95% ethanol, and a stirring bar. The 50-mL, pressure-equalizing dropping funnel is charged with 100 mg (0.19 mmol) of $Ru(bpy)_2Cl_2 \cdot 2H_2O$ and 15 mL of deaerated 95% ethanol. The flask is placed in an oil bath on a stirrer/hotplate, and the stirred ethanolic suspension of 2,3-dpp is heated to reflux. The ethanolic suspension of $Ru(bpy)_2Cl_2 \cdot 2H_2O$, stirred by means of the nitrogen flow, is then dropwise added (in ~ 1 h) to the refluxing solution of the ligand. The sparging needle may also be used to prevent funnel clogging. During this step the N_2 is provided only by the inlet and the N_2 line stopcock is closed. At the end of the addition, the N_2 line stopcock is open and the funnel is substituted with a ground-glass stopper. The reaction mixture is refluxed for 12 h, then cooled to room temperature. The reaction can be followed by TLC: neutral alumina, $CH_2Cl_2/MeOH$ 9 : 1 (v/v); under these conditions the R_f values decrease in the following order: (1) the ligand (the spot is visible under a UV lamp), (2) the solvated metallic precursor (bordeaux spot), (3) the product (orange spot), (4) the binuclear complex $[(bpy)_2Ru(\mu-2,3\text{-dpp})Ru(bpy)_2](PF_6)_4$ sideproduct (purple spot). After 12 h, some solvated metallic precursor can still be present, but it is not possible to achieve a higher conversion with a longer reaction time.

From now on, the reaction mixture can be handled under air. It is transferred in one-necked, round-bottomed flask and the solvent evaporated to dryness on a rotary

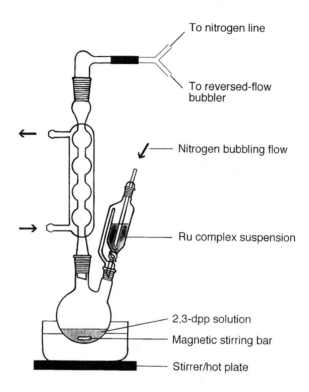

Figure 1. Apparatus for the preparation of $[Ru(bpy)_2(2,3\text{-dpp})]^{2+}$.

evaporator. The crude product so obtained is purified as follows. It is dissolved in the minimum quantity of a mixture H_2O/acetone 5 : 3 (v/v) 0.05 M in NaCl and chromatographed on a Sephadex-CM C-25 (Aldrich Catalog 27,124-1. Column dimensions: 2.5×20 cm) using the same solvent mixture as eluant. The product is contained in the dark orange band. The eluate is rotary-evaporated to remove acetone, then the product is precipitated from the aqueous solution by adding an excess of solid NH_4PF_6. The dark-orange solid so obtained is collected on a sintered-glass filter by suction and dried overnight under vacuum in a desiccator.

   If a spectrofluorimeter is available, it is possible to check the purity of the sample and the effectiveness of the purification by comparing the emission spectra of the crude product to that of the purified product. In the spectrum of the former, the band of the product [$\lambda_{max} = 670$ nm, in acetonitrile at room temperature (RT)] and the band of the binuclear sideproduct ($\lambda_{max} = 756$ nm, in acetonitrile at RT) can be observed; after purification, the emission band of the sideproduct is no longer present (excitation wavelength, 450 nm).[7]

In order to remove possible traces of inorganic salts, the product is dissolved in the minimum quantity of acetonitrile and chromatographed on Sephadex G-10 (Aldrich Catalog 27,103-9. Column dimensions: 2.5×5 cm) by using the same solvent. The eluate is rotary-evaporated to the small volume and then Et_2O is added to induce the precipitation of the product that is collected on a sintered-glass filter by suction, washed three times with Et_2O and dried under vacuum. Yield: 140 mg (78%).

(*Note*: It is necessary to swell the Sephadex-CM C-25 resin in excess eluant mixture, at room temperature without stirring, for about 24 h before use. Attention must be paid to the quantity of Sephadex-CM C-25 resin used, since its volume increases about twice on swelling. It is recommended that the Sephadex G-10 resin be allowed to swell in excess acetonitrile for at least 3 h prior to use.)

Anal. Calcd. for $C_{34}H_{26}N_8F_{12}P_2Ru \cdot H_2O$: C, 42.73; H, 2.95; N, 11.73. Found: C, 42.80; H, 2.75; N, 11.28.

B. TETRAKIS-HEXAFLUOROPHOSPHATE-BIS-BISBIPYRIDYL-2,3-BIS(2-PYRIDYLPYRAZINE)RUTHENIUM(II)-DICHLORORUTHENIUM(II) ({Cl₂Ru[(μ-2,3-dpp)Ru(bpy)₂]₂}(PF₆)₄)

$$2[\text{Ru(bpy)}_2(2,3\text{-dpp})](\text{PF}_6)_2 + \text{RuCl}_3 \xrightarrow{\text{Cl}^-} \{\text{Cl}_2\text{Ru}[(\mu\text{-}2,3\text{-dpp})\text{Ru(bpy)}_2]_2\}(\text{PF}_6)_4$$

This synthesis must be repeated twice in order to get the amount of product necessary for the subsequent synthesis of $(\text{Ru}\{(\mu\text{-}2,3\text{-dpp})\text{Ru}[(\mu\text{-}2,3\text{-dpp})\text{Ru}(\text{bpy})_2]_2\}_3)(\text{PF}_6)_{20}$.

A nitrogen flushed 50-mL, two-necked, round-bottomed flask fitted with a condenser is charged with 11.5 mg (0.04 mmol) of $RuCl_3 \cdot xH_2O$, 80 mg (0.08 mmol) of $[Ru(bpy)_2(2,3\text{-dpp})](PF_6)_2$, 11.9 mg (0.28 mmol) of LiCl, 15 mL of 95% ethanol, and a stirring bar. The flask is placed in an oil bath on a stirrer/hot-plate, and the stirred reaction mixture is heated to reflux. After 7 h the reaction mixture is cooled to room temperature. The reaction can be followed by TLC: neutral alumina, $CH_2Cl_2/MeOH$ 9 : 1 (v/v); in these conditions the R_f values decrease in the following order: (1) mononuclear precursor (orange spot), (2) trinuclear product (green spot).

From now on, the reaction mixture can be handled under air. An excess of solid NH_4PF_6 is added causing the formation of a dark green solid that is collected on a sintered-glass filter by suction, washed with Et_2O. The crude product is dissolved in the minimum quantity of a 1 : 1 (v/v) CH_3CN/toluene mixture and purified by column chromatography on neutral aluminum oxide (diameter 2.5 cm, length 20 cm; aluminum oxide activity: 1) by using the same solvent mixture [1 : 1 (v/v) CH_3CN/toluene] as eluant. The elution of the column is performed slowly. In these experimental conditions, the first band eluted (orange) contains the unreacted mononuclear $[Ru(bpy)_2(2,3\text{-dpp})](PF_6)_2$ complex. After the collection of this first band, the quantity of CH_3CN in the eluition mixture is increased progressively [up to 3 : 1 (v/v) CH_3CN/toluene] in order to obtain the green-blue band that contains the trinuclear product. At times, after the elution of the trinuclear complex, a dark byproduct (maybe a tetranuclear complex) stays on the top of the column. The green-blue band is eluted and the solution containing the desired product is rotary-evaporated to dryness. The product is dissolved in a very small amount ($\sim$2 mL) of CH_3CN and precipitated by adding diethyl ether. The column yield is 70%. Total yield of the reaction (after purification) 55%.[*]

If a spectrofluorimeter is available, it is possible to check the purity of the sample and the effectiveness of the purification by comparing the emission spectra of the crude product and the purified product. In the former the band of the mononuclear complex can be present ($\lambda_{max} = 670$ nm in acetonitrile at RT);[7] after purification, no emission should be detectable (excitation wavelength 450 nm).

Anal. Calcd. for $C_{68}H_{52}N_{16}Cl_2F_{24}P_4Ru_3 \cdot 3H_2O$: C, 38.87; H, 2.78; N, 10.67; P, 5.90; Ru, 14.43. Found: C, 38.89; H, 2.66; N, 10.56; P, 6.38; Ru, 14.21. FAB^+ (*m*-NBA). Calcd. for $(MPF_6)^-$: 1903.0. Found: 1903.

C. ICOSAKISHEXAFLUOROPHOSPHATE-TRIS-BIS-BISBIPYRIDYL-2,3-BIS(2-PYRIDYLPYRAZINE)RUTHENIUM(II)-2,3-BIS (2-PYRIDYLPYRAZINE) RUTHENIUM(II) [({Ru[(μ-2,3-DPP)Ru (μ-2,3-DPP)Ru(BPY)$_2$]$_2$}$_3$)(PF$_6$)$_{20}$]

$$[Ru(2,3\text{-dpp})_3](PF_6)_2 + 3\{Cl_2Ru[(\mu\text{-}2,3\text{-dpp})Ru(bpy)_2]_2\}(PF_6)_4 \xrightarrow{PF_6^-}$$
$$\{Ru[(\mu\text{-}2,3\text{-dpp})Ru[(\mu\text{-}2,3\text{-dpp})Ru(bpy)_2]_2]_3\}(PF_6)_{20}$$

A nitrogen-flushed 50-mL, two-necked, round-bottomed flask fitted with a condenser is charged with 116 mg (0.06 mmol) of $\{Cl_2Ru[(\mu\text{-}2,3\text{-dpp})Ru$ $(bpy)_2]_2\}(PF_6)_4$, 19.2 mg (0.12 mmol) of $AgNO_3$, 4 mL of a $MeOH/H_2O$ 2 : 1 (v/v) mixture, and a stirring bar. After 2 h stirring at room temperature 20.6 mg (0.02 mmol) of $[Ru(2,3\text{-dpp})_3](PF_6)_2$ and 4 mL of ethylene glycol are added, and the reaction mixture is refluxed for 48 h. The reaction can be followed by TLC: neutral alumina, CH_2Cl_2/MeOH 9 : 1 (v/v); in these conditions the R_f values decrease in the following order: (1) mononuclear core (orange spot), (2) solvated trinuclear precursor (green-blue spot), (3) possible impurities of the tetra- and heptanuclear complexes in which one or two free chelating sites are still present on the core (purple spots), (4) decanuclear complex (purple spot).

---

[*] The checkers point out that the activity grade of the alumina can be important; a higher activity grade of alumina (i.e., freshly opened bottle of alumina activity 1) requires a higher CH_3CN/toluene ratio for release all the product. They suggest that the optimal CH_3CN/toluene ratio be determined with a preliminary "disposable pipette column" study.

After cooling to room temperature the reaction mixture can be handled under air. The AgCl formed is removed by repeated centrifugations. The reaction mixture is rotary-evaporated to remove the low-boiling solvents. An excess of solid NH_4PF_6 is added to the remaining solution (mainly ethylene glycol) and the mixture is stirred for 20 min. After addition of an equal volume of Et_2O, a two-phase system is formed. Methanol is added dropwise, under stirring, until a single solvent phase is obtained, and this solution is stirred for 30 min. A purple precipitate forms and is collected on a sintered glass filter by suction. This crude product is dissolved in the minimum volume of acetonitrile and chromatographed on Sephadex G-25 (Aldrich Catalog 27,109-8. Column dimensions: 2.5×5 cm) by using the same solvent. An amount of 95% ethanol ($\sim$6 mL) is added to the eluate and the solution is rotary-evaporated until the product precipitates. The solid is collected on a sintered-glass filter by suction, washed 3 times with Et_2O and dried under vacuum. Yield: 118 mg (80%).

If a spectrofluorimeter is available, it is possible to check the purity of the sample by looking at the emission spectrum of the product; only the band of the decanuclear complex ($\lambda_{max} = 785$ nm in acetonitrile at RT)[7] must be present. The band of the mononuclear core ($\lambda_{max} = 635$ nm, in acetonitrile at RT)[7] should be absent (excitation wavelength 450 nm). (*Note*: For the swelling of Sephadex G-25 resin, see the indications given for the Sephadex G-10.)

Anal. Calcd. for $C_{246}H_{186}N_{60}F_{120}P_{20}Ru_{10} \cdot 10H_2O$: C, 36.60; H, 2.57; N, 10.41; P, 7.67; Ru, 12.51. Found: C, 36.19; H, 2.70; N, 10.83; P, 8.30; Ru, 12.01.

Properties

The complexes dealt with in this contribution are soluble in polar solvents (like acetonitrile and acetone). They are air-stable in both the solid state and in

TABLE I. Properties of Complexes 1, 2, 3, and 4[a]

Compound	IR Frequencies (KBr Pellets), cm^{-1}	Absorption λ_{max}, nm (ε, M^{-1} cm^{-1})	E_{ox}, V. vs SCE
1	1576(w), 1556(w), 1544(m), 1462(w), 1428(w), 990(w)	455 (13,000) 281 (52,300)	+1.60
2	1590(m), 1555(m), 1578(s), 1436(s), 1402(m), 1386(m), 990(w)	475(sh) (11,500) 285 (72,100)	+1.31
3	1607(m), 1559(w,br), 1470(s), 1450(s), 1423(m), 1393(s)	615 (24,300) 285 (107,000)	+0.72 +1.45
4	1605(m), 1560(w,br), 1470(s), 1450(s), 1420(s), 1399(s)	541 (125,000) 282 (312,000)	+1.53

[a] The numbering of the complexes is identical to that defined in the introduction and also shown in Eqs. (1) and (2).

solution. Their solutions are strongly colored. Relevant spectroscopic and electrochemical data are gathered in Table I. The weak band at 990 cm^{-1} in the IR spectra is diagnostic of the presence of unbridged 2,3-dpp ligand.

Acknowledgments

We thank C. Mingazzini for the elemental analyses and G. Gubellini for technical assistance.

References and Notes

1. S. Campagna, G. Denti, S. Serroni, M. Ciano, and V. Balzani, *Inorg. Chem.* **30**, 3728 (1991); S. Serroni and G. Denti, *Inorg. Chem.* **31**, 4251 (1992); S. Campagna, G. Denti, S. Serroni, A. Juris, M. Venturi, V. Ricevuto, and V. Balzani, *Chem. Eur. J.* **1**, 211 (1995); V. Balzani, A. Juris, M. Venturi, S. Campagna, and S. Serroni, *Chem. Rev.* **96**, 759 (1996).
2. J. M. J. Fréchet, *Science* **263**, 1710 (1994).
3. K. J. Brewer, W. R. Murphy, Jr., S. R. Spurlin, and J. D. Petersen, *Inorg. Chem.* **25**, 882 (1986).
4. W. R. Murphy Jr., K. J. Brewer, G. Gettliffe, and J. D. Petersen, *Inorg. Chem.* **28**, 81 (1989); S. Campagna, G. Denti, L. Sabatino, S. Serroni, M. Ciano, and V. Balzani, *J. Chem. Soc., Chem. Commun.* 1500 (1989); G. Denti, S. Campagna, L. Sabatino, S. Serroni, M. Ciano, and V. Balzani, *Inorg. Chem.* **29**, 4750 (1990); G. Denti, S. Serroni, S. Campagna, V. Ricevuto, and V. Balzani, *Coord. Chem. Rev.* **111**, 227 (1991).
5. (a) P. A. Lay, A. M. Sargeson, and H. Taube, *Inorg. Synth.*, **24**, 292 (1986); (b) P. Belser and A. von Zelewsky, *Helv. Chim. Acta* **63**, 1675 (1980); (c) D. A. Buckingham, F. P. Dwyer, H. A. Goodwin, and A. M. Sargeson, *Aust. J. Chem.* **55**, 325 (1964); (d) G. Denti, S. Serroni, L. Sabatino, M. Ciano, V. Ricevuto, and S. Campagna, *Gazz. Chim. Ital.* **121**, 37 (1991).
6. H. A. Goodwin and F. Lions, *J. Am. Chem. Soc.* **81**, 6415 (1959).
7. Emission spectra have been obtained with a Perkin-Elmer LS-50 spectrofluorimeter equipped with a Hamamatsu R958 phototube. Data are uncorrected for spectral sensitivity of the phototube.

3. DODECATUNGSTOALUMINIC ACID AND ITS MONOLACUNARY AND MIXED-ADDENDUM DERIVATIVES

Submitted by JENNIFER J. COWAN,[*] CRAIG L. HILL,[*]
RICHARD S. REINER,[†] and IRA A. WEINSTOCK[†]
Checked by WALT KLEMPERER[‡] and KEITH MAREK

Heteropolyoxometalates of the Keggin structural class, including their transition-metal-substituted derivatives, are effective homogeneous catalysts for selective oxidations of organic and inorganic substrates by a variety of oxidants.[1,2]

[*] Department of Chemistry, Emory University, Atlanta, GA 30322.
[†] USDA Forest Service, Forest Products Laboratory, One Gifford Pinchot Dr., Madison, WI 53705.
[‡] Department of Chemistry, University of Illinois, Urbana-Champagne 600 South Matthews Ave., Urbana, IL 61801.

While many polyoxometalates (POMs) of this class are remarkably stable to oxidizing conditions, they are often unstable with respect to hydrolysis, limiting the use of POMs in water to acid pH values.[3] In light of the increasing drive toward more environmentally benign chemical systems, it would be advantageous to develop POMs that are stable in water over a wider range of pH values. Compared to the Si-centered and P-centered analogs[4] of the Keggin anion, an Al-centered Keggin anion possesses a more negative charge, which correlates with greater base stability.[5,6] The $[AlW_{12}O_{40}]^{5-}$ anion is known, but comparatively little has been published about it; this may be attributable to the low yield of the published synthetic method.[7–10] The shortcoming of this method is that it fails to account for the ability of Al(III) to function as both a heteroatom and an addendum atom, which results in the competitive formation of $[Al(AlOH_2)W_{11}O_{40}]^{6-}$ during aluminum addition.[8–11]

This contribution details a new, high-yield route to $H_5[AlW_{12}O_{40}]$. By deliberately adjusting the reaction stoichiometry to 2Al:11W and choosing the proper pH value, $[Al(AlOH_2)W_{11}O_{40}]^{6-}$ is formed selectively and then is cleanly converted to $[AlW_{12}O_{40}]^{5-}$ by refluxing at low pH.[12,13] As this procedure generates a mixture of the α and β isomers, the separation of the β isomer and the conversion to the α-isomer are described. Also included is the first reported synthesis of its monolacunary derivative α-$[AlW_{11}O_{40}]^{9-}$ which is an obvious precursor to a variety of transition-metal-substituted Keggin anions. The preparation of one of these, α-$[AlVW_{11}O_{40}]^{7-}$, is provided here by way of example. The preparation of the oxidized derivative of the vanadium-substituted anion, α-$[AlVW_{11}O_{40}]^{6-}$, is also presented as an example of the use of ozone as an oxidant that introduces no exogenous elements.

A useful property of the aluminum heteroatom is that the ^{27}Al isotope, present in 100% abundance and possessing a nuclear spin quantum number of $\frac{5}{2}$, is easily and rapidly observed by nuclear magnetic resonance (NMR) spectroscopy. The chemical shift and linewidth of ^{27}Al NMR signals provide readily accessible information regarding both the coordination number of the aluminum atom and the chemical symmetry of its environment.[8] However, care must be taken in collection of samples to allow a sufficient delay between pulses, as narrow resonances of quadrupolar nuclei relax more slowly (a 5-s delay is usually sufficient). An additional caveat regards the use of dilute samples or those with broad signals; in some cases a broad resonance has been observed (from ∼105 to 35 ppm) in addition to those expected from the sample (it is also observed in the absence of a glass sample tube in the probe). This broad signal is due to aluminum present in glass in the NMR probe itself (not all NMR instruments possess Al-containing glassy compounds). In either case, the relative intensity of the resonance can be minimized by using concentrated samples. The effectiveness of this technique improves with decreasing peak width. There are other (less satisfactory) methods of dealing with this problem, which are addressed in Ref. 12.

Materials and General Procedures

Reagent-grade chemicals (sodium tungstate dihydrate was Folin reagent grade) were obtained from commercial sources. Deionized water is used throughout. All NMR experiments were carried out in deuterium oxide and referenced externally: ^{27}Al to 0.1 M $AlCl_3$ in deuterium oxide, ^{183}W to saturated Na_2WO_4 in deuterium oxide, and ^{51}V to neat $VOCl_3$.

A. DODECATUNGSTOALUMINIC ACID $H_5[AlW_{12}O_{40}]$

Step 1: $11[WO_4]^{2-} + 2Al^{3+} + 10H^+ \rightarrow [Al(AlOH_2)W_{11}O_{39}]^{6-} + 5H_2O$

Step 2: $12[Al(AlOH_2)W_{11}O_{39}]^{6-} + 56H^+ \rightarrow 11[AlW_{12}O_{40}]^{5-} + 13Al^{3+} + 28H_2O$

Procedure

■ **Caution.** *Hydrochloric acid is corrosive. Precautions should be taken to avoid coming into direct contact or breathing HCl fumes.*

Step 1. Sodium tungstate dihydrate ($Na_2WO_4 \cdot 2H_2O$, 100 g, 0.304 mol) is dissolved in 400 mL of H_2O in a 1000-mL, three-necked, round-bottomed flask containing a magnetic stirring bar and fitted with an addition funnel and condenser. Hydrochloric acid ($\sim$23.0 mL, 0.276 mol) is added to the solution dropwise with vigorous stirring to pH 7.7 (use of a calibrated pH meter in the solution during this procedure is necessary). After every several drops the addition is momentarily stopped to allow the local precipitate of tungstic acid to dissolve. The solution is then heated to reflux, and aluminum chloride hexahydrate ($AlCl_3 \cdot 6H_2O$, 13.32 g, 0.0552 mol), dissolved in 80 mL deionized water, is added dropwise by means of the addition funnel over $\sim$90 min ($\sim$5–6 drops/min) with constant stirring. During the addition, the solution becomes slightly cloudy. However, addition should be kept at a slow enough rate to prevent the mixture from becoming opaque. If this should occur, however, the addition must be stopped and the solution stirred until it becomes clearer. After all the $AlCl_3$, has been added, the solution is kept at reflux for 1 h, cooled to room temperature, and filtered through a 0.5-in.-thick pad of Celite (diatomaceous earth) on a medium sintered-glass-fritted funnel. The final pH should be approximately 7.

Step 2. The solution, now containing $Na_6[Al(AlOH_2)W_{11})O_{39}]$ (^{27}Al NMR: 73 ppm, $\Delta\nu_{1/2} = 89$ Hz; 8 ppm, $\Delta\nu_{1/2} = 256$ Hz), is transferred to a 1000-mL, round-bottomed flask fitted with a reflux condenser. The solution is acidified to pH 0 by careful dropwise addition of concentrated sulfuric acid ($\sim$20 mL, 0.376 mol). After the pH has reached 0, an additional 3 mL of conc. sulfuric acid is added and the solution is heated to reflux. The solution becomes cloudy and slightly yellow as the acid is added, but should clear within 16 h of the beginning of reflux. To ensure complete conversion to product, the solution should be kept at reflux for 6 days (monitoring the progress by ^{27}Al NMR is also recommended). Then, after cooling to room temperature, the solution may be filtered (if cloudy) using a medium-porosity glass-fritted funnel {this is seldom necessary if the two steps are performed in quick succession, that is, provided the $[Al(AlOH_2)W_{11}O_{39}]^{6-}$ is not allowed to stand in solution for more than 24 h}.

Workup. The solution, which now contains $H_5[AlW_{12}O_{40}]$ and 1.08 (i.e., $\frac{13}{12}$) equiv of soluble Al(III) salts, is transferred to a 1000-mL beaker and cooled to 0°C.

■ **Caution.** *The following acidification and extraction procedure should be performed in a ventilated hood, while wearing appropriate safety clothing including splash goggles. Sulfuric acid is extremely corrosive, and during the ether extraction, due to possible excessive pressure generation, there is a danger of splashing the sulfuric acid. Careful, frequent ventilation of the separatory funnel must be applied.*

Cold (0°C) conc. sulfuric acid (147 mL) is added carefully to avoid excessive heating. The solution is then cooled in an ice-water bath to 0°C and transferred to a 2000-mL separatory funnel. Diethyl ether (500 mL) is added and the mixture shaken very gently with frequent ventilation until rapid evaporation of diethyl ether subsides. Then, the mixture is shaken more vigorously, still with frequent venting, and allowed to settle until three layers separate. The top clear colorless layer is diethyl ether, the middle somewhat cloudy layer is the aqueous phase, and the bottom layer (a dense, pale yellow, viscous liquid) is the etherate of $H_5[AlW_{12}O_{40}]$. The bottom (etherate) layer is collected and the shaking and venting procedure is repeated until the etherate layer no longer forms (the aqueous layer appears clearer as the extraction nears completion). The combined etherate layers ($\sim$20 mL) are concentrated to dryness by rotary evaporation. The crude product (69.2 g, 95%) is reprecipitated by dissolving in 20 mL of hot water, concentrating to a volume of 23 mL by gentle heating, and then cooling to 0°C for 16 h. Yield: 50.46 g, 64%.[*]

---

[*] The product prepared as described is a mixture of α- and β-isomers, with β the dominant isomer (typically 85–90%). However, when $K_9[AlW_{11}O_{39}]$ is prepared from this mixture as described below, the product is 100% α-$K_9[AlW_{11}O_{39}]$.

Separation of the β isomer of $H_5[AlW_{12}O_{40}]$ from the α isomer may be accomplished by careful fractional crystallization—this procedure is not required for the preparation of the lacunary species, α-$K_9[AlW_{11}O_{39}]$; see footnote below. [Conversion to and collection of the pure α isomer, i.e., α-$Na_5[AlW_{12}O_{40}]$, is described in the following section.] Monitoring of the products by ^{27}Al NMR is necessary to observe whether the separation is complete. The product from the procedure described above is dissolved in minimum warm H_2O in a 250-mL beaker and allowed to stand in a drafty location (the back of a fume hood, for example) until crystals form, 3 days or more.

If the crystallization takes place slowly enough, the first crop (9.56 g) contains a high percentage of α-$H_5[AlW_{12}O_{40}]$ (77% α isomer in one case, determined by ^{27}Al NMR), due to its lower solubility. The subsequent mother liquor is treated in a similar fashion—the product is allowed to crystallize slowly and then is removed by filtration once a few grams of crystals form. The second crop (7.31 g) collected in this manner has a smaller (19%), but still substantial, amount of the α isomer. The mother liquor is then allowed to stand again to give a larger third crop (26.42 g) that contains a small percentage of α isomer (4.4%). Very slow recrystallization of the third crop from a minimum of H_2O (perhaps over one week) gives pure β-$H_5[AlW_{12}O_{40}]$ (7.76 g, 9.8%).

Properties

The mixture of α- and β-dodecatungstoaluminic acid prepared as described in step 2 is a water-soluble slightly yellow amorphous solid. The pure β isomer is pale yellow and crystalline, but becomes amorphous on drying. To avoid reduction of the free acid $H_5[AlW_{12}O_{40}]$, metal implements should not be used in handling it as a solid, and especially while in solution. The two polyanions are stable in water below pH 6 and characterized by ^{27}Al NMR: β isomer, 71.6 ppm ($\Delta\nu_{1/2} = 4.6$ Hz); α isomer, 72.1 ppm ($\Delta\nu_{1/2} = 1.3$ Hz); ^{183}W NMR β isomer (integration), $-110.8(1)$, $-118.7(2)$, and $-136.8(1)$ ppm; α isomer, -110.1 ppm. IR (2–5 wt% KBr pellet, cm^{-1}): 972(s), 899(s), 795(broad, s), 747(broad, s), 538(m), and 477(m).

Anal. Calcd. (found) for $H_5[AlW_{12}O_{40}]$ $15H_2O$: H, 1.12 (1.15); W, 70.07 (70.23); Al, 0.86 (0.89).

B. α-PENTASODIUM DODECATUNGSTOALUMINATE
α-Na₅[AlW₁₂O₄₀]

This procedure converts a mixture of α- and β-$H_5[AlW_{12}O_{40}]$ to $\sim 95\%$ α-$Na_5[AlW_{12}O_{40}]$. The α isomer, because of its lower solubility, is easily separated from the residual amount of β isomer. A 31.9-g sample of $H_5[AlW_{12}O_{40}]$

is dissolved in 65 mL of water and filtered if cloudy. The pH of the solution is adjusted to ~6 using a 0.75 M solution of Na_2CO_3 (~42.5 mL) and is then heated at reflux for 3 days. At this point, the solution typically contains about 95% α-$[AlW_{12}O_{40}]^{5-}$ (with 5% β, determined by ^{27}Al NMR). The solution may be heated longer if the proportion of α isomer is less than 95%. After cooling, the solution is concentrated by rotary evaporation until precipitate begins to form, and is then refrigerated at 5°C. The product is collected on a coarse glass-fritted funnel and air-dried. Yield (2 crops): 20.07 g (60.8%).

Properties

^{183}W NMR, 8: −112.8 ppm. ^{27}Al NMR, 8: 72.1 ppm ($\Delta\nu_{1/2} = 1.1$ Hz). IR (KBr pellet, cm^{-1}): 955(m), 883(s), 799(s), 758(s), 534(w), 498(w).

Anal. Calcd. (found) for $Na_5[AlW_{12}O_{40}]\cdot13H_2O$: H, 0.81 (0.78); W, 68.47 (68.22); Al, 0.84 (0.88); Na, 3.57 (3.39).

C. α-NONAPOTASSIUM UNDECATUNGSTOALUMINATE α-K₉[AlW₁₁O₃₉]

$$H_5[AlW_{12}O_{40}] + 7.5K_2CO_3 + 0.5H_2O \rightarrow$$
$$\alpha\text{-}K_9[AlW_{11}O_{39}] + KHWO_4 + 5KHCO_3 + 2.5CO_2(g)$$

Procedure

The isomeric composition of the starting $H_5[AlW_{12}O_{40}]$ does not affect that of the lacunary product; under the conditions described here, the α isomer is always obtained in near-quantitative yield. The typical starting material used in this procedure is the isomeric mixture of $H_5[AlW_{12}O_{40}]$ prepared and collected as described in procedure A, step 2 (without separation of the β isomer by fractional crystallization). Dodecatungstoaluminic acid $H_5[AlW_{12}O_{40}]\cdot13H_2O$ (43.76 g, 14.1 mmol) is dissolved in 100 mL of H_2O and heated with stirring to 60°C. Three equivalents of potassium carbonate (1.5 hydrate) ($K_2CO_3\cdot1.5$ H_2O, 6.97 g, 42.3 mmol) are added gradually as a solid. The pH should rise to about 2. Another 5 equiv of potassium carbonate (1.5 hydrate) (11.62 g, 70.5 mmol, dissolved in 20 mL H_2O) are added dropwise carefully over about 60 min. The pH of the solution should not be allowed to rise above 8 until at least 75% of the potassium carbonate solution has been added, and should at all times be kept below 8.5. A white precipitate begins to form as addition of the potassium carbonate solution proceeds. The final pH of the mixture should be near 8.25. After addition of the potassium carbonate solution is complete, the mixture is cooled to 5°C for several hours. The product, a fine white precipitate, is then collected, washed 3 times with H_2O, and dried on a medium-porosity glass frit. Yield 41.8 g, 92%.

Properties

The potassium salt of the lacunary anion is a white amorphous solid, which is slightly soluble in water (2 g/100 mL at 22°C). ^{27}Al NMR (47,000 scans at 30°C): 63.3 ppm ($\Delta\nu_{1/2} = 784$ Hz). IR (2–5 wt% KBr pellet, cm^{-1}): 937(m), 868(s), 789(s), 756(sh), 704(m), 524(w), 493(w).

Anal. Calcd. (found) for $K_9[AlW_{11}O_{39}] \cdot 12H_2O$: H, 0.75 (0.79); W, 62.39 (62.05); Al, 0.83 (0.92); K, 10.86 (10.80).

D. α-HEPTAPOTASSIUM VANADOUNDECATUNGSTOALUMINATE α-$K_7[AlV^{IV}W_{11}O_{40}]$

$$\alpha\text{-}K_9[AlW_{11}O_{39}] + VOSO_4 \rightarrow \alpha\text{-}K_7[AlVW_{11}O_{40}] + K_2SO_4$$

Procedure

To a well-stirred slurry of $\alpha\text{-}K_9[AlW_{11}O_{39}]$ (5.43 g, 1.80 mmol) in 10 mL of H_2O in a 50-mL beaker, vanadyl sulfate trihydrate ($VOSO_4 \cdot 3H_2O$, 0.39 g, 1.80 mmol), dissolved in 5 mL of H_2O, is added dropwise rapidly at room temperature. The color changes immediately to dark purple. The mixture is stirred for 30 min and filtered on a medium-porosity glass frit, then the dark purple filtrate is cooled to 5°C for 2 h. The resulting dark purple crystals are collected on a coarse fritted funnel and recrystallized from a minimum of warm (60°C) H_2O. Recrystallized yield: 3.5 g (61%).

Properties

Dark purple crystalline $\alpha\text{-}K_7[AlVW_{11}O_{40}] \cdot 15H_2O$ is characterized in the solid state by IR: (2–5 wt% KBr pellet, cm^{-1}): 942(m), 871(m), 793(s), 761(m), 697(w), 537(w), 492(w), 473(w).

Anal. Calcd. (found) for $K_7[AlVW_{11}O_{40}] \cdot 15H_2O$: H, 0.92 (0.84); W, 61.58 (61.66); Al, 0.82 (0.97); V, 1.55 (1.32); K, 8.33 (8.17).

The vanadyl ($V^{IV} = O)^{2+}$ containing anion, $[AlVW_{11}O_{40}]^{7-}$, is paramagnetic. However, solutions of $[AlVW_{11}O_{40}]^{7-}$ are readily oxidized to bright yellow solutions of $[AlVW_{11}O_{40}]^{6-}$ by addition of elemental bromine. Diamagnetic $[AlVW_{11}O_{40}]^{6-}$, prepared in situ, can be observed by ^{27}Al, ^{51}V, and ^{183}W NMR. ^{27}Al: 72.5 ppm ($\Delta\nu_{1/2} = 175$ Hz); ^{51}V: -535.5 ppm ($\Delta\nu_{1/2} = 220$ Hz); ^{183}W (integration): $-79.7(2)$, $-96.0(2)$, $-116.2(2)$, $-119.5(1)$, $-121.4(2)$, $-141.2(2)$ ppm.

E. α-HEXAPOTASSIUM VANADOUNDECATUNGSTOALUMINATE α-K₆[AlV^V W₁₁O₄₀]

The fully oxidized anion, $[AlV^VW_{11}O_{40}]^{6-}$, is conveniently prepared from the reduced species $[AlV^{IV}W_{11}O_{40}]^{7-}$, prepared in situ from $K_9[AlW_{11}O_{39}]$ and vanadyl sulfate as described in procedure D, by passing a stream of ozone through its acidic solution (acid must be present during the ozonation to consume the hydroxide generated by the reduction of ozone in water). If ozone is not available, other oxidants, such as Br_2 or sodium hypochlorite, may be used. A 0.5 M solution of vanadyl sulfate trihydrate ($VOSO_4 \cdot 3H_2O$, 12.20 mL, 6.1 mmol) is added dropwise to a well-stirred slurry of α-$K_9[AlW_{11}O_{39}]$ (20 g, 6.1 mmol in 50 mL H_2O), resulting in a dark purple solution of $[AlVW_{11}O_{40}]^{7-}$. Then 2 equiv of hydrochloric acid (4 mL of a 3 M solution) are added to the polyanion solution and a stream of ozone is bubbled through the solution until the dark purple color is discharged and a bright yellow solution obtained (the extinction coefficient of the reduced species is large enough that even very small amounts can be detected by eye in the presence of the oxidized species). Then oxygen is bubbled through the solution for several minutes to flush out unreacted ozone, and the solution is concentrated by rotary evaporation to approximately half its volume. The solution is refrigerated overnight at 5°C; yellow crystals of $K_6[AlV^VW_{11}O_{40}]$ are collected and dried on a medium frit, and recrystallized from a minimum of hot H_2O. Yield: 12.95 g (66%).

Properties

The potassium salt of α-$K_6[AlV^VW_{11}O_{40}]$ is a water-soluble yellow crystalline solid. ^{27}Al NMR, δ: 72.5 ppm ($\Delta\nu_{1/2} = 175$ Hz); ^{51}V NMR, δ: -535.5 ppm ($\Delta\nu_{1/2} = 220$ Hz); ^{183}W (relative intensities): $-83.1(2)$, $-99.1(2)$, $-119.5(2)$, $-123.0(1)$, $-124.0(2)$, $-144.4(2)$ ppm. IR (2–5 wt% KBr pellet, cm^{-1}): 950(m), 878(s), 794(s), 756(s), 542(w), 487(w).

Anal. Calcd. (found) for $K_6[AlVW_{11}O_{40}] \cdot 13H_2O$: H, 0.82 (0.78); W, 63.02 (62.97); Al, 0.84 (0.88); V, 1.59 (1.88); K, 7.31 (7.29).

References and Notes

1. M. T. Pope and A. Müller, *Angew. Chem., Int. Ed. Engl.* **30**, 34 (1991).
2. C. L. Hill and C. M. Prosser-McCartha, *Coord. Chem. Rev.* **143**, 407 (1995).
3. T. Okuhara, N. Mizuno, and M. Misono, *Adv. Catal.* **41**, 113 (1996).
4. M. T. Pope, *Heteropoly and Isopolyoxometalates,* Springer-Verlag, New York, 1993.
5. Increasing x in the series $[PV_xW_{(12-x)}O_{40}]^{(3+x)-}$ results in an increase in the negative charge on the anion and greater hydrolytic stability; see D. P. Smith and M. T. Pope, *Inorg. Chem.* **12**, 331 (1973).
6. While $[PVW_{11}O_{39}]^{4-}$ is stable in water at pH values between 2 and 3, the silicon analog, $[SiVW_{11}O_{39}]^{5-}$, is stable indefinitely at reflux in water at pH values between 3 and 4. See I. A. Weinstock, R. A. Atalla, R. S. Reiner, M. A. Moen, K. E. Hammel, C. L. Hill, and M. K. Harrup, *J. Mol. Catal. A, Chemical* **116**, 59 (1997).
7. Preparation of $H_5[A1W_{12}O_{40}]$ (mistakenly described as 11-tungstoaluminic acid): J. A. Mair and J. L. T. Waugh, *J. Chem. Soc.*, 2372 (1950).
8. Preparation and ^{27}Al NMR spectrum, of $H_5[A1W_{12}O_{40}]$ (mixture of α and β isomers): J. W. Akitt and A. Farthing, *J. Chem. Soc. Dalton Trans.*, 1615 (1981).
9. X-ray powder diffractometry of $Cs_5[A1W_{12}O_{40}]$: D. H. Brown, *J. Chem. Soc.*, 3281 (1962).
10. IR spectroscopy of $H[(C_4H_9)_4N]_4[A1W_{12}O_{40}]$: K. Nomiya and M. Miwa, *Polyhedron* **2**, 955 (1983).
11. Solution ^{27}Al NMR spectroscopy of reaction mixtures: M. A. Fedotov and L. P. Kazanskii, Izv. *Akad. Nauk SSSR,* Ser. Khim. 9, 2000 (1988) (Engl. trans., p. 1789).
12. I. A. Weinstock, J. J. Cowan, E. M. G. Barbuzzi, H. Zeng, and C. L. Hill, *J. Am. Chem. Soc.* **121**, 4608 (1999).
13. J. J. Cowan, A. J. Bailey, R. A. Heintz, B. T. Do, K. I. Hardcastle, C. L. Hill, and I. A. Weinstock, *Inorg. Chem.* **40** (in press).

4. SUPRAMOLECULAR COMPLEXES OF BIS(2,2′-BIPYRIDINE)OSMIUM(II) AND RUTHENIUM(II)

Submitted by **JEFF A. CLARK, MARK M. RICHTER,**[†] and **KAREN J. BREWER**[*]
Checked by **D. SENIVERATNE** and **J. F. ENDICOTT**[‡]

Interest in osmium(II) and ruthenium(II) polypyridyl complexes originally developed because of their extreme stability and intense colors.[1–3] Later they were shown to possess very interesting excited-state properties and undergo facile energy and electron transfer.[4–7] A more recent trend is this area of chemistry is to attach a variety of polyazine ligands capable of bridging to allow for the construction of supramolecular systems.[8–13] One useful series of polyazine bridging ligands is that of dpp [2,3-bis(2-pyridyl)pyrazine], dpq [2,3-bis(2-pyridyl)quinoxaline], and dpb [2,3-bis(2-pyridyl)benzoquinoxaline][8,14,15] (Fig. 1).

[†] Department of Chemistry, Southwest Missouri State University, Springfield, MO 65804.
[*] Department of Chemistry, Virginia Polytechnic Institute and State University, Blacksburg, VA 24061-0212 (kbrewer@vt.edu).
[‡] Department of Chemistry, Wayne State University, Detroit, MI 48202-3489.

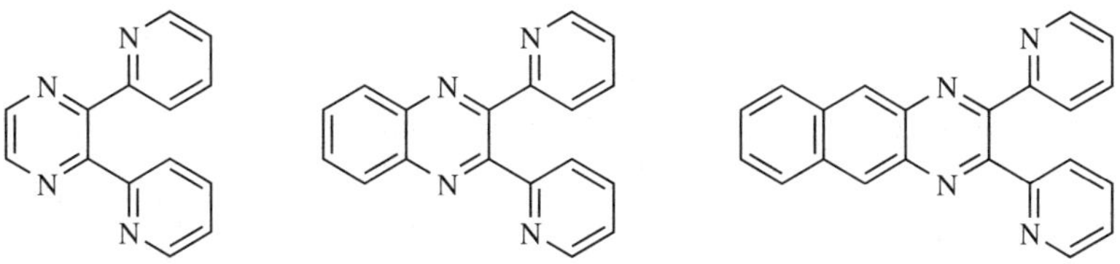

dpp   dpq   dpb

Figure 1

This procedure describes the preparation of the dpq and dpb ligand and a series of general synthetic methods for the synthesis of osmium(II) and mixed-metal osmium(II) and ruthenium(II) bipyridine complexes of these polyazine bridging ligands. The procedures for the preparation of the starting complexes [M(bpy)$_2$ Cl$_2$], M = Os^{II} [2,17] or Ru^{II} [1,16,17] have been reported in previous *Inorganic Syntheses* volumes. The dpq[11,14] and dpb[11,15] ligands are prepared by a modification of the earlier literature preparations and purified by adsorption chromatography. The metal complexes are prepared by the reaction of the appropriate [M(bpy)$_2$Cl$_2$] with either the free BL (BL = bridging ligand), to prepare the monometallic[11] and symmetric bimetallic systems[12] or BL already bound to the osmium, [Os(bpy)$_2$(BL)]$^{2+}$, to prepare the following mixed-metal bimetallic complexes:[13]

$$[Os(bpy)_2Cl_2] + BL \rightarrow [Os(bpy)_2(BL)]^{2+} + 2Cl^-$$
$$[Os(bpy)_2(BL)]^{2+} + [Ru(bpy)_2Cl_2] \rightarrow [(bpy)_2Os(BL)Ru(bpy)_2]^{4+} + 2Cl^-$$
$$2[Os(bpy)_2Cl_2] + BL \rightarrow [(bpy)_2Os(BL)Os(bpy)_2]^{4+} + 4Cl^-$$

General Chromatographic Procedures

All the ligands and complexes reported herein are purified by column chromatography using Fisher Scientific adsorption alumina. Most other types of alumina and sources of alumina possess more active surfaces, and this often results in either the need for more polar solvents to elute the system of interest or an irreversible adsorption of the compounds to the chromatographic support. Columns are typically 2–3 cm in diameter and 40–50 cm in length. The columns are prepared by first loading with the solvent of interest and then adding the dry adsorption alumina with the solvent flowing through the column. This enhances the packing of the column and eliminates air bubbles within the alumina base.

Sufficient adsorption alumina is added to fill the column approximately two-thirds full, allowing for dead volume on the top to load solutions of compounds and for solvent addition as the compound passes through the column. The solvent used varies with each system and is specified in the synthetic procedure. The total volume of solvent needed depends on many factors, including water content of solvents and humidity of the laboratory and vary from $\sim 100\text{--}500$ mL for the purifications described below. The compound is dissolved in a minimal amount of the solvent used for the column preparation, filtered, and loaded on the column. The solvent is then added for the chromatographic step and the products collected as they elute from the column. Most of the systems studied are colored and visual inspection is used to separate product bands. For the purification of the colorless dpq ligand a handheld UV lamp makes it possible to see the blue emission of the dpq ligand. Alternatively, the ligand can be detected by electronic absorption spectroscopy as dpq absorbs highly in the UV region. In all chromatographic separations the product band is collected and the solvent is removed by rotary evaporation. The product is then redissolved in a minimal amount of methylene chloride for dpq and dpb and CH_3CN for the metal complexes and precipitated by the addition to ~ 250 mL of diethyl ether while stirring. The product is removed by vacuum filtration and dried under vacuum. Specific solvent conditions and visual observations of product colors are described below.

■ **Caution.** *2,2'-pyridil is toxic and an irritant. 1,2-diaminobenzene is toxic and an irritant. 2,3-diaminonapthalene is highly toxic and a suspected carcinogen. 2,3-bis(2-pyridyl)pyrazine is toxic and an irritant. 2,3-bis(2-pyridyl) quinoxaline is toxic. 2,3-bis(2-pyridyl)benzoquinoxaline is toxic. Potassium hexafluorophosphate is toxic and corrosive. Avoid contact with skin or inhalation for all of these chemicals. Ethanol is flammable. This procedure should be performed in a fume hood. Ethylene glycol is toxic and flammable. Toluene is toxic and flammable. Acetonitrile is toxic and flammable. Diethyl ether is toxic and highly flammable. Avoid contact with skin, inhalation, and breathing vapors for all chemicals. All procedures should be performed in a fume hood. Refluxing ethylene glycol is very hot, use a heat-resistant glove to remove the reaction mixture from the heat.*

A. 2,3-BIS(2-PYRIDYL)QUINOXALINE (dpq)

Procedure

The total time for the preparation and purification of this product is about 10 h. The ligand dpq is prepared by the method of Goodwin and Lions as described below.[14]

Solid 2,2'-pyridyl (available from Aldrich Chemicals) (1.09 g, 5.12 mmol) is combined with 0.56 g (5.2 mmol) of 1,2-diaminobenzene in a 100-mL round-bottomed flask containing a magnetic stir bar. To this is added 30 mL of absolute ethanol. The flask is equipped with a reflux condenser and the solution is stirred and heated at reflux for 5 h. The flask is removed from the heating mantle and the solution is allowed to cool to room temperature, at which time solid colorless crystals of the dpq product form. The product is removed by vacuum filtration. Purification can be achieved by recrystallization from hot ethanol or by chromatography on adsorption alumina using methylene chloride as the eluent as described above. The chromatographic purification of the crude dpq ligand yields the desired colorless band containing the purified product as the first band that elutes prior to the unreacted starting materials. The chromatographic procedure can be monitored by UV spectroscopy or with the use of a UV lamp as the dpq ligand displays a blue emission. Purification by recrystallization is also possible. The crude product is dissolved in a minimum amount of boiling EtOH. The EtOH is allowed to cool to RT and the product removed by vacuum filtration. Two recrystallizations are typically required to produce the pure, colorless dpq ligand. The purified product is washed with 50 mL of diethylether and dried under vacuum for 4 h. Yield: 0.96 g, 66%.

Anal. Calc. for $C_{18}H_{12}N_4$: C, 76.03; H, 4.26; N, 19.71. Found: C, 75.11; H, 4.06; N, 19.52. ^{13}C NMR d(CDCl$_3$): 157.4, 152.5, 148.6, 141.1, 136.6, 130.5, 129.4, 124.2, 123.0. $E_{1/2}^{red}$ (0.1 M Bu$_4$NPF$_6$ in CH$_3$CN) $= -1.43$ V versus Ag/AgCl.

B. 2,3-BIS(2-PYRIDYL)BENZOQUINOXALINE (dpb)

Procedure

The total time for the preparation and purification of this product is about 10 h. The ligand dpb was prepared by a modification of the method of Buu-Hoi and Saint-Ruf.[15] (*Note:* This procedure should be performed in a fume hood.)

Solid 2,2'-pyridil (1.08 g 5.10 mmol) is combined with 0.82 g (5.2 mmol) of 2,3-diaminonaphthalene in a 100-mL round-bottomed flask containing a magnetic stir bar. To this is added 30 mL of absolute ethanol. The flask is equipped with a reflux condenser. This solution is then stirred and heated at reflux for 5 h. The round bottom is removed from the heat and the solution is allowed to cool to room temperature, at which time solid yellow crystals of the dpb product form. The product is removed by vacuum filtration. Purification is achieved by chromatography on adsorption alumina using methylene chloride as the eluent as described above. The desired yellow band containing the dpb product elutes prior to the unreacted starting materials. The purified product is then washed with 50 mL of diethylether and dried under vacuum for 4 h. Yield: 1.14 g, 67%.

Anal. Calc. for $C_{22}H_{14}N_4$: C, 79.02; H, 4.22; N, 16.76. Found: C, 77.74; H, 4.20; N, 16.08. ^{13}C NMR d(CDCl$_3$) 157.5, 153.1, 148.4, 137.7, 136.7, 134.3, 128.6, 127.9, 127.0, 124.0, 123.0. $E_{1/2}^{red}$ (0.1 M Bu$_4$NPF$_6$ in CH$_3$CN) $= -1.14$ V versus Ag/AgCl.

C. *cis*-BIS(2,2'-BIPYRIDINE)2,3-BIS(2-PYRIDYL)PYRAZINEOSMIUM (II)HEXAFLUOROPHOSPHATE HYDRATE $\{[\text{Os(bpy)}_2(\text{dpp})](\text{PF}_6)_2 \cdot \text{H}_2\text{O}\}$

$$[\text{Os(bpy)}_2\text{Cl}_2] + \text{dpp} \rightarrow [\text{Os(bpy)}_2(\text{dpp})]^{2+} + 2\text{Cl}^-$$

Procedure

The total time required for the preparation and purification of this product is about 3 h. The procedure is a modification of a published synthesis by Kalyanasundaram.[9]

The ligand dpp (0.23 g, 1.0 mmol) and [Os(bpy)$_2$Cl$_2$] (0.28 g, 0.48 mmol) are added to a 100-mL round-bottomed flask. To this is added 10 mL of ethylene glycol. The round-bottomed flask is equipped with a reflux condenser and a magnetic stir bar and stirrer and a heating mantle. While the reaction mixture is being stirred, it is heated to reflux for 60 min. The heat is removed and the reaction mixture is allowed to cool to room temperature. A saturated, aqueous solution of KPF$_6$ is prepared by adding an excess of KPF$_6$ to 30 mL of deionized water and stirring to ensure saturation. The solution is allowed to sit for 5 min and then 20 mL of the supernatant is decanted into a 100-mL beaker. The reaction mixture is added dropwise to this saturated, aqueous KPF$_6$ to induce precipitation of the brown crude product. The solid is removed by vacuum filtration and washed with two 30 mL portions of diethyl ether. The crude product is purified by chromatography on adsorption-type alumina using a 2 : 1 v/v toluene/acetonitrile eluent as described above. The first visible band to elute is the desired brown product. It is important to discard any solvent that elutes prior to this product as it contains some unreacted dpp ligand that is not visible to the eye. A small amount of the purple bimetallic species can be seen to follow the desired brown product on the alumina column. The product obtained from this first column is rechromatographed on a new adsorption alumina column using the same procedure and 2 : 1 toluene/acetonitrile eluent. Again the brown band is collected. The purified product is washed with two 30-mL portions of diethylether and dried under vacuum overnight. Yield 0.42 g, 85%.

Anal. Calc. for $C_{34}H_{28}N_8OP_2F_{12}Os$: C, 39.08; H, 2.70; N, 10.73. Found: C, 39.21; H, 2.60; N, 10.75.

D. *cis*-BIS(2,2'-BIPYRIDINE)2,3-BIS(2-PYRIDYL) QUINOXALINEOSMIUM(II) HEXAFLUOROPHOSPHATE {[Os(bpy)$_2$(dpq)](PF$_6$)$_2$}

$$[Os(bpy)_2Cl_2] + dpq \rightarrow [Os(bpy)_2(dpq)]^{2+} + 2Cl^-$$

Procedure

The total time required for the preparation and purification of this product is about 3 h. This complex is prepared as above for [Os(bpy)$_2$(dpp)](PF$_6$)$_2 \cdot$H$_2$O substituting dpq (0.29 g, 1.0 mmol) for the dpp used above.

The crude product is purified by chromatography on adsorption alumina using a 2 : 1 toluene/acetonitrile solvent mixture as described above. The first visible band to elute from the column is the desired red-purple product band. It is important to discard any solvent that elutes prior to this product as it contains some unreacted dpq ligand that is not visible to the eye. A small amount of the dark purple bimetallic species can be seen to follow the desired red-purple product on the alumina column. This red-purple solid obtained from this chromatographic separation is rechromatographed on a new adsorption alumina column using the same procedure and 2 : 1 toluene/acetonitrile solvent mixture. Again the red-purple band is collected. The purified product is washed with two 30-mL portions of diethylether and dried under vacuum overnight. Yield: 0.42 g, 80%.

Anal. Calc. for $C_{38}H_{28}N_8P_2F_{12}Os$: C, 42.38; H, 2.62; N, 10.41. Found: C, 42.99; H, 2.86; N, 10.65.

E. *cis*-BIS(2,2′-BIPYRIDINE)2,3-BIS(2-PYRIDYL) BENZOQUINOXALINEOSMIUM(II) HEXAFLUOROPHOSPHATE {[Os(bpy)₂(dpb)](PF₆)₂}

$$[Os(bpy)_2Cl_2] + dpb \rightarrow [Os(bpy)_2(dpb)]^{2+} + 2Cl^-$$

Procedure

The total time required for the preparation and purification of this product is about 3 h. This complex is prepared as described above for [Os(bpy)₂(dpp)] (PF₆)₂·H₂O by substituting dpb (0.34 g, 1.0 mmol) for dpp. Purification is achieved again by chromatography on adsorption alumina as described above using a 2 : 1 toluene/acetonitrile eluent. The first visible band to elute from the column is typically unreacted dpb ligand that is yellow in color and elutes with the solvent front. This should be discarded. The desired product comes next and is purple. A small amount of the green bimetallic species can be seen to follow the desired purple product on the alumina column. The desired purple product is collected. This product is rechromatographed using the same procedure and 2 : 1 toluene/acetonitrile solvent mixture. Again the purple band is collected. The purified product is washed with two 30-mL portions of diethylether and dried under vacuum overnight. Yield: 0.41 g, 75%.

Anal. Calc. for $C_{42}H_{30}N_8P_2F_{12}Os$: C, 44.76; H, 2.68; N, 9.95. Found: C, 44.74; H, 2.88; N, 9.87.

F. *cis*-TETRAKIS(2,2′-BIPYRIDINE)-μ-[2,3-BIS(2-PYRIDYL) PYRAZINE]DIOSMIUM(II) HEXAFLUOROPHOSPHATE DIHYDRATE {[(bpy)₂Os(dpp)Os(bpy)₂](PF₆)₄·2H₂O}

$$2[Os(bpy)_2Cl_2] + dpp \rightarrow [(bpy)_2Os(dpp)Os(bpy)_2]^{4+} + 4Cl^-$$

Procedure

The total time required for the preparation and purification of this product is about 3 h, and the product is prepared by a modification of the published procedure of Campagna.[10] The ligand dpp (0.056 g, 0.24 mmol) and [Os(bpy)₂Cl₂] (0.34 g, 0.60 mmol) are added to a 100-mL round-bottomed flask. To this is added 30 mL of ethylene glycol. The round-bottomed flask is equipped with a reflux condenser and a magnetic stir bar and stirrer. While the reaction mixture is being stirred it is heated at reflux for 45 min.

The heat is removed and reaction allowed to cool to room temperature. A saturated, aqueous solution of KPF_6 is prepared by adding an excess of KPF_6 to 90 mL of deionized water and stirring to ensure saturation. The solution is allowed to sit for 5 min and then 75 mL of the supernatant is decanted into a 150-mL beaker. The reaction mixture is added dropwise to this saturated, aqueous KPF_6 to induce precipitation of the purple-brown crude product. The solid is removed by vacuum filtration and washed with two 30-mL portions of diethyl ether. The crude product is purified by chromatography on adsorption alumina as described above using a 2 : 1 toluene/acetonitrile eluent. The first visible band to elute is a small amount of the brown monometallic. Occasionally a small amount of unreacted $[Os(bpy)_2Cl_2]$ will elute prior to the brown monometallic. After the monometallic band is eluted, the solvent is changed to 1 : 2 toluene/acetonitrile to elute the desired purple product. The purple product band is collected. This product is then rechromatographed on a new alumina column using the same procedure and solvent mixtures. Again the purple product band is collected. The purified product is washed with two 30-mL portions of diethylether and dried under vacuum overnight. Yield: 0.25 g, 56%.

Anal. Calc. for $C_{54}H_{46}N_{12}O_2P_4F_{24}Os_2$: C, 34.96; H, 2.50; N, 9.06. Found: C, 35.05; H, 2.58; N, 9.00.

G. *cis*-TETRAKIS(2,2′-BIPYRIDINE)-μ-[2,3-BIS(2-PYRIDYL) QUINOXALINE]DIOSMIUM(II) HEXAFLUOROPHOSPHATE DIHYDRATE {[(bpy)₂Os(dpq)Os(bpy)₂](PF₆)₄·2H₂O}

$$2[Os(bpy)_2Cl_2] + dpq \rightarrow [(bpy)_2Os(dpq)Os(bpy)_2]^{4+} + 4Cl^-$$

Procedure

The total time required for the preparation and purification of this product is about 3 h. This product is prepared as above for $[(bpy)_2Os(dpp)Os(bpy)_2]$ $(PF_6)_4 \cdot 2H_2O$ substituting dpq (0.068 g, 0.24 mmol) for the dpp above.

This crude product is purified by chromatography using adsorption alumina as described above initially using a 2 : 1 toluene/acetonitrile solvent mixture and changing solvents as described below. The first visible band to elute from this column is a small amount of the red-purple monometallic complex. After this band is eluted the solvent is changed to 1 : 2 toluene/acetonitrile to elute the desired purple product. This band elutes second. This purple product is rechromatographed on a new adsorption alumina column using the same 2 : 1 toluene/acetonitrile solvent mixture followed by the change to 1 : 2 toluene/acetonitrile after elution of the red-purple monometallic. The purified product is washed with two 30-mL portions of diethylether and dried under vacuum overnight. Yield: 0.46 g, 58%.

Anal. Calc. for $C_{58}H_{48}N_{12}O_2P_4F_{24}Os_2$: C, 36.56; H, 2.54; N, 8.82. Found: C, 36.61; H, 2.65; N, 8.98.

H. *cis*-TETRAKIS(2,2′-BIPYRIDINE)-μ-[2,3-BIS(2-PYRIDYL)BENZO-QUINOXALINE]DIOSMIUM(II) HEXAFLUOROPHOSPHATE TRIHYDRATE{[(bpy)$_2$Os(dpb)Os(bpy)$_2$] (PF$_6$)$_4$·3H$_2$O}

$$2[Os(bpy)_2Cl_2] + dpb \rightarrow [(bpy)_2Os(dpb)Os(bpy)_2]^{4+} + 4Cl^-$$

Procedure

The total time required for the preparation and purification of this product is about 3 h. This product is prepared as above for [(bpy)$_2$Os(dpp)Os(bpy)$_2$] (PF$_6$)$_4$·2H$_2$O substituting dpb (0.081 g, 0.24 mmol) for the dpp described above.

The product is purified by chromatography on adsorption alumina using the procedure described above and a 2 : 1 toluene/acetonitrile solvent mixture. The first visible band to elute from this column is unreacted dpb that is yellow in color and elutes with the solvent front. Next a small amount of the purple monometallic elutes. After this band is eluted, the solvent is changed to 1 : 2 toluene/acetonitrile to elute the desired green product. The green product is rechromatographed on a new alumina column using the same procedure and solvent mixtures. The purified product is washed with two 30-mL portions of diethylether and dried under vacuum overnight. Yield: 0.20 g, 43%.

Anal. Calc. for $C_{62}H_{52}N_{12}O_3P_4F_{24}Os_2$: C, 37.73; H, 2.65; N, 8.52. Found: C, 37.60; H, 2.56; N, 8.32.

I. *cis*-TETRAKIS(2,2′-BIPYRIDINE)-μ-[2,3-BIS(2-PYRIDYL) PYRAZINE]RUTHENIUM(II)-OSMIUM(II) HEXAFLUORO-PHOSPHATE DIHYDRATE {[(bpy)$_2$Os(dpp)Ru(bpy)$_2$](PF$_6$)$_4$·2H$_2$O}

$$[Os(bpy)_2(dpp)]^{2+} + [Ru(bpy)_2Cl_2] \rightarrow [(bpy)_2Os(dpp)Ru(bpy)_2]^{4+} + 2Cl^-$$

Procedure

The total time required for the preparation and purification of this product is about 3 days, and the product is prepared by a modification of the published procedure of Kalyanasundaram.[9]

The osmium complex $[Os(bpy)_2(dpp)](PF_6)_2 \cdot H_2O$ (0.11 g, 0.10 mmol) and $[Ru(bpy)_2Cl_2]$ (0.24 g, 0.50 mmol) are added to a 500-mL round-bottomed flask. To this is added 300 mL of 2 : 1 v/v ethanol/water. The round-bottom flask is equipped with a reflux condenser and a magnetic stir bar and stirrer. While the reaction mixture is being stirred it is heated at reflux for 48 h. The heat is removed and reaction allowed to cool to room temperature. The volume is reduced to 75 mL by rotary evaporation. A saturated, aqueous solution of KPF_6 is prepared by adding an excess of KPF_6 to 80 mL of deionized water and stirring to ensure saturation. The solution is allowed to sit for 5 min, and then 60 mL of the supernatant is decanted into a 250-mL beaker. The reaction mixture is added dropwise to this saturated, aqueous KPF_6 to induce precipitation of the purple-brown crude product. The solid is removed by vacuum filtration and washed with two 30-mL portions of diethyl ether. The crude product is purified by chromatography on adsorption alumina as described above using a 1 : 1 v/v toluene/acetonitrile eluant. The first visible band to elute is a small amount of $[Ru(bpy)_2Cl_2]$. Next a small amount of the brown osmium monometallic is eluted. After this band is eluted, the solvent is changed to 1 : 3 toluene/acetonitrile to elute the desired purple product. The purple product band is collected. This product is then rechromatographed on a new alumina column using the same procedure and solvent mixtures. The purified product is washed with two 30-mL portions of diethylether and dried under vacuum overnight. Yield: 0.13 g, 70%.

Anal. Calc. for $C_{54}H_{46}N_{12}O_2P_4F_{24}OsRu$: C, 36.72; H, 2.62; N, 9.52. Found: C, 36.66; H, 2.73; N, 9.52.

J. *cis*-TETRAKIS(2,2'-BIPYRIDINE)-μ-[2,3-BIS(2-PYRIDYL)QUINOXA-LINE]RUTHENIUM(II)-OSMIUM(II) HEXAFLUOROPHOSPHATE TRIHYDRATE {$[(bpy)_2Os(dpq)Ru(bpy)_2](PF_6)_4 \cdot 3H_2O$}

$$[Os(bpy)_2(dpq)]^{2+} + [Ru(bpy)_2Cl_2] \rightarrow [(bpy)_2Os(dpq)Ru(bpy)_2]^{4+} + 2Cl^-$$

Procedure

The total time required for the preparation and purification of this product is about 3 days. This product is prepared as described above for $[(bpy)_2Os(dpp)$

Ru(bpy)$_2$](PF$_6$)$_4$·2H$_2$O substituting [Os(bpy)$_2$(dpq)](PF$_6$)$_2$ (0.12 g, 0.11 mmol) for the [Os(bpy)$_2$(dpp)](PF$_6$)$_2$·H$_2$O.

The product is purified by chromatography on adsorption alumina using the procedure described above and a 2 : 1 toluene/acetonitrile solvent mixture initially. The first visible band to elute from this column is a small amount of [Ru(bpy)$_2$Cl$_2$]. Next a small amount of the red-purple osmium monometallic is eluted. After this band is eluted, the solvent is changed to 1 : 3 toluene/acetonitrile to elute the desired blue-green product. The blue-green product is rechromatographed on a new alumina column using the same procedure and solvent mixtures. The purified product is washed with two 30-mL portions of diethylether and dried under vacuum overnight. Yield: 0.18 g, 85%.

Anal. Calc. for C$_{58}$H$_{50}$N$_{12}$O$_3$P$_4$F$_{24}$OsRu: C, 37.98; H, 2.75; N, 9.17. Found: C, 37.82; H, 2.94; N, 9.46.

K. *cis*-TETRAKIS(2,2′-BIPYRIDINE)-μ-[2,3-BIS(2-PYRIDYL) BENZOQUINOXALINE]RUTHENIUM(II)-OSMIUM(II) HEXAFLUOROPHOSPHATE DIHYDRATE {[(bpy)$_2$Os(dpb)Ru(bpy)$_2$](PF$_6$)$_4$·2H$_2$O}

$$[Os(bpy)_2(dpb)]^{2+} + [Ru(bpy)_2Cl_2] \rightarrow [(bpy)_2Os(dpb)Ru(bpy)_2]^{4+} + 2Cl^-$$

Procedure

The total time required for the preparation and purification of this product is about 3 days. This product is prepared as described above for [(bpy)$_2$Os(dpp)Ru (bpy)$_2$](PF$_6$)$_4$·2H$_2$O substituting [Os(bpy)$_2$(dpb)](PF$_6$)$_2$ (0.11 g, 0.10 mmol) for the [Os(bpy)$_2$(dpp)](PF$_6$)$_2$·H$_2$O.

The product is purified by chromatography on adsorption alumina using the procedure described above and an initial 2 : 1 toluene/acetonitrile solvent mixture. The first visible band to elute from this column is a small amount of [Ru(bpy)$_2$Cl$_2$]. Next a small amount of the purple osmium monometallic is eluted. After this band is eluted the solvent is changed to 1 : 3 toluene/acetonitrile to elute the desired green product. The green product is rechromatographed on a new alumina column using the same procedures and solvent mixtures. The purified product is washed with two 30-mL portions of diethylether and dried under vacuum overnight. Yield: 0.16 g, 87%.

Anal. Calc. for C$_{62}$H$_{50}$N$_{12}$O$_2$P$_4$F$_{24}$OsRu: C, 39.90; H, 2.70; N, 9.01. Found: C, 40.25; H, 2.95; N, 9.18.

TABLE I. Electronic Spectral Properties^a **of [Os(bpy)$_2$(BL)]$^{2+}$,[11]**
[(bpy)$_2$Os(BL)Os(bpy)$_2$]$^{4+}$,[12] and [(bpy)$_2$Os(BL)Ru(bpy)$_2$]$^{4+}$[13]
(BL = dpp, dpq or dpb)^b

Complex	λ_{max}^{abs}, nm (ε, M^{-1} cm^{-1})
[Os(bpy)$_2$(dpp)]$^{2+}$	290 (67,300), 432 (11,000), 486 (12,700)
[Os(bpy)$_2$(dpq)]$^{2+}$	286 (73,200), 346 (25,100), 428 (11,000), 536 (13,400)
[Os(bpy)$_2$(dpb)]$^{2+}$	290 (70,000), 366 (24,400), 408 (17,300), 446 (12,800), 570 (13,800)
[(bpy)$_2$Os(dpp)Os(bpy)$_2$]$^{4+}$	286 (96,700), 356 (26,400), 432 (19,800), 552 (25,200)
[(bpy)$_2$Os(dpq)Os(bpy)$_2$]$^{4+}$	286 (90,500), 356 (21,700), 398 (19,600), 428 (14,300), 630 (20,900)
[(bpy)$_2$Os(dpb)Os(bpy)$_2$]$^{4+}$	286 (90,700), 354 (35,300), 422(16,600), 510 (8,160), 610 (12,900), 670 (16,800)
[(bpy)$_2$Os(dpp)Ru(bpy)$_2$]$^{4+}$	286 (102,700), 348 (31,300), 430 (22,300), 542 (28,500)
[(bpy)$_2$Os(dpq)Ru(bpy)$_2$]$^{4+}$	286 (118,600), 398 (28,400), 424 (19,100), 616 (28,500)
[(bpy)$_2$Os(dpb)Ru(bpy)$_2$]$^{4+}$	286 (100,300), 344 (40,700), 408 (20,600), 424 (20,500), 658 (18,500)

^{a} CH$_3$CN solution at RT.
^{b} *Abbreviations*: dpp = 2,3-bis(2-pyridyl)pyrazine, dpq = 2,3-bis(2-pyridyl)quinoxaline; dpb = 2,3-bis (2-pyridyl)benzoquinoxaline.

Properties

Spectral and electrochemical properties are listed in Tables I and II. The complexes are all quite stable as solids and as acetonitrile solutions. They all display intense electronic transitions in the ultraviolet that represent bpy and BL-based $\pi \rightarrow \pi^*$ transitions. In the visible spectrum they display metal-to-ligand charge transfer (MLCT) transitions associated with each metal center and its coordinated ligands. The lowest-lying electronic transition in all the complexes reported is a Os $\rightarrow$ BL CT transition.

All the complexes display reversible metal-based oxidations M$^{II/III}$ with the osmium oxidizing prior to the ruthenium in the mixed-metal bimetallic complexes. They also possess reversible ligand-based reductions with the monometallic complexes displaying BL$^{0/-}$, bpy$^{0/-}$, and bpy$^{0/-}$ couples and the bimetallics displaying BL$^{0/-}$, BL$^{-/2-}$, and four bpy$^{0/-}$ couples.

TABLE II. Electrochemical Data^a **for [Os(bpy)$_2$(BL)]$^{2+}$,^{11} [(bpy)$_2$Os (BL)Os(bpy)$_2$]$^{4+}$,^{12} and [(bpy)$_2$Os(BL)Ru(bpy)$_2$]$^{4+}$ ^{13} (BL = dpp, dpq or dpb)^b**

Complex	$E_{1/2}$ (V vs. Ag/AgCl)
[Os(bpy)$_2$(dpp)]$^{2+}$	0.94, -1.02, -1.38, -1.58
[Os(bpy)$_2$(dpq)]$^{2+}$	0.99, -0.76, -1.26, -1.57
[Os(bpy)$_2$(dpb)]$^{2+}$	1.00, -0.61, -1.16, -1.55
[(bpy)$_2$Os(dpp)Os(bpy)$_2$]$^{4+}$	1.22, 0.91, -0.61, -1.00, -1.28, -1.38, -1.58, -1.76
[(bpy)$_2$Os(dpq)Os(bpy)$_2$]$^{4+}$	1.28, 0.98, -0.34, -0.90, -1.26, -1.36, -1.54, -1.72
[(bpy)$_2$Os(dpb)Os(bpy)$_2$]$^{4+}$	1.30, 0.98, -0.25, -0.76, -1.31, -1.39, -1.58, -1.74
[(bpy)$_2$Os(dpp)Ru(bpy)$_2$]$^{4+}$	1.56, 1.01, -0.62, -1.03, -1.34, -1.41, -1.61, -1.74
[(bpy)$_2$Os(dpq)Ru(bpy)$_2$]$^{4+}$	1.61, 1.09, -0.33, -0.94, -1.32, -1.42, -1.56, -1.73
[(bpy)$_2$Os(dpb)Ru(bpy)$_2$]$^{4+}$	1.62, 1.09, -0.21, -0.76, -1.34, -1.42, -1.59, -1.75

^{a} Recorded in CH$_3$CN for monometallics and oxidative couples and DMF for bimetallic reductive couples with 0.1 MBu$_4$NPF$_6$ supporting electrolyte and potentials are reported versus Ag/AgCl (0.27 V vs. NHE).
^{b} Abbreviations: dpp = 2,3-bis(2-pyridyl)pyrazine, dpq = 2,3-bis(2-pyridyl)quinoxaline; dpb = 2,3-bis (2-pyridyl)benzoquinoxaline.

Acknowledgments

The authors acknowledge the generous support of this work by the National Science Foundation (CHE-9632713). We thank Johnson Matthey, an Alfa Aesar Company, for the loan of the ruthenium trichloride used in this study. We also thank the checkers for their timely assistance in testing this series of preparations and to the editor for his patience and useful comments.

References

1. F. P. Dwyer, H. A. Goodwin, and E. C. Gyarfas, *Aust. J. Chem.* **16**, 544 (1963).
2. D. A. Buckingham, F. P. Dwyer, and A. M. Sargeson, *Aust. J. Chem.* **17**, 325 (1964).
3. F. P. Dwyer, N. K. King, and M. E. Winfield, *Aust. J. Chem.* **12**, 139 (1959).
4. A. Juris, V. Balzani, F. Barigelletti, S. Campagna, P. Belser, and A. von Zelewsky, *Coord. Chem. Rev.* **84**, 85 (1988).
5. K. Kalyanasundaram, *Coord. Chem. Rev.* **46**, 159 (1982).
6. B. Durham, J. V. Caspar, J. K. Nagle, and T. J. Meyer, *J. Am. Chem. Soc.* **104**, 4803 (1982).
7. S. R. Johnson, T. D. Westmoreland, J. V. Caspar, K. R. Barqawi, and T. J. Meyer, *Inorg. Chem.* **27**, 3195 (1988).

8. For a review, see V. Balzani, A. Juris, M. Venturi, S. Campagna, S. Serroni, *Chem. Rev.* **96**, 759 (1996).
9. K. Kalyanasundaram and M. K. Nazeeruddin, *Chem. Phys. Lett.* **158**, 45 (1989).
10. G. Denti, S. Serroni, L. Sabatino, M. Ciano, V. Ricevuto, and S. Campagna, *Gazz. Chim. Ital.* **121**, 37 (1991).
11. M. M. Richter and K. J. Brewer, *Inorg. Chim. Acta.* **180**, 125 (1991).
12. M. M. Richter and K. J. Brewer, *Inorg. Chem.* **32**, 2827 (1993).
13. M. M. Richter and K. J. Brewer, *Inorg. Chem.* **31**, 1594 (1992).
14. H. A. Goodwin and F. Lions, *J. Am. Chem. Soc.* **81**, 6415 (1959).
15. N. P. Buu-Hoi and G. Saint-Ruf, *J. Chem. Soc.* 2257 (1961).
16. B. P. Sullivan, D. J. Salmon, and T. J. Meyer, *Inorg. Chem.* **17**, 3334 (1978).
17. P. A. Lay, A. M. Sargeson, and H. Taube, *Inorg. Synth.* **24**, 291 (1986).

5. BINUCLEAR OXOMOLYBDENUM–METALLOPORPHYRIN COMPLEXES

Submitted by PARTHA BASU,[*] MICHAEL VALEK,[†] and JOHN H. ENEMARK[†]
Checked by H. TETSUO UYEDA and M. J. THERIEN[‡]

In this section, syntheses of {5-(m,n-catecholato[hydrotris(3,5-dimethyl-1-pyrazolyl)borato]oxomolybdenum(V)}-10,15,20-tri-p-tolylporphinatometal [ML(m,n-Mo-TTP)], [ML = Zn(II), Fe(III)Cl, Cu; m,n = 2,3 or 3,4] are described.

Supramolecular assemblies of metal complexes are of considerable interest because of their versatile chemistry that can be adapted to produce materials with applications in several areas such as separation, electron transfer, catalysis, magnetic devices, and optical devices.[1,2] Transition metal complexes with the ability to shuttle between multiple redox states are playing a pivotal role in developing such materials. Moreover, some of these complexes are important in understanding the complex clusters in biological macromolecules.

The multinuclear assemblies described here were developed as models for understanding the interprosthetic group interaction between the molybdenum and heme centers in sulfite oxidase.

The ligand system of Fig. 1 is designed to coordinate two metal centers at constrained distances and is well suited for studying weak magnetic interactions

[*] Current address: Department of Chemistry and Biochemistry, Duquesne University, Pittsburgh, PA 15282.
[†] Department of Chemistry, University of Arizona, Tucson, AZ 85721-0041.
[‡] Department of Chemistry, University of Pennsylvania, Philadelphia, PA 19104-6323.

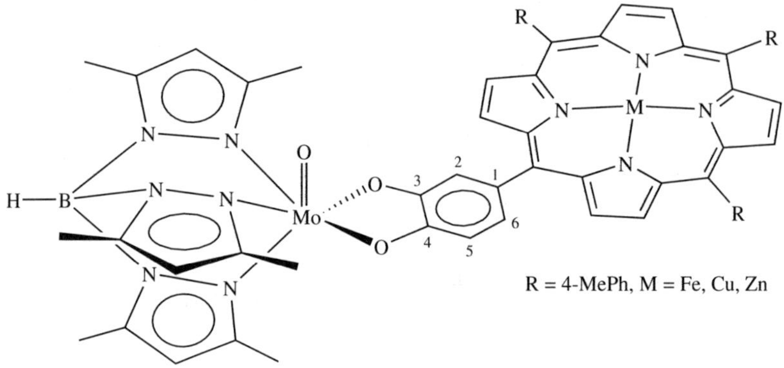

R = 4-MePh, M = Fe, Cu, Zn

Figure 1

between Mo(V) and a paramagnetic metalloporphyrin center.[3–6] The corresponding Zn(II) derivatives exhibit interesting photoinduced electron transfer processes.[7] The synthetic approach to preparation of these complexes is outlined
in Scheme 1.

The free-base porphyrins 3,4-OH-TTP and 2,3-OH-TTP are synthesized by
demethylating the corresponding dimethoxy porphyrins.[8] The dimethoxy porphyrins are prepared by direct condensation of pyrrole and an appropriate mixture of aldehydes following Adler's method.[9]

A. 5-(3,4-DIMETHOXYPHENYL)-10,15,20-TRI-*p*-TOLYLPORPHYRIN (3,4-OME-TTP)

3,4-$(OMe)_2$-benzaldehyde + 3*p*-tolualdehyde + 4pyrrole $\rightarrow$

$$3,4\text{-OMe-TTP} + \text{TTP} + \text{other products} \quad (1)$$

Procedure

3,4-Dimethoxybenzaldehyde (8.3 g, 50 mmol) and *p*-tolualdehyde (18 g, 150 mmol)
are dissolved in propionic acid (500 mL) and brought to reflux. Freshly distilled
pyrrole (13.4 g, 200 mmol) is added to the solution and reflux continued for
45 min. The reaction mixture is cooled overnight at 11°C and filtered, and the
purple-black precipitate washed with cold absolute ethanol. The crude product
is purified in 500-mg batches. Tar and other insoluble impurities are removed
by addition of 50 mL of dichloromethane and rapid elution through a 4×5-cm
silica column with dichloromethane until the eluted product is no longer purple.
Evaporation of the solvent gives a purple solid that is purified by chromatography
on a 4×15-cm column using 230–400-mesh 60 Å silica gel. Up to 500 mg of
crude product can be loaded on the column, and the major TTP component

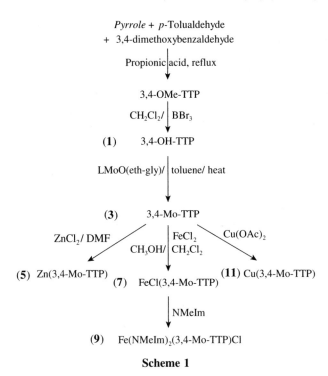

Scheme 1

rapidly elutes with ∼1 L of dichloromethane-hexane (2 : 3) containing 0.1%
triethylamine. After the TTP has been flushed from the column the desired
3,4-OMe-TTP can be eluted with 0.5–1 L of dichloromethane/hexane (5 : 1) containing 0.1% triethylamine. The purple 3,4-OMe-TPP band is collected and the
solvent removed under vacuum. The product can be recrystallized by dissolving
it in a minimum amount of dichloromethane (1–2 mL) in a test tube and slowly
layering ∼10 mL of heptane over the surface with a pipette. Allowing the solvents to diffuse together slowly in a refrigerator over about 3 days yields shiny
purple crystals. Yield: 0.70 g (2%).

Anal. Calcd. for 3,4-OMe-TPP: C, 82.07; H, 5.62; N, 7.85. Found: C, 80.90; H,
5.75; N, 7.72. Mass spectrum: calcd. molecular ion: 716.32 amu; found (FAB,
nitrobenzyl alcohol): 717.00 amu. Proton NMR in CDCl$_3$ at 22°C: δ 2.69(s)
(9H); δ 3.97(s) (3H); δ 4.16(s) (3H); δ 7.23(d) = 8 Hz (1H); δ 7.54(d) = 8 Hz
(6H); δ 7.74(d) = 8 Hz (1H); δ 7.76(s) (1H); δ 8.1(d) = 8 Hz (6H); δ 8.85(m)
(8H). Electronic spectrum in toluene at 25°C (λ in nm, ε in L mol^{-1} cm^{-1}):
516 (17,400), 552 (9330), 594 (5010), 649 (4370).

B. 5-(2,3-DIMETHOXYPHENYL)-10,15,20-TRI-*p*-TOLYLPORPHYRIN (2,3-OME-TTP)

This compound is prepared from 2,3-dimethoxybenzaldehyde (8.3 g, 50 mmol) with 3 equiv of *p*-tolualdehyde (10.8 g, 90 mmol) and 4 equiv of freshly distilled pyrrole (8.1 g, 120 mmol) by the procedure used for 3,4-OMe-TTP. The crude product is purified by adsorption chromatography on silica gel with dichloromethane–hexane as the elutant, as described for 3,4-OMe-TTP. The product is recrystallized from dichloromethane–heptane. Yield: 0.55 g (2.6%). Mass spectrum: calcd. molecular ion: 716.32 amu; found (FAB, nitrobenzyl alcohol): 716.60. Proton NMR in $CDCl_3$ at 22°C: δ 2.70(s) (9H); δ 3.18(s) (3H); δ 4.13(s) (3H); δ 7.38(m) (2H); δ 7.55(d) = 8 Hz (6H); δ 7.65(m) (1H); δ 8.1(m) (6H); δ 8.3(m) (8H). Electronic spectrum in toluene at 25°C (λ in nm, ε in L mol^{-1} cm^{-1}): 516 (19,500), 550 (8510), 592 (5370), 652 (5620).

C. 5-(3,4-DIHYDROXYPHENYL)-10,15,20-TRI-*p*-TOLYLPORPHYRIN (3,4-OH-TTP)

This compound is synthesized from 3,4-OMe-TTP by demethylating with boron tribromide.[10]

$$3,4\text{-OMe-TTP} + BBr_3 \rightarrow (3,4\text{-O-TTP})BBr + 2MeBr$$
$$(3,4\text{-O-TTP})BBr + 3H_2O \rightarrow (3,4\text{-OH-TTP}) + H_3BO_3 + HBr \tag{2}$$

Procedure

First, 0.2 g (2.8×10^{-4} mol) of 3,4-OMe-TTP dissolved in a minimum volume of freshly distilled dry dichloromethane is added dropwise to a BBr_3 solution (2mL BBr_3 in 5 mL dichloromethane) at -80°C. The solution is stirred for 1 h, slowly brought to room temperature, and then stirred for an additional 24 h. Water is added slowly to the green solution to hydrolyze any excess BBr_3 present, and the solution is neutralized with triethylamine to precipitate the target compounds as a purple solid. The solid mass is filtered, washed with water, dried, and recrystallized from dichloromethane-heptane solution.[11] Yield: 0.17 g (90%).

Anal. Calcd. for 3,4-OH-TTP (**1**): C, 81.92; H, 5.27; N, 8.17. Found: C, 81.49; H, 5.56; N, 7.92. Mass spectrum: calcd. molecular ion 688.28 amu; found (FAB, nitrobenzyl alcohol): 688.20 amu. Proton NMR in CD_2Cl_2 at 22°C: δ 2.70(s) (9H); 7.12(d) = 8 Hz (1H); 7.49(s) (1H), 7.57(d) = 8 Hz (7H); 8.09(d) = 8 Hz (6H); 8.87(s) (8H). Electronic spectrum in toluene at 25°C (λ in nm, ε in L mol^{-1} cm^{-1}): 516 (17,800), 551 (9550), 594 (5250), 652 (7590).

D. 5-(2,3-DIHYDROXYPHENYL)-10,15,20-TRI-*p*-TOLYLPORPHYRIN (2,3-OH-TTP)

This compound is synthesized from 2,3-OMe-TTP by the method described above. Yield: (88%).

Anal. Calcd. for 2,3-OH-TTP (**2**): C, 81.92; H, 5.27; N, 8.17. Found: C, 80.86; H, 5.62; N, 7.87. Mass spectrum: calcd. molecular ion 688.28 amu; found (FAB, nitrobenzyl alcohol): 689.45 amu. Proton NMR in CD_2Cl_2 at 22°C: δ 2.71(s) (9H); 7.23(t) = 8 Hz (1H); 7.33(d) = 8 Hz (1H); 7.59(d) = 8 Hz (7H); 8.10(d) = 8 Hz (6H); 8.89(m) (8H). Electronic spectrum in toluene at 25°C (λ in nm; ε in L mol^{-1} cm^{-1}): 515 (18,600), 550 (8710), 592 (6030), 651 (9550).

Properties

Compounds 3,4-OMe-TTP; 2,3-OMe-TTP; 3,4-OH-TTP, and 2,3-OH-TTP are bright purple crystalline substances that are stable in air in the solid state. However, their solutions, which are purple in color, are light-sensitive. The solids almost always contain solvents of crystallization that interfere with the elemental analyses. They are highly soluble in organic solvents such as chloroform and dichloromethane. They are also slightly to moderately soluble in methanol and ethanol.

Materials and General Procedures

A convenient source of the molybdenum center is $LMoO(eth\text{-}gly)$[12] (where L = 3,5-dimethyl-1-pyrazolylborate and eth-gly = ethylene glycolate). The low acidity of ethylene glycol makes it a good leaving group from $LMoO(eth\text{-}gly)$ that can be easily replaced by stronger acids, such as the pendant catechol function of the porphyrin catecholate ligands.

E. [HYDROTRIS(3,5-DIMETHYL-1-PYRAZOLYL)BORATO]OXOMOLYBDENUM(V)DICHLORIDE (LMoOCl₂)

$$MoCl_5 + K(C_5H_7N_2)_3BH \rightarrow Cl_2MoO(C_5H_7N_2)_3BH$$

Procedure

To 16.4 g (60 mmol) of $MoCl_5$ in a 200-mL airless flask at −77°C is slowly added 120 mL of tetrahydrofuran (also at −77°C) with vigorous stirring. The reaction mixture is gradually brought to room temperature with continuous stirring. Near room temperature an exothermic reaction begins and the color changes from dark red-brown to green, and a green precipitate subsequently forms. To the resulting slurry is added 20 g (59.5 mmol) of potassium hydrotris(3,5-dimethyl-1-pyrazolyl)borate[13,14] (KL), and the mixture is heated at 50°C for 12 h with stirring. The resultant green precipitate is separated from the dark red supernatant by filtration, washed several times with acetonitrile, and dried in vacuo. The crude product is dissolved in $\sim$1 L of refluxing 1,2-dichloroethane; the solution is filtered to remove potassium chloride and evaporated to dryness in vacuo. The green product, LMoOCl$_2$, is washed several times with acetonitrile to remove a red impurity. This green material is suitable for subsequent syntheses; additional purification can be effected by recrystallization from 1,2-dichloroethane. Yield of LMoOCl$_2$ (C$_{15}$H$_{22}$N$_6$OCl$_2$BMo): 20 g (70%). IR (KBr pellet): ν (MoO) 960 cm^{-1}. Electronic spectrum in dichloroethane (λ in nm; ε in L mol^{-1} cm^{-1}): 337 (5200), 435 (1000), 705 (50).

F. [HYDROTRIS(3,5-DIMETHYL-1-PYRAZOLYL)BORATO]OXOMOLYBDENUM(V) ETHYLENEGLYCOLATE [LMoO(Eth-Gly)]

$$Cl_2MoO(C_5H_7N_2)_3BH + H_2(\text{eth-gly}) + Et_3N \rightarrow$$
$$(\text{eth-gly})MoO(C_5H_7N_2)_3BH + Et_3NHCl$$

Procedure

To a stirring slurry of 2 g (4.2 mmol) of LMoOCl$_2$ in 50 mL of toluene at 70°C is added a mixture of 0.6 mL (8.4 mmol) of triethylamine and 0.5 mL (8.4 mmol) of ethylene glycol in 5 mL of toluene. The progress of the reaction can be monitored by thin-layer chromatography (TLC). After the reaction is complete ($\sim$1.5 h) the mixture is cooled to room temperature, filtered, and evaporated to dryness in vacuo. The crude product is dissolved in toluene and the solution filtered to remove Et$_3$NHCl. Evaporation of the filtrate gives a blue powder that is dissolved in a minimum volume of dichloromethane and purified by column chromatography on a neutral alumina column (mesh 80–200) with dichloromethane as eluant. The solvent is removed under vacuum to give blue LMoO(eth-gly). Yield of LMoO(eth-gly) (C$_{17}$H$_{26}$N$_6$O$_3$BMo): 1.0 g (50%). IR: ν (MoO) 938 cm^{-1}. Electronic spectrum in 1,2-dichloroethane (λ in nm; ε in L mol^{-1} cm^{-1}): 348 (3940), 525 (200), 640 (200).

G. 5-{3,4-CATECHOLATO[HYDROTRIS(3,5-DIMETHYL-1-PYRAZOLYL)BORATO]OXOMOLYBDENUM(V)}-10,15,20-TRI-*p*-TOLYLPORPHYRIN (3,4-Mo-TTP)

$$\text{LMoO(eth-gly)} + \text{3,4-OH-TTP} \rightarrow \text{3,4-Mo-TTP} + \text{H}_2\text{(eth-gly)}$$

Attachment of the molybdenum center to the pendant catecholato group is described in this section and Section H.

Procedure

Amounts of 3,4-OH-TTP (0.5 g, 0.7 mmol) and LMo^VO(eth-gly) (0.7 g, 1.5 mmol) are placed in an evacuated Schlenk flask (250 mL). Degassed dry toluene (100 mL) is added and the mixture is stirred at 70°C. Heating the solution to ~100°C lowers the yield. Heating is continued until the reaction is complete as evidenced by TLC; the organic solvent is removed under vacuum. Dichloromethane/toluene (1 : 1) (25 mL) is added to yield a purple-brown solution. The solution is purified by chromatography on a silica gel column (3×30 cm) with dichloromethane/toluene (1 : 1) as eluant. The solvent is removed under vacuum. Recrystallization from dichloromethane–heptane gives dark purple microcrystals. Yield: 0.65 g (82%).

Anal. Calcd. for 3,4-Mo-TTP (**3**) ($C_{62}H_{56}N_{10}O_3BMo$): C, 67.95; H, 5.15; N, 12.78. Found: C, 67.88; H, 5.64; N, 12.24. Mass spectrum: calculated molecular ion: 1097.37 amu; found (FAB, nitrobenzyl alcohol): 1098.41 amu. IR (KBr): ν_{B-H} 2545(w); $\nu_{Mo=O}$ 941(s) cm^{-1}. Electronic spectrum in toluene at 25°C (λ in nm; ε in L mol^{-1} cm^{-1}): 518 (19,100), 554 (12,600), 595 (5890), 653 (7590).

H. 5-{2,3-CATECHOLATO[HYDROTRIS(3,5-DIMETHYL-1-PYRAZOLYL)BORATO]OXOMOLYBDENUM(V)}-10,15,20-TRI-*p*-TOLYLPORPHYRIN (2,3-Mo-TTP)

This synthesis follows the procedure described above for 3,4-Mo-TTP but utilizing the 2,3-OH-TTP porphyrin. Yield: 0.70 g (88%).

Anal. Calcd. for 2,3-Mo-TTP ($C_{62}H_{56}N_{10}O_3BMo$): C, 67.95; H, 5.15; N, 12.78. Found: C, 68.10; H, 5.33; N, 12.62. Mass spectrum: calculated molecular ion: 1097.37 amu; found (FAB, nitrobenzyl alcohol): 1096.52 amu. IR (KBr): ν_{B-H} 2547(w); $\nu_{Mo=O}$ 939(s) cm^{-1}. Electronic spectrum in toluene at 25°C (λ in nm; ε in L mol^{-1} cm^{-1}): 516 (18,200), 551 (9550), 594 (5370), 652 (7240).

Properties

Both 3,4-Mo-TTP and 2,3-Mo-TTP are purple in color in the solid state as well as in solution. Both compounds are moderately sensitive to acid and show a strong Mo=O and a weak B—H stretch in the IR. The electronic spectra show strong bands due to the porphyrin center that obscure weak absorptions originating from the molybdenyl fragment. Because of the long electronic relaxation time of the molybdenyl center, the ^{1}H NMR spectra of the molybdenyl fragment is too broad to be meaningfully interpreted.

We now discuss metalation of the porphyrin. Insertion of a metal ion into the porphyrin generates binuclear metal complexes. Metalation of a free-base porphyrin generates 2 equiv of protons per mole of porphyrin. The hydrolytic sensitivity of the molybdenum center necessitates removal of the acid from the reaction mixture as it is generated. This is achieved by bubbling argon or dinitrogen through the solution during the course of the reaction.[15]

I. 5-{3,4-CATECHOLATO[HYDROTRIS(3,5-DIMETHYL-1-PYRAZOLYL)BORATO]OXOMOLYBDENUM(V)}-10,15,20-TRI-*p*-TOLYLPORPHINATOZINC(II) [Zn(3,4-Mo-TTP)]

$$3,4\text{-Mo-TTP} + ZnCl_2 \rightarrow Zn(3,4\text{-Mo-TTP}) + 2HCl$$

In this section we describe insertion of zinc to produce binuclear zinc–molybdenum compounds.

Procedure

The compound 3,4-Mo-TTP (0.175 g, 0.16 mmol) is dissolved in dimethylformamide (100 mL) and brought to reflux. $ZnCl_2$ (0.30 g, 2.2 mmol) is added to the solution with vigorous stirring. Reflux is maintained until zinc insertion is complete as evidenced by the loss of the band at 652 nm from the electronic spectrum of **3**. The reaction mixture is diluted with 200 mL of water and extracted with dichloromethane (3 × 50 mL) to yield a pink-purple solution. This solution is washed with 3 × 200-mL portions of water to remove inorganic zinc salts and the organic solvent removed by rotary evaporation. The purple residue is dissolved in a minimum amount of dichloromethane and purified by chromatography on silica gel with dichloromethane/methanol (20 : 1) as eluant. Removal of the solvent in vacuo gives a purple solid. This solid can be recrystallized by dissolving in a minimum amount of toluene at ambient temperature and adding heptane until incipient crystallization. Yield: 0.12 g (65%).

Anal. Calcd. for Zn(3,4-Mo-TTP) (**5**) ($C_{62}H_{54}N_{10}O_3BMoZn$): C, 64.20; H, 4.69; N, 12.13. Found: C, 65.73; H, 5.18; N, 11.17. Mass spectrum: calculated molecular ion: 1159.29 amu; found (FAB, nitrobenzyl alcohol): 1159.43 amu. IR (KBr): ν_{B-H} 2545(w); $\nu_{Mo=O}$ 940(s) cm^{-1}. Electronic spectrum in toluene at 25°C: (λ in nm; ε in L mol^{-1} cm^{-1}): 515 (4470), 552 (24,000), 592 (6920).

J. 5-{2,3-CATECHOLATO[HYDROTRIS(3,5-DIMETHYL-1-PYRAZOLYL)BORATO]OXOMOLYBDENUM(V)}-10,15,20-TRI-*p*-TOLYLPORPHINATOZINC(II) [Zn(2,3-Mo-TTP)]

The compound Zn(2,3-Mo-TTP) is synthesized from 2,3-Mo-TTP following the procedure outlined above for Zn(3,4-Mo-TTP). Yield: 0.10 g (60%).

Anal. Calcd. for Zn(2,3-Mo-TTP) (**6**) ($C_{62}H_{54}N_{10}O_3BMoZn$): C, 64.20; H, 4.69; N, 12.13. Found: C, 65.28; H, 5.03; N, 11.28. Mass spectrum: calculated molecular ion: 1159.27 amu; found (FAB, nitrobenzyl alcohol): 1159.43 amu. IR (KBr): ν_{B-H} 2542(w); $\nu_{Mo=O}$ 940(s) cm^{-1}. Electronic spectrum in toluene at 25°C: (λ in nm; ε in L mol^{-1} cm^{-1}): 513 (3020), 551 (22,900), 590 (4900).

Properties

Compounds Zn(3,4-Mo-TTP) and Zn(2,3-Mo-TTP) are purple in the solid state and readily dissolve in organic solvents such as dichloromethane, chloroform, and toluene to give pink solutions. Both compounds are photosensitive, and syntheses and manipulations should be performed in low light or with reaction vessels and columns individually wrapped to minimize exposure to ambient light. Their dimethylformamide and dichloromethane solutions are unstable on prolonged heating. The compounds should be stored in an inert atmosphere and in the dark.

Next, we discuss insertion of iron to synthesize binuclear high-spin iron(III) ($S = \frac{5}{2}$)-molybdenum compounds. Iron insertion into the porphyrin core is more complicated and requires strict exclusion of HCl from the reaction mixture and product complexes to avoid the acid hydrolysis of the catecholato ligand from the molybdenyl group. The μ-oxo dimer generated during purification is cleaved by aqueous NaCl in a biphasic reaction.

K. 5-{3,4-CATECHOLATO[HYDROTRIS(3,5-DIMETHYL-1-PYRAZOLYL)BORATO]OXOMOLYBDENUM(V)}-10,15,20-TRI-*p*-TOLYLPORPHINATOIRON(III)CHLORIDE [FeCl(3,4-Mo-TTP)]

$$3,4\text{-Mo-TTP} + FeCl_2 \rightarrow FeCl(3,4\text{-Mo-TTP}) + 2HCl$$

$$FeCl(3,4\text{-Mo-TTP}) \rightarrow (FeCl(3,4\text{-Mo-TTP})_2O$$

$$(FeCl(3,4\text{-Mo-TTP})_2O + 2H^+ + 2NaCl \rightarrow 2FeCl(3,4\text{-Mo-TTP}) + 2Na^+ + H_2O$$

An amount of 3,4-Mo-TTP (0.2 g, 1.82×10^{-4} mol) is dissolved in 20 mL of dichloromethane, and $FeCl_2 \cdot 4H_2O$ (0.2 g, 1.08×10^{-3} mol) is dissolved in a 15-mL methanol/dichloromethane mixture. After the addition of the iron solution to the porphyrin solution, the mixture is refluxed until iron insertion is complete as evidenced by the electronic spectra. The solution is evaporated to dryness under vacuo and dissolved in dichloromethane (50 mL) to yield a brown solution. After washing with three 100-mL portions of water, the organic phase is purified by chromatography on a silica gel column (2.5×50 cm, 70–230 mesh) using 400 mL dichloromethane/MeOH (10 : 1) as eluant. This procedure converts the chloro iron species to a μ-oxo dimer that is greenish brown in solution. The greenish brown solution is evaporated, redissolved in dichloromethane (25 mL), and reacted with two 200-mL portions of aqueous NaCl (1 M) with vigorous stirring [addition of 3–5 drops of 6(N) HCl in each liter of NaCl solution facilitates the reaction]. The remaining dichloromethane solution is washed with water (100 mL), and the solvent is removed in vacuo. Recrystallization from dichloromethane–heptane gives dark black microcrystals. Yield: 0.14 g (65%).

Anal. Calcd. for FeCl(3,4-Mo-TTP) ($C_{62}H_{54}N_{10}O_3BMoFeCl$): C, 62.79; H, 4.59; N, 11.86; Cl, 2.99. Found: C, 63.97; H, 5.40; N, 11.00; Cl, 2.82. Mass spectrum: calculated molecular ion: 1186.26 amu; found (FAB, nitrobenzyl alcohol): 1150.63 amu (M $-$ Cl $= 1151.29$ amu). IR (KBr): ν_{B-H} 2545(w); $\nu_{Mo}=O$ 940(s) cm^{-1}. Electronic spectrum in toluene at 25°C: (λ in nm; ε in L mol^{-1} cm^{-1}): 509 (14,100), 571 (5500), 693 (3720).

L. 5-(2,3-CATECHOLATO[HYDROTRIS(3,5-DIMETHYL-1-PYRAZOLYL)BORATO]OXOMOLYBDENUM(V)}-10,15,20-TRI-*p*-TOLYLPORPHINATOIRON(III)CHLORIDE [FeCl(2,3-Mo-TTP)]

This compound is synthesized by the same procedure used for FeCl(3,4-Mo-TTP, utilizing 2,3-Mo-TTP as the molybdenum-containing porphyrin. Recrystallization from dichloromethane–heptane gives a purple powder. Yield: (75%).

Anal. Calcd. for FeCl(2,3-Mo-TTP) ($C_{62}H_{54}N_{10}O_3BMoFeCl$): C, 62.79; H, 4.59; N, 11.86; Cl, 2.99. Found: C, 62.69; H, 4.78; N, 11.88; Cl, 3.06. Mass spectrum: calculated molecular ion: 1186.26 amu; found (FAB, nitrobenzyl alcohol): 1149.51 amu (M $-$ HCl $= 1150.28$ amu). IR (KBr): ν_{B-H} 2550(w); $\nu_{Mo}=O$ 943(s) cm^{-1}. Electronic spectrum in toluene at 25°C: (λ in nm; ε in L mol^{-1} cm^{-1}): 509 (15,100), 573 (4170), 690 (3630).

Properties

Compounds (FeCl(3,4-Mo-TTP) and (FeCl(2,3-Mo-TTP) are dark-colored solids that are stable in the solid state. They are soluble in most organic solvents, giving brown-colored solutions. Both compounds are acid/alkali-sensitive.

Next we discuss preparation of binuclear low-spin iron(III) ($S = \frac{1}{2}$)-molybdenum compounds. The low-spin iron(III) compounds are generated in solution by adding excess base to the corresponding high-spin complexes.[16]

M. 5-{3,4-CATECHOLATO[HYDROTRIS(3,5-DIMETHYL-1-PYRAZOLYL)BORATO]OXOMOLYBDENUM(V)}-10,15,20-TRI-*p*-TOLYLPORPHINATOIRON(III)BIS(*N*-METHYLIMIDAZOLE) ADDUCT {Fe(*N*-MeIm)$_2$[(3,4-Mo-TTP)]CL}

$$FeCl(3,4\text{-Mo-TTP}) + 2N\text{-MeIm} \leftrightarrow [Fe(N\text{-MeIm})_2(3,4\text{-Mo-TTP})]^+ Cl^-$$

Procedure

[Fe(*N*-MeIm)$_2$(3,4-Mo-TTP)]Cl ($\sim 1 \times 10^{-5}$ mol) is generated in solution by addition of 0.4 mL (0.41 g, d = 1.03) of *N*-methylimidazole (5×10^{-3} mol) to 10 mg of FeCl(3,4-Mo-TTP) (7.41×10^{-6} mol), and making the final volume of the solution to 10 mL. The solvents are either dimethylformamide/methylene chloride or methylene chloride/toluene (1 : 1). Mass spectrum: calculated molecular ion for the cation [Fe(*N*-MeIm)$_2$(3,4-Mo-TTP)]$^+$: 1314 amu; found (ESI in dimethylformamide): 1313.75. Electronic spectrum in dimethylformamide (λ in nm; ε in L mol^{-1} cm^{-1}): 556 (7240), 581(sh) (6610), 644(sh) (2500), [where (sh) represents a shoulder].

N. 5-{2,3-CATECHOLATO[HYDROTRIS(3,5-DIMETHYL-1-PYRAZOLYL)BORATO]OXOMOLYBDENUM(V)}-10,15,20-TRI-*p*-TOLYLPORPHINATOIRON(III)BIS(*N*-METHYLIMIDAZOLE) ADDUCT {Fe(*N*-MeIm)$_2$[(2,3-Mo-TTP)]CL}

[Fe(*N*-MeIm)$_2$(2,3-Mo-TTP)]Cl is prepared in solution following the procedure for [Fe(*N*-MeIm)$_2$(3,4-Mo-TTP)]Cl. Electronic spectrum in dimethylformamide (λ in nm; ε in L mol^{-1} cm^{-1}): 555 (7240), 579 (6170), 647(sh) (1580).

Properties

The solutions of [Fe(*N*-MeIm)$_2$(3,4-Mo-TTP)]Cl and [Fe(*N*-MeIm)$_2$(2,3-Mo-TTP)]Cl are red in color and show distinctive paramagnetically shifted pyrrole ^{1}H resonances centered about -17 ppm at 23°C.[17] At 500 MHz [Fe(*N*-MeIm)$_2$(3,4-Mo-TTP)]Cl shows four peaks in the -15- to -18-ppm range, whereas [Fe(N-MeIm)$_2$(2,3-Mo-TTP)]Cl shows eight peaks in the -10- to -24-ppm range.

O. 5-{3,4-CATECHOLATO[HYDROTRIS(3,5-DIMETHYL-1-PYRAZOLYL)BORATO]OXOMOLYBDENUM(V)}-10,15,20-TRI-*p*-TOLYLPORPHINATOCOPPER(II) [Cu(3,4-Mo-TTP)]

$$3,4\text{-Mo-TTP} + \text{Cu(OAc)}_2 \rightarrow \text{Cu(3,4-Mo-TTP)} + 2\text{AcOH}$$

This synthesis and the procedure described in Section P involve insertion of copper to produce bimetallic copper–molybdenum compounds.

Procedure

An amount of 3,4-Mo-TTP (58 mg, 0.05 mmol) is dissolved in 20 mL of dichloromethane and the solution is degassed thoroughly with argon. This solution is brought to reflux, and a dichloromethane/methanol (3 : 1) solution (10 mL) of copper acetate (35 mg, 0.16 mmol) is added [use of copper chloride instead of copper acetate leads to demolybdated product(s)]. Refluxing is continued until the electronic spectrum indicates complete insertion of the copper ($\sim$30 min). After removal of the solvents under vacuum, the remaining solid mass is dissolved in dichloromethane and washed with water in order to remove excess inorganic material. The organic layer is evaporated to dryness, the solid redissolved in dichloromethane, and chromatographed on a silica gel column. The target compound is eluted as a reddish brown band with dichloromethane. The compound is isolated by the evaporation of the organic solvent and further purified by recrystallization from dichloromethane–heptane. Yield: 45 mg (80%). Mass spectrum: calculated molecular ion, 1158.3 amu; found (FAB, nitrobenzyl alcohol): 1156.7 amu. IR (KBr): ν_{B-H} 2542(w); $\nu_{Mo=O}$ 940(s). Electronic spectrum in toluene at 25°C (λ in nm; ε in L mol^{-1} cm^{-1}): 501 (4300), 542 (18,300), 580 (3700), 617 (2200).

P. 5-{2,3-CATECHOLATO[HYDROTRIS(3,5-DIMETHYL-1-PYRAZOLYL)BORATO]OXOMOLYBDENUM(V)}-10,15,20-TRI-*p*-TOLYLPORPHINATOCOPPER(II) [Cu(2,3-Mo-TTP)]

This compound is synthesized according to the procedure for Cu(3,4-Mo-TTP) described above but using 2,3-Mo-TTP instead of 3,4-Mo-TTP.

Mass spectrum: calculated molecular ion, 1158.3 amu; found (FAB, nitrobenzyl alcohol), 1158 amu. IR (KBr): ν_{B-H} 2540(w); $\nu_{Mo=O}$ 940(s). Electronic spectrum in toluene at 25°C (λ in nm; ε in L mol^{-1} cm^{-1}): 507 (2300), 540 (10,800), 575 (2200), 617 (3500).

Properties

Both Cu(3,4-Mo-TTP) and Cu(2,3-Mo-TTP) are dark brown in the solid state and dissolve in most organic solvents to give brown solutions.

Acknowledgments

We thank Mr. J. Weibrecht for experimental assistance and Dr. J. McMaster for helpful discussions. Partial support of this research by the National Institutes of Health (GM-37773) and the Materials Characterization Program of the University of Arizona is gratefully acknowledged. MV thanks the DuPont-Arizona Research Experience (DARE) program for a summer undergraduate research fellowship.

References and Notes

1. V. Balzani and L. De Cola (Eds.), *Supramolecular Chemistry*, NATO ASI Series (Series C: Mathematical and Physical Sciences), Vol. 371, Kluwer, Dordrecht, 1992.
2. F. Ciardelli, E. Tsuchida, and D. Wöhrle (Eds.), *Macromolecule-Metal Complexes*, Springer, Berlin, 1996.
3. P. Basu, A. M. Raitsimring, M. J. LaBarre, I. K. Dhawan, J. L. Weibrecht, and J. H. Enemark, *J. Am. Chem. Soc.* **116**, 7166 (1994).
4. M. J. LaBarre, A. M. Raitsimring, and J. H. Enemark, in *Molybdenum Enzymes, Cofactors and Model Systems*, E. I. Stiefel, D. Coucouvanis, and W. E. Newton, (Eds.), ACS Symp. Series 535; American Chemical Society, Washington, DC, 1993, p. 130.
5. A. M. Raitsimring, P. Basu, N. V. Shokhirev, and J. H. Enemark, *Appl. Mag. Res.* **9**, 173 (1995).
6. P. Basu, A. M. Raitsimring, and J. H. Enemark, 207th National Meeting of the American Chemical Society, San Diego, CA, March 1994; INOR 309.
7. M. H. Wall, Jr., P. Basu, T. Buranda, B. S. Wicks, E. W. Findsen, M. Ondrias, J. H. Enemark, and M. L. Kirk, *Inorg. Chem.* **36**, 5676–5677 (1997).
8. D. Gust, T. A. Moore, R. V. Bensasson, P. Mathes, E. J. Land, C. Chachaty, A. L. Moore, and P. A. Liddle, *J. Am. Chem. Soc.* **107**, 3631–3640 (1985).
9. A. D. Adler, F. R. Longo, J. D. Finarelli, J. Goldmacher, J. Assour, and L. Korsakoff, *J. Org. Chem.* **32**, 476 (1957); A. D. Adler, F. R. Longo, F. Kampas, and J. Kim, *J. Inorg. Nucl. Chem.* **32**, 2443–2445 (1970).
10. J. F. W. McOmie, M. L. Watts, and D. E. West, *Tetrahedron* **24**, 2289–2292 (1968).
11. The checkers isolated **1** and **2** by extracting the neutralized reaction mixture into 2×100 mL of dichloromethane, rather than by filtration of the purple solid. Removal of the solvent gives a purple powder that is purified by column chromatography (SiO_2, 3 : 1 dichloromethane/hexane).
12. W. E. Cleland, Jr., K. M. Barnhart, K. Yamanouchi, D. Collision, F. E. Mabbs, R. B. Ortega, and J. H. Enemark, *Inorg. Chem.* **26**, 1017–1025 (1987).
13. S. J. Trofimenko, *J. Am. Chem. Soc.* **89**, 6288–6294 (1967).
14. S. J. Trofimenko, *Inorg. Synth.* **12**, 99 (1970). See also Chapter 4, Synthesis 37 in this volume.
15. The checkers obtained similar yields without purging the solution with argon to remove stoichiometric amounts of HCl formed in the reaction.
16. (a) Low-spin complexes were prepared in proper concentration according to the equilibrium constant for individual complexes: P. Basu, A. M. Raitsimring, J. H. Enemark, and F. A. Walker, *Inorg. Chem.* **36**, 1088–1094 (1997); (b) F. A. Walker, V. L. Balke, and J. T. West, *J. Am. Chem. Soc.* **107**, 1226–1233 (1985). For the equation that follows Synthesis M.
17. P. Basu, N. V. Shokhirev, F. A. Walker, and J. H. Enemark, *J. Am. Chem. Soc.* **117**, 9042 (1995).

6. IRON SANDWICH POLYOXOANION COMPOUNDS

Submitted by XUAN ZHANG, DEAN C. DUNCAN,
QIN CHEN, and CRAIG L. HILL[*]
Checked by H. WEINER and R. FINKE[†]

The title complexes are the first well-characterized multiiron sandwich compounds prepared via rational synthesis from lacunary or defect species. They have possible applications as catalysts for the oxidation of organic compounds with the environmentally friendly oxidant H_2O_2[1–3] and constitute a well-defined multiiron species to facilitate the further investigation and delineation of magnetic interactions in such systems.[4,5]

The most prevalent and studied sandwich polyoxoanions are those derived from trivacant forms of the Keggin and Wells-Dawson parent polyoxoanion structures, in which two such d^0 polyanion fragments sandwich four d electron-containing transition metal ions. The first family of formula $[(M)_4(PW_9O_{34})_2]^{10-}$, initially prepared by Weakley et al.[6] and the second family, of formula $[(M)_4(P_2W_{15}O_{56})_2]^{16-}$, first prepared by Finke et al.,[7] are known only for some divalent first-row transition metal ions ($M = Co^{II}$, Mn^{II}, Ni^{II}, Cu^{II}, Zn^{II}), however. The literature procedure does not work well with Fe^{3+} because of the acidity of the solution. Related iron-containing sandwich compounds that have been reported include $[(Fe_2Cu_2)(FeW_9O_{34})_2H_2]^{10-}$[8] and $[(Zn_xFe_{3-x}W)(ZnW_9O_{34})_2]^{(9+x)-}$ ($n = 0,1$).[9] The procedure reported here provides routes to both Keggin and Wells-Dawson–derived Fe_4 sandwich polyanions.

A. HEXAKISTETRABUTYLAMMONIUM OCTADECATUNGSTOTETRAIRONDIPHOSPHATE {(TBA)$_6$[Fe$^{III}_4$(H$_2$O)$_2$(PW$_9$O$_{34}$)$_2$]}

$$4Fe^{2+} + 2PW_9O_{34}{}^{9-} + O_2 + 4H^+ \rightarrow [Fe^{III}_4(H_2O)_2(PW_9O_{34})_2]^{6-}$$

Procedure

Ferrous chloride tetrahydrate ($FeCl_2 \cdot 4H_2O$, 0.48 g) is dissolved in 25 mL of H_2O in a 60-mL beaker, and 3.0 g (1.09 mmol) of Δ-$Na_8HPW_9O_{34} \cdot 19H_2O$[10] is added

[*] Department of Chemistry, Emory University, 1515 Pierce Dr. Atlanta, GA 30322.
[†] Department of Chemistry, Colorado State University, Fort Collins, CO 80523-1872.

slowly as a solid at room temperature with vigorous stirring. The solution is heated to 60°C for 5–10 min and filtered hot through a medium-porosity sintered-glass frit. Potassium chloride (3.6 g; 48.29 mmol) is added and an immediate dark precipitate forms. The solution is stirred as it cools to ambient temperature. The precipitate is collected on a 50-mL medium-porosity sintered-glass frit, washed with 3×25 mL (i.e., three 25-mL portions) of ethanol and dried at 40°C in vacuum for 1 h. About 3 g of this precipitate is dissolved in 100 mL of H_2O (pH$\sim$7 without adjustments). Tetrabutylammonium (TBA) chloride (2.4 g; 8.64 mmol) dissolved in 90 mL of CH_2Cl_2 is added to the aqueous solution. The resulting two-layer mixture is transferred to and shaken in a separatory funnel to produce a dark-chocolate-colored upper layer and yellowish-green bottom layer. The bottom layer is removed, an additional 90 mL of CH_2Cl_2 is added, and the mixture is shaken again. A yellow precipitate forms in the upper layer while the bottom layer remains a dark greenish-brown solution. The bottom organic layer is removed, placed in a 400-mL recrystallizing dish, covered with parafilm containing several holes, and allowed to oxidize in the air overnight (the color changes from dark brown to light brown). The next day the remaining oil organic phase is washed with 150 mL of H_2O to give a greenish yellow solid. This solid is collected on a medium-porosity frit, washed with 3×25 mL of H_2O and then dried at 40°C under vacuum overnight. The powder is dissolved in a minimum amount ($\sim$1–3 mL) of CH_3CN in a 250-mL beaker to form a dark brownish-green solution. Diethyl ether (100 mL) is added and a light yellow powder precipitates. After the mixture is stirred for 2–3 h, the precipitate is collected on a 50-mL medium porosity sintered-glass frit and dried at 40°C under vacuum overnight. The yield is 0.8–1.2 g. The compound can be recrystallized by dissolving 0.05 g (8.1×10^{-3} mmol) of the crude product in a 10-mL vial containing 4 mL of a 5 : 1 v/v mixture of CH_2Cl_2 and CH_3CN, then placing this vial into a 20-mL vial containing 5 mL of diethyl ether. This two-vial assembly is then capped and placed in a freezer at -20°C for several days. For X-ray-quality crystals, the compound is recrystallized by dissolving 0.02 g (3.2×10^{-3} mmol) of the crude product in a 20-mL vial containing 10 mL of a 9 : 1 v/v mixture of CH_2Cl_2 and CH_3CN, then placing this vial in a 50-ml vial or jar containing 5 mL of diethyl ether. This two-vial assembly is then capped and placed in a freezer at -20°C for approximately one week.

Properties

The tetrabutylammonium salt, $(TBA)_6[Fe^{III}_4(H_2O)_2(PW_9O_{34})_2]$, is a light yellow solid. It is soluble in acetonitrile, *N,N*-dimethylformamide, and dimethylsulfoxide. It is slightly soluble in acetone and methylene chloride. It is characterized in the solid state by its IR spectrum (2% KBr pellet, 1300–400 cm^{-1}): 1066(m), 1014(w), 970(m), 957(m), 932(m, sh), 868(s), 823(s), 769(vs), 699(s), 623(w),

589(w,sh), 520(w), 496(w,sh), and 454(w). No IR bands arising from the presence of solvent molecules are visible as the crystals lose solvent molecules of crystallization quickly on exposure to the air. The crystal data for $(TBA)_6[Fe^{III}_4$ $(H_2O)_2(PW_9O_{34})_2]\cdot 4CH_3CN\cdot 2CH_2Cl_2\cdot 2H_2O$ are orthorhombic, space group *Pbca*, with $a = 25.3330(3)$ Å, $b = 24.9326(4)$ Å, $c = 28.4164(3)$ Å, $V = 17948.30$ Å^3, $Z = 4$.

Anal. Calcd. for $C_{96}H_{220}Fe_4N_6O_{70}P_2W_{18}$ or $(TBA)_6[Fe^{III}_4(H_2O)_2(PW_9O_{34})_2]$: C, 18.68; H, 3.59; Fe, 3.62; N, 1.36; P, 1.00; W, 53.60. Found: C, 18.73; H, 3.55; Fe, 3.35; N, 1.33; P, 0.89; W, 53.24.

B. DODECASODIUM TRIACONTATUNGSTOTETRAIRONTETRA PHOSPHATE HYDRATE $\{Na_{12}[Fe^{III}_4(H_2O)_2(P_2W_{15}O_{56})_2]\cdot 58H_2O\}$

$$4Fe^{3+} + 2P_2W_{15}O_{56}^{12-} + 2H_2O \rightarrow [Fe^{III}_4(H_2O)_2(P_2W_{15}O_{56})_2]^{12-}$$

Procedure

To a solution of 0.82 g (3 mmol) of $FeCl_3\cdot 6H_2O$ in 30 mL of a 1 M NaCl solution in a 60-mL beaker is added slowly with vigorous stirring 6.0 g (1.4 mmol) of solid α-$Na_{12}P_2W_{15}O_{56}\cdot 18H_2O$.[10] The solution is heated to 80°C for 5–10 min and filtered hot. The solution is left to stand. After several days, yellow crystalline solid precipitates (1.8 g, $\sim$30% yield); the filtrate is discarded. X-ray-quality crystals can be obtained by recrystallizing 1 g of this yellow crystalline solid in 5 mL of a 2 M NaCl solution. The needle-shaped crystals are dichroic: yellow along the thin axis and brown along the long axis.

Properties

The sodium salt of this Wells-Dawson-derived sandwich polyoxoanion is a yellow crystalline solid that is soluble in water. It is characterized in solution by ^{31}P NMR (9 mM solution in H_2O, D_2O in a capillary insert): one resonance for the distal P atoms at -11.1 ppm ($\Delta\nu_{1/2} = 70$ Hz). In the solid state the compound always contains NaCl as it must be crystallized from aqueous NaCl solution. If NaBr is substituted for NaCl in the aqueous crystallization process, crystals of the compound still form but now contain NaBr in place of NaCl. This sandwich complex is characterized in the solid state by IR (2% KBr pellet, 1300–400 cm^{-1}): 1091(s), 1017(w), 951(s,sh), 917(m), 826(s), 757(s), 695(s), 630(m,sh), 526(w). The crystal data for $Na_{12}[Fe^{III}_4(H_2O)_2(P_2W_{15}O_{56})_2]\cdot 58H_2O$

are triclinic, space group $P\bar{1}$, with $a = 12.536$ Å, $b = 16.150$, $c = 18.980$ Å, $\alpha = 87.618°$, $\beta = 79.895°$, $\gamma = 74.926°$, $V = 3633.9$ Å^3, $Z = 1$.

Anal. Calcd. for $H_{120}ClFe_4Na_{13}P_4O_{172}W_{30}$ or $Na_{12}[Fe^{III}_4(H_2O)_2 (P_2W_{15}O_{56})_2] \cdot$ $NaCl \cdot 58H_2O$: Na, 3.30; Fe 2.46; P, 1.37; W, 60.81; Cl, 0.39. Found: Na, 3.30; Fe, 2.35; P, 1.24; W, 60.9ɔ; Cl, 0.48.

References

1. R. Neumann and M. Gara, *J. Am. Chem. Soc.* **116**, 5509 (1994).
2. R. Neumann and M. Gara, *J. Am. Chem. Soc.* **117**, 5066 (1995).
3. A. M. Khenkin and C. L. Hill, *Mendeleev Commun.*, 140 (1993).
4. E. Coronado and C. J. Gómez-García, *Comm. Inorg. Chem.* **17**, 255 (1995).
5. C. J. Gómez-García, J. J. Borrasalmenar, E. Coronado, and L. Ouahab, *Inorg. Chem.* **33**, 4016 (1994).
6. T. J. R. Weakley, H. T. Evans, Jr., J. S. Showell, G. F. Tourné, and C. M. Tourné, *J. Chem. Soc. Chem. Commun.*, 139 (1973).
7. R. G. Finke, M. Droege, J. R. Hutchinson, and O. Gansow, *J. Am. Chem. Soc.* **103**, 1587 (1981).
8. S. H. Wasfi, A. L. Rheingold, G. F. Kokoszka, and A. S. Goldstein, *Inorg. Chem.* **26**, 2934 (1987).
9. C. M. Tourné, G. F. Tourné, and F. Zonnevijlle, *J. Chem. Soc., Dalton Trans.*, 143 (1991).
10. R. G. Finke, M. W. Droege, and P. J. Domaille, *Inorg. Chem.* **26**, 3886 (1987).

7. DIPYRRYL AND PORPHYRINIC PRECURSORS TO SUPRAMOLECULAR CONJUGATED (PORPHINATO)METAL ARRAYS: SYNTHESES OF DIPYRRYLMETHANE AND (5,15-DIPHENYLPORPHINATO)ZINC(II)

Submitted by VICTOR S.-Y. LIN, PETER M. IOVINE,
STEPHEN G. DiMAGNO, and MICHAEL J. THERIEN*
Checked by STEVE MALINAK and DIMITRI COUCOUVANIS

Metal-mediated cross-coupling reactions involving *meso*-haloporphyrins enable the fabrication of porphyrin arrays that exhibit exceptional electronic interactions between their constituent porphyrinic building blocks.[1-3] Because *meso*-haloporphyrins derive from direct halogenation of the porphyrinic aromatic macrocycle, porphyrins bearing unsubstituted *meso* positions are important synthetic precursors to these supramolecular, multichromophoric systems.

*Department of Chemistry, University of Pennsylvania, Philadelphia, PA 19104-6323.

Archetypal examples of these strongly coupled (porphinato)metal assemblies are highlighted by a *meso*-to-*meso* ethynyl or butadiynyl linkage topology between the porphyrin units of the array and feature a linear arrangement of chromophores, typified by structures **I** and **II** (Fig. 1).

Similar design elements can be incorporated into more elaborate supramolecular structures (**III** and **IV**, Fig. 1),[4] which serve as models in which to probe ground- and excited-state electronic interactions along multiple conjugation

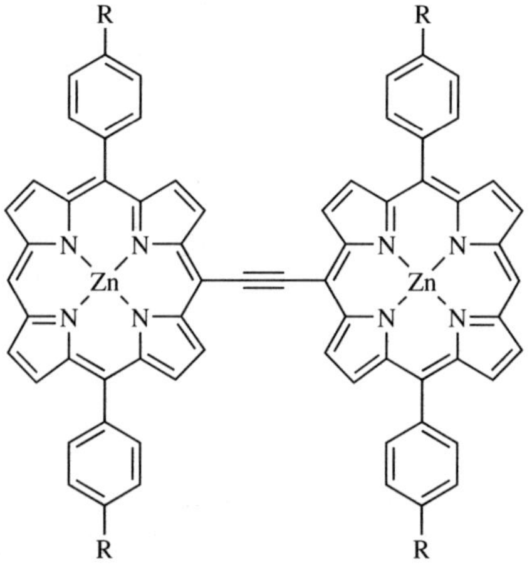

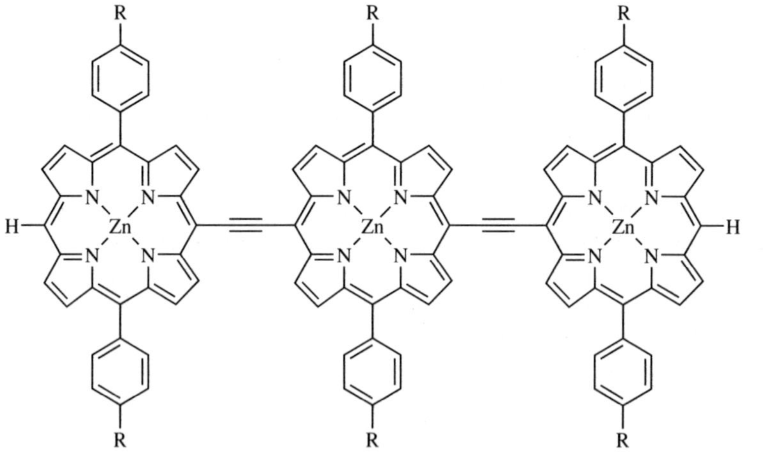

Figure 1. Highly conjugated ethyne–bridged (porphinato)zinc(II) arrays **I– IV**.

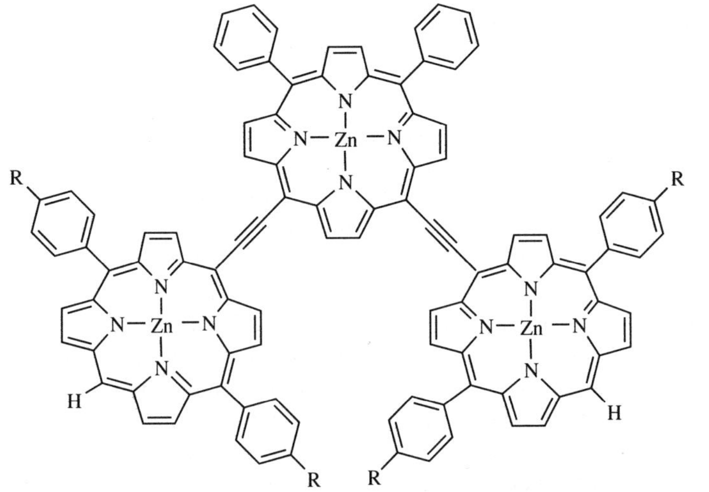

Figure 1   (*Continued*)

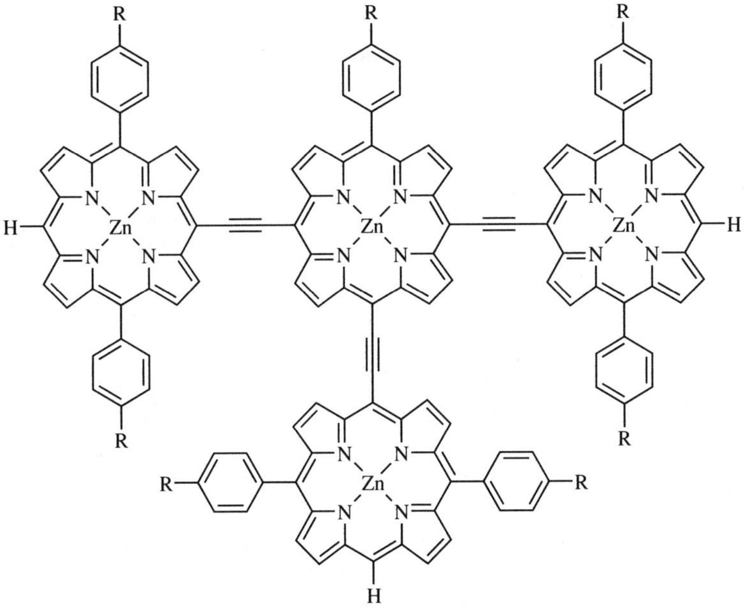

pathways as well as important oligochromophoric precursors to higher-order structures.

While the 5-phenylporphyrin and 5,10-diphenylporphyrin components of conjugated porphyrin arrays **III** and **IV** can be prepared by a number of methods,

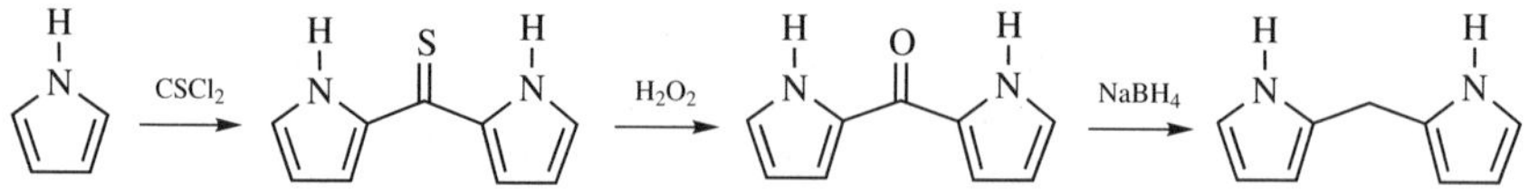

Scheme 1. Clezy's synthetic route to dipyrrylmethane.[6]

routes exploiting McDonald-type $2+2$ acid-catalyzed condensations of dipyrryl precursors to such asymmetrically *meso*-substituted parent porphyrin complexes offer numerous advantages with respect to syntheses that rely on the direct reaction of organic aldehydes with monopyrroles, the most important of which is significantly simplified chromatographic purification of products. As such, dipyrrylmethane along with its α-substituted and α,α'-disubstituted derivatives, are key building blocks for these highly conjugated multiporphyrin structures.

Although a variety of synthetic methods for β-unsubstituted-*meso*-substituted, β-substituted-*meso*-substituted, and β-substituted-*meso*-unsubstituted dipyrrylmethanes have been reported in the past,[5] the preparation of the parent compound dipyrrylmethane has generally been accomplished using a three-step synthesis developed by Clezy (Scheme 1).[6] Drawbacks to this method include the toxicity of thiophosgene as well as the difficult isolation of the thioketone and its tendency to polymerize; such difficulties make large-scale preparations of dipyrrylmethane problematic. Recently, Bruce published an improved procedure for dipyrrylmethane synthesis involving a one-step reaction of excess pyrrole and paraformaldehyde in a mixture of methanol and acetic acid.[7] The poor solubility of paraformaldehyde in most organic solvents coupled with the required chromatographic purification of the product, however, likewise precluded large-scale preparations of this compound.

We report herein a significantly improved and simplified preparation of dipyrrylmethane, which is based on Lindsey's route to *meso*-substituted dipyrryl porphyrin precursors,[8] as well as the syntheses of (5,15-diphenylporphinato)-zinc(II), a key building block of conjugated porphyrin arrays **I–IV** (Fig. 1).

Experimental Procedure

Standard Schlenk techniques were employed to manipulate air-sensitive solutions. All manipulations involving air-sensitive materials were carried out under nitrogen previously passed through an O_2 scrubbing tower (Schweizerhall R3-11 catalyst) and a drying tower (Linde 3-Å molecular sieves). All solvents utilized in this work were obtained from Fisher Scientific (HPLC Grade). Methylene chloride was distilled from calcium hydride under N_2. Chromatographic purification (Silica Gel 60, 230–400 mesh, EM Science) of products, when required, was accomplished on the bench top. Reagents were obtained from Aldrich Chemical Company.

A. 2,2'-DIPYRRYLMETHANE (C$_9$H$_{10}$N$_2$)

Procedure

Pyrrole-2-carboxaldehyde (10 g, 0.1 mol) was dissolved in 300 mL of water in an Erlenmeyer flask. NaBH$_4$ (11 g, 0.3 mol) was dissolved in 100 mL of water and added to the pyrrole-2-carboxaldehyde solution dropwise over a 10-min period. The mixture was allowed to stir at room temperature for 1 h. The reaction solution was then extracted with diethylether (3 × 100 mL), washed with satd Na$_2$CO$_3$ (3 × 100 mL), and dried over MgSO$_4$ (anhydrous). The solvent was evaporated to yield 9 g of 2-(hydroxymethyl)pyrrole, isolated as a colorless oil (90% yield based on 10 g of the pyrrole-2-carboxaldehyde starting material).

The 2-(hydroxymethyl)pyrrole was added to a Schlenk flask containing 70 mL of pyrrole which served as the solvent for the reaction. The solution was immediately purged with N$_2$ for 5 min, following which BF$_3$·Et$_2$O (1.23 mL) was slowly added via syringe. The color of the solution changed from light yellow to dark brown over a period of 2 h. Stirring at room temperature under N$_2$ was continued for 12 h; the reaction mixture was then diluted with 200 mL of CH$_2$Cl$_2$, thrice washed with 100-mL of 0.1 N NaOH(aq), and neutralized with saturated aqueous Na$_2$CO$_3$. The organic layer was isolated, dried over MgSO$_4$, and evaporated to give a dark brown oil which contained both the dipyrryl-methane product and unreacted pyrrole. The pyrrole was removed by vacuum transfer (10 μm Hg, 50°C) to give an oily mixture which was placed under high vacuum for an additional 3 h to completely remove any remaining pyrrole. Hexane (600 mL) was added and the solution was refluxed for 15 min prior to carrying out a hot filtration on the bench top. The filtrate was evaporated to give a light yellow pale solid product. The hexane extraction procedure was repeated 3 times, giving a total 7.5 g of pure dipyrrylmethane [56% yield based on 9 g of the 2-(hydroxymethyl)pyrrole starting material]. The dipyrrylmethane is stored at −40°C under inert conditions. The compound, under these conditions, is stable in excess of 6 months. ^{1}H NMR (250 MHz, CDCl$_3$): δ 7.89 ppm (bs, 2H, NH), 6.65 (q, 2H, β-H), 6.13 (q, 2H, β-H), 6.02 (q, 2H, β-H), 3.97 (s, 2H, *meso*-H). All other characterization data was consistent with that previously reported.[6,7]

B. 5,15-DIPHENYLPORPHYRIN[9,10]

Procedure

A flame-dried 1000-mL flask equipped with a magnetic stirring bar was charged with 2,2'-dipyrrylmethane (458 mg, 3.1 mmol), benzaldehyde (315 µL, 3.1 mmol), and 600 mL of freshly distilled methylene chloride. The solution was degassed with a stream of dry nitrogen for 10 min. Trifluoroacetic acid (150 µL, 1.95 mmol) was added via syringe, the flask was shielded from light with aluminum foil, and the solution was stirred for 3 h at room temperature. The reaction was quenched by the addition of 900 mg (3.96 mmol) of 2,3-dichloro-5,6-dicyanobenzoquinone (DDQ), and the solution was stirred for an additional 30 min. Pyridine (3 mL) was then added, which serves two purposes: (i) it neutralizes excess acid and 1,4-dihydroxy-2,3-dichloro-5,6-dicyanobenzene, and (ii) it reacts with excess DDQ to form an insoluble precipitate which can be filtered from the solution prior to chromatographic work up. After reducing the volume of the filtrate, it was poured directly on top of a silica gel column (20×2 cm) packed in CH_2Cl_2. The product was eluted in 700 mL of CH_2Cl_2. The solvent was evaporated leaving purple crystals that were washed once with hexane, filtered, and dried. This compound (518 mg, 1.12 mmol, 72.2%) was sufficiently pure for further reactions.

C. (5,15-DIPHENYLPORPHINATO)ZINC(II)

Procedure

A 500-mL round-bottom flask was charged with 5,15-diphenylporphyrin (518 mg, 1.12 mmol), zinc acetate (1 g, 5.5 mmol), and 300 ml of a $9:1$ $CHCl_3$: THF solution and refluxed. The porphyrin metalation reaction was monitored by optical spectroscopy and was completed within 2 h. The solution was washed with water ($3 \times$), dried over calcium chloride, and filtered. Evaporation of the solvent gave 562 mg of 5,15-(diphenylporphinato)zinc(II) (95% yield, based on 518 mg of the porphyrin starting material). No further purification was necessary.

Anal. ^{1}H NMR (500 MHz, CDCl$_3$): δ 10.17 ppm (s, 2H, *meso*-H), δ 9.32 (d, 4H, J = 4.4 Hz, β-H), δ 9.04 (d, 4H, J = 4.4 Hz, β-H), δ 8.22 (m, 4H, *o*-phenyl), 7.72 (m, 6H, *m,p*-phenyl). ^{13}C NMR (500 MHz, CDCl$_3$): 105.61 ppm, 119.35, 126.30, 127.04, 131.33, 132.02, 134.74, 143.29, 149.32, 149.90. HRMS: Calcd. 524.0979; Found: 524.0973 [MH$^+$].

References and Notes

1. V. S.-Y. Lin, S. G. DiMagno, and M. J. Therien, *Science* **264**, 1105–1111 (1994).
2. V. S.-Y. Lin and M. J. Therien, *Chem. Eur. J.* **1**, 645–651 (1995); P. J. Angiolillo, V. S.-Y. Lin, J. M. Vanderkooi, and M. J. Therien, *J. Am. Chem. Soc.* **117**, 12514–12527 (1995).
3. For other examples of conjugated porphyrin arrays, see D. P. Arnold and L. J. Nitschinsk, *Tetrahedron* **48**, 8781–8792 (1992); D. P. Arnold and G. A. Heath, *J. Am. Chem. Soc.* **115**, 12197–12198 (1993); H. L. Anderson, S. J. Martin, and D. D. C. Bradley, *Angew. Chem. Int. Ed. Engl.* **33**, 655–657 (1994); H. L. Anderson, *Inorg. Chem.* **33**, 972–981 (1994); S. Prathapan, T. E. Johnson, and J. S. Lindsey, *J. Am. Chem. Soc.* **115**, 7519–7520 (1993); R. W. Wagner and J. S. Lindsey, *J. Am. Chem. Soc.* **116**, 9759–9760 (1994).
4. P. J. Angiolillo, K. Susumu, H. T. Uyeda, V. S.-Y. Lin, R. Shediac, and M. J. Therien, *Synthetic Metals* **116**, 247–253 (2001); V. S.-Y. Lin, R. Kumble, and M. J. Therien. Manuscript in preparation.
5. R. B. Woodard, *Angew. Chem.* **72**, 651–662 (1960); G. P. Arsenault, E. Bullock, and S. F. MacDonald, *J. Am. Chem. Soc.* **82**, 4384–4389 (1960); D. M. Wallace, S. H. Leung, M. O. Senge, and K. M. Smith, *J. Org. Chem.* **58**, 7245–7257 (1993); D. M. Wallace and K. M. Smith, *Tetrahedron Lett.* **31**, 7265–7268 (1990); T. Ema, Y. Kuroda, and H. Ogoshi, *Tetrahedron Lett.* **32**, 4529–4532 (1991); T. Mizutani, T. Ema, T. Tomita, Y. Kuroda, and H. Ogoshi, *J. Chem. Soc., Chem. Commun.* 520–522 (1993); C.-H. Lee, F. Li, K. Iwamoto, J. Dadok, A. A. Bothner-By, and J. S. Lindsey, *Tetrahedron* **51**, 11645–11672 (1995); T. Mizutani, T. Ema, T. Tomita, Y. Kuroda, and H. Ogoshi, *J. Am. Chem. Soc.* **116**, 4240–4250 (1994).
6. P. S. Clezy, G. A. Smith, *Aust. J. Chem.* **22**, 239 (1969); R. Chong, P. S. Clezy, A. J. Liepa, and A. W. Nichol, *Aust. J. Chem.* **22**, 229 (1969).
7. Q. M. Wang and D. W. Bruce, *Synlett.* 1267–1268 (1995).
8. C.-H. Lee and J. S. Lindsey, *Tetrahedron* **50**, 11427–11440 (1994).
9. J. S. Manka and D. S. Lawrence, *Tetrahedron Lett.* **30**, 7341–7344 (1989).
10. S. G. DiMagno, V. S.-Y. Lin, and M. J. Therien, *J. Org. Chem.* **58**, 5983–5993 (1993).

8. SYNTHESIS OF DODECAOXOHEXADECACARBOXY-LATOTETRAAQUO-DODECAMANGANESE $[Mn_{12}O_{12}(O_2CR)_{16}(H_2O)_4]$ (R = Me,Et,Ph,Cr) COMPLEXES

Submitted by HILARY J. EPPLEY and GEORGE CHRISTOU*
Checked by NEIL A. LAW and VINCENT L. PECORARO†

Great interest and excitement have developed since the late 1980s among members of the chemistry and physics communities about complexes with the formula $[Mn_{12}O_{12}(O_2CR)_{16}(H_2O)_4]$.[1-7] In these dodecanuclear complexes, a central Mn_4^{IV} cubane moiety is surrounded by a nonplanar ring of alternating Mn^{III} and oxide ions (Fig. 1). These complexes have shown unusual bulklike magnetic

* Department of Chemistry, University of Florida, Gainesville, FL 32611–7200.
† Department of Chemistry, University of Michigan, Ann Arbor, MI 48109.

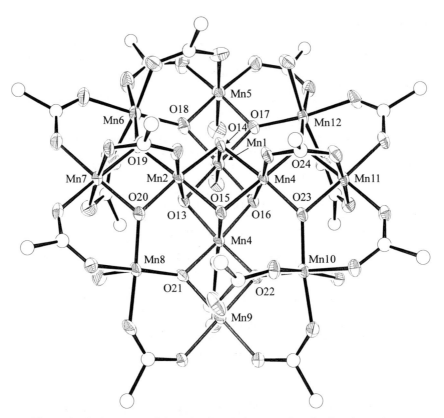

Figure 1. The structure of the $[Mn_{12}O_{12}(O_2CR)_{16}(H_2O)_x]$ family of complexes.

properties and have been a source of novel discoveries at the interface of the quantum and classical regimes. The Mn_{12} complexes are well-characterized, single-molecule models for superparamagnets and have exhibited abnormally slow magnetic relaxation for a molecular species, resulting in hysteresis of magnetization and out-of-phase components in the AC magnetic susceptibility.[1,2,7] Rather than resulting from a magnetic phase transition, these unusual magnetic properties result from a large ground-state spin value of $S = 9$ or 10 and a high degree of molecular anisotropy in the complexes, the latter due to near-parallel alignment of the Jahn–Teller elongation axes of the eight Mn^{III} ions. In addition to these fascinating properties, the Mn_{12} molecules can also serve as models for the investigation of quantum tunneling of magnetization, which is of particular interest to physicists.[4,5] These complexes, despite their structural complexity, are nevertheless quite easy to synthesize.[1,6,7] Syntheses of four examples of the dodecanuclear complexes are given herein.

Procedure

All reagents were used as purchased. All reactions were carried out under ambient atmospheric conditions. The procedure is a slightly modified version of the original synthesis reported by T. Lis.[6]

A. $[Mn_{12}O_{12}(O_2CMe)_{16}(H_2O)_4] \cdot 4H_2O \cdot 2HO_2CMe$

Distilled water (16 mL) and glacial acetic acid (24 mL) are mixed in a 125-mL Erlenmeyer flask at room temperature (20°C). To this solution 4.04 g (16.5 mmol) $Mn(O_2CMe)_2 \cdot 4H_2O$ is added with vigorous stirring using a very large stir bar (5 cm). When the $Mn(O_2CMe)_2 \cdot 4H_2O$ has dissolved, 1.00 g (6.33 mmol) of finely crushed $KMnO_4$ solid is added all at once with continued vigorous stirring. A gradual color change occurs from light pink to red-brown, and finally to a very dark brown. The solution is stirred vigorously until all the $KMnO_4$ has dissolved. It is quite difficult to observe whether complete dissolution has occurred because the solution is so dark, and accurate evaluation of the reaction's progress requires inspection of the bottom of the flask with a flashlight. The reaction should not stir longer than 20 min; longer reaction times produce a powder that has nearly the same IR spectrum as the desired product but is insoluble. The stirring is halted when the $KMnO_4$ has all reacted, and the solution is allowed to stand at room temperature for 24–72 h. Black rectangular crystals are collected by filtration through a Büchner funnel with a coarse fritted glass disk, and excess powder and reaction solution is washed away with copious amounts of acetone (100 mL). Yield is 3.12 g ($\sim$80% based on Mn).

Anal. Calcd. for $C_{36}H_{72}O_{56}Mn_{12}$: C, 21.0; H, 3.5; Mn 32.0. Found: C, 21.3; H, 3.2; Mn, 31.2

Properties

Solubility: $[Mn_{12}O_{12}(O_2CMe)_{16}(H_2O)_4] \cdot 4H_2O \cdot 2HO_2CMe$ is soluble in CH_3CN. Although the bulk material is analytically pure, a small amount of material will remain insoluble. The proportion of this insoluble form increases with an increased rate of crystallization of **A**, and appears to be a different solvate form. The complex is stable indefinitely in air and atmospheric moisture but decomposes into amorphous manganese oxides when exposed to a large amount of H_2O. The lattice $MeCO_2H$ cannot be easily removed by drying in vacuo, but it can be removed by recrystallization from CH_3CN/Et_2O. Selected IR peaks (KBr, cm^{-1}): 3650(w), 3300(w,b), 1709(w). 1586(s), 1561(s), 1527(s), 1450(s), 1389(s), 1334(m), 1258(s), 1049(w), 1026(w), 959(w), 935(w), 714(m), 673(m), 640(m), 611(m), 563(m), 553(m), 518(w). ^1H NMR in CD_3CN (ppm):

48.2, 41.8, 13.9 (plus a peak due to an average of coordinated and free H_2O, which varies significantly in broadness and shift depending on sample concentration). All peaks are very broad because of the presence of unpaired electrons in the complex.

B. $[Mn_{12}O_{12}(O_2CEt)_{16}(H_2O)_3](H_2O)$

Procedure

The synthesis is an optimized version of the published synthesis of B.[7] To a slurry containing 1.00 g (0.49 mmol) $[Mn_{12}O_{12}(O_2CMe)_{16}(H_2O)_4] \cdot 4H_2O \cdot 2HO_2CMe$ (**A**) and 50 mL of toluene in a 250-mL round-bottomed flask is added 3 mL (40 mmol) of $EtCO_2H$. The slurry is concentrated using a rotary evaporator at reduced pressure (water aspiration) with gentle warming (40°C) to remove acetic acid as the toluene azeotrope. The synthesis may also be performed successfully using a vacuum pump to remove the solvent. The slurry of Mn_{12} dissolves as ligand substitution occurs, forming a dark brown solution. When the solution has been concentrated to only a few milliliters (2–4 mL), 50 mL of toluene is added, and removed at reduced pressure as stated previously, and then another 50 mL of toluene is added to the flask and the solution is filtered to remove insoluble material (~0.2 g). Another 3 mL of HO_2CEt is added to the dark brown oil. Toluene (15 mL) is added to the oil and the solution filtered. To the dark brown filtrate 400 mL of hexanes is added with mixing and the solution allowed to stand overnight, whereupon small black crystals form. The solution is filtered to isolate the crystals, which are then washed with three 30-mL portions of hexane. The product is dried in vacuo, giving 850 mg of compound (~84% based on Mn).

Anal. Calcd. for $C_{48}H_{88}O_{48}Mn_{12}$: C, 27.63; H, 4.27; Mn, 30.52. Found: C, 27.90; H, 4.27; Mn, 31.51.

Properties

Solubility: $[Mn_{12}O_{12}(O_2CEt)_{16}(H_2O)_3](H_2O)$ is soluble in a variety of organic solvents, including toluene, diethyl ether, dichloromethane, acetonitrile, benzonitrile, dichloroethane, and others, making it well suited for solution-state characterization. The product can be recrystallized from CH_2Cl_2/hexanes to give well-formed crystals containing no lattice H_2O. Selected IR bands (KBr, cm^{-1}): 3370(m,b), 1586(s), 1575(s), 1466(s), 1432(vs), 1379(s), 812(m), 719(m), 650(s,b), 559(m,b). Conductivity (25°C): 4.1 and 0.3S cm^2 mol^{-1} in CH_3CN and CH_2Cl_2, respectively. ^1H NMR in CD_3CN (ppm): 52.7, 47.3, 46.7, 12.2, 11.1, 3.7, −4.9 (plus a peak due to an average of coordinated and free H_2O, which varies significantly in broadness and shift depending on sample concentration).

All peaks are broadened because of the presence of unpaired electrons in the complex.

C. [Mn$_{12}$O$_{12}$(O$_2$CPh)$_{16}$(H$_2$O)$_4$]

Procedure

This procedure is based on a published preparation.[1] To a slurry of crystalline [Mn$_{12}$O$_{12}$(O$_2$CMe)$_{16}$(H$_2$O)$_4$] from **A** (3.00 g, 1.5 mmol) in CH$_2$Cl$_2$ (90 mL) is added 6.00 g (48 mmol) of HO$_2$CPh, and the reaction is stirred for 24 h. As the slurry is stirred and ligand substitution progresses, the insoluble starting material dissolves. After 24 h, the solution is filtered by gravity through filter paper to remove the unreacted starting material. To the filtrate 2 : 1 Et$_2$O/hexanes (v/v) is added to precipitate the crude product, which is collected by filtration and washed with two 15-mL portions of hexane. In our syntheses, this requires 400 mL of solvent and results in precipitation of the product within 30 min. However, more solvent may be required to precipitate the product depending on the exact volume of dichloromethane remaining at this stage. The crude product is redissolved in 100 mL of CH$_2$Cl$_2$ containing an additional 6.00 g of benzoic acid and the solution is stirred overnight. The solution is filtered and then 400 mL (or more) 2 : 1 Et$_2$O/hexanes is added carefully, until a slight cloudiness in the solution indicates saturation. The solution is then allowed to stand overnight, during which time the product precipitates from solution. The brown-black microcrystals are filtered, washed with two 40-mL portions of 2 : 1 Et$_2$O/hexanes, and dried in vacuo. The yield is 2.63 g ($\sim$57% based on Mn).

Anal. Calcd. for C$_{112}$H$_{88}$O$_{48}$Mn$_{12}$: C, 47.0; H, 3.1; Mn, 21.5. Found; C, 46.5; H, 3.4; Mn, 21.2.

Properties

Solubility: Mn$_{12}$O$_{12}$(O$_2$CPh)$_{16}$(H$_2$O)$_4$ is soluble in dichloromethane and similar solvents, but relatively insoluble in CH$_3$CN. Selected IR peaks (KBr, cm^{-1}): 1599(m), 1561(m), 1525(m), 1493(m), 1448(m), 1418(s), 1350(m), 1308(w), 1179(w), 1159(w), 1140(w), 1070(w), 1026(w), 718(m), 677(m), 652(m), 615(m), 550(w), 513(m).

D. [Mn$_{12}$O$_{12}$(O$_2$CCH=CHCH$_3$)$_{16}$(H$_2$O)$_3$]

Procedure

To a slurry containing 1.00 g of **A** (0.49 mmol) in 50 mL of toluene in a 250-mL round-bottomed flask is added 1.00 g (11.6 mmol) of crotonic acid. The solution is concentrated using rotary evaporation at reduced pressure (water aspiration) with gentle warming ($\sim 40°C$). The solid slowly dissolves as ligand substitution occurs, giving a dark brown solution. This solution is evaporated to an oil, and then another 50 mL of toluene is added. The solvent removal process is repeated twice more with additional 50-mL aliquots of toluene. To the final oil, 15 mL of toluene is added and the dark brown solution is filtered to remove any insoluble material. To the filtrate, 125 mL of hexane is added to precipitate the product as a dark brown solid. This is collected by filtration and washed with five 15-mL hexane aliquots. The yield of the product is 780 mg ($\sim 70\%$).

Anal. Calcd. for $C_{64}H_{86}O_{47}Mn_{12}$: C, 33.9; H, 3.8. Found: C, 34.1; H, 4.1.

Properties

Solubility: $Mn_{12}O_{12}(O_2CCH{=}CHCH_3)_{16}(H_2O)_3$ is soluble in a variety of organic solvents, including CH_2Cl_2, CH_3CN, toluene, Et_2O, and benzonitrile. Excess ligand can be easily separated from the product by washing with hexanes (crotonic acid is quite soluble in hexanes). Selected IR bands (KBr, cm^{-1}): 3044(w), 2967(w), 2938(w), 2913(w), 2851(w), 1657(s), 1568(s), 1495(s), 1425(vs), 1354(s), 1296(m), 1256(m), 1233(m), 1103(w), 1020(w), 966(m), 916(w), 849(w), 731(w), 702(m), 654(m), 612(m), 552(w), 517(w). ^1H NMR in CD_3CN (ppm): 51.8, 27.8, 23.8, 14.2, 6.6, 4.9, 3.1, -5.6, -12.5 (plus a peak due to an average of coordinated and free H_2O, which varies significantly in broadness and chemical shift depending on sample concentration). All peaks are broad because of the unpaired electrons in the complex.

All the reactions (**A–D**) may be scaled up, but the yields are somewhat reduced. If reactions **B–D** do not yield completely substituted product (as evidenced by the presence of peaks from **A** in the ^1H NMR spectrum), the ligand substitution process can be repeated until the product is pure. All products are stable in air for extended periods of time.

References

1. R. Sessoli, H.-L. Tsai, A. R. Schake, S. Wang, J. B. Vincent, K. Folting, D. Gatteschi, G. Christou, and D. N. Hendrickson, *J. Am. Chem. Soc.* **115**, 1804 (1993).
2. A. Caneschi, D. Gatteschi, and R. Sessoli, *Nature* **365**, 141 (1993).
3. J. Villain, F. Hartman-Boutron, R. Sessoli, and A. Rettori, *Europhys. Lett.* **27**, 159 (1994).
4. L. Thomas, F. Lionti, R. Ballou, D. Gatteschi, R. Sessoli, and B. Barbara, *Nature* **383**, 145 (1996).
5. J. R. Friedman, M. P. Sarachik, J. Tejada, J. Maciejewski, and R. Ziolo, *J. Appl. Phys.* **79**, 6031 (1996).
6. T. Lis, *Acta Crystallography* **B36**, 2042 (1980).
7. J. Eppley, H.-L Tsai, K. Folting, D. N. Hendrickson, and G. Christou, *J. Am. Chem. Soc.* **117**, 301 (1995).

9. USE OF PICOLINEHYDROXIMATE COMPLEXES IN PREPARATION OF URANYL AND LANTHANIDE 15-METALLACROWN-5: UO₂(NO₃)₂[15-MCCu(II)N(PIC)-15] AND {Eu(NO₃)₂[15-MCCu(II)N(PIC)-15]}NO₃

Submitted by ANN J. STEMMLER and VINCENT L. PECORARO[*]
Checked by MINSHENG HUANG and D. COUCOUVANIS[*]

Metallacrowns are a new class of molecular recognition agents that selectively bind cations and/or anions in structures that are similar to organic crown ethers.[1-15] Metallacrowns in the 9-MC-3, 12-MC-4, and 15-MC-5 structure motifs were prepared using variants of the ligand salicylhydroxamic acid. This and related ligands form five- and six-membered chelate rings to the ring metals of the metallacrown. While nonplanar 15-metallacrown-5[8] structure types were prepared with these ligands, planar structures that would form a pentagonal coordination environment around the captured metal could not be realized by this approach.

The strategy we used to prepare planar 15-MC-5 chelating agents required the use of a ligand such as picoline hydroximate (H₂picHA), which forms two 5 membered chelate rings. Five of these ligands can be accommodated into a planar structure to achieve the desired structure. In these new metallacrowns, metals with higher preferred coordination geometries can be sequestered. Herein we provide syntheses for uranyl[-13] and lanthanide[-10] encapsulated 15-MC-5 (Fig. 1).

Materials and General Procedures

Planar 15-metallacrown-5 molecules can be synthesized in a one-step reaction, by the self-assembly of a simple ligand and metal salts. The ligand that is used is picoline hydroxamic acid, H₂PicHA. This ligand was easily prepared by the reaction of hydroxylamine hydrochloride with ethyl picolinate in ethanol using a well-documented procedure for the preparation of hydroxamic acids from esters.[16,17] Reagents used are ethyl picolinate, copper(II) acetate, uranyl(II) nitrate, hydroxyl amine hydrochloride, and potassium hydroxide, which were obtained from Aldrich Chemical Co. All other chemicals and solvents were reagent-grade.

[*] Department of Chemistry, The Willard H. Dow Chemical Laboratories, University of Michigan, Ann Arbor, MI 48109–1055.

A. UO$_2$(NO$_3$)$_2$[15-MC$_{Cu(II)N(pic)}$-15]

Procedure

The metallacrown complex UO$_2$(NO$_3$)$_2$[15-MC$_{Cu(II)N(picHA)}$-15] is synthesized by dissolving Cu(OAc)$_2$·H$_2$O (0.40 g, 2 mmol) in 50 mL of DMF in a 500-mL Erlenmeyer flask. To this solution is added H$_2$picHA (0.28g, 2 mmol) dissolved

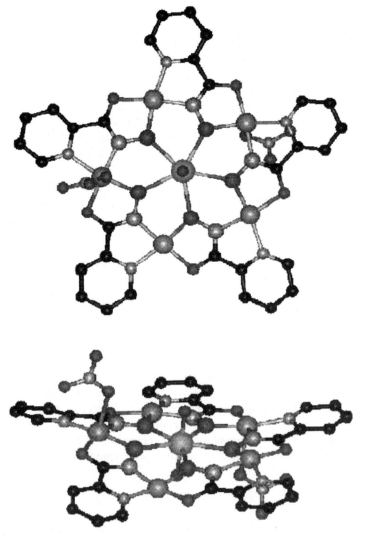

Figure 1. The structure of the Mn$_{12}$O$_{12}$(O$_2$CR)$_{16}$(H$_2$O)$_4$ family of complexes.

in 50 mL of DMF. The solution is allowed to stir for 1 hour, at which point solid $UO_2(NO_3)_2 \cdot 6H_2O$ (0.20 g, 2/5 mmol) is added and the solution is stirred for several hours. Vapor diffusion of ether into the DMF solution resulted in a 69% yield of dark green, rhombic crystals.

Characterization: elemental analysis calculated for Cu_5 $C_{33}H_{29}N_{13}O_{20}U$: Cu, 21.1; C, 26.92; H, 2.31; N, 11.89. Found: Cu, 21.4; C,26.72; H, 1.97; N, 12.27. ESI+MS in methanol gave a molecular ion of 1330 *m/e* and 634 *m/e*. The ^1H NMR spectrum had four resonances at 108, 41.87, 40.97, and 13.21 ppm in d^4 methanol. An infrared spectrum (KBr) exhibits the characteristic $O=U=O$ stretch at 920 cm^{-1}.

B. $\{Eu(NO_3)_2[15\text{-}MC_{Cu(II)N(picHA)}\text{-}15]\}NO_3$

Procedure

The metallacrown complex $\{Eu(NO_3)_2[15\text{-}MC_{Cu(II)N(pic)}\text{-}15]\}NO_3$ is synthesized by dissolving $Cu(OAc)_2 \cdot H_2O$ (0.40 g, 2 mmol) in 50 mL of DMF in a 500 mL Erlenmeyer flask. To this solution is added H_2picHA (0.28g, 2 mrnol) dissolved in 50 mL of DMF. The solution is allowed to stir for one hour, at which point solid $Eu(NO_3)_3 \cdot 6H_2O$ (0.18 g, 0.4 mmol) is added, and the solution is stirred for several hours. Slow evaporation of the solvent results in a 79% yield of dark green, rhombic crystals. (The checkers found it necessary to reduce the volume of the solution to a third of the original volume before crystals formed.) These crystals, obtained in 71% yield, were washed with ether, were dried, and gave satisfactory elemental analysis.

Characterization: Elemental analysis calculated for $Cu_5C_{45}H_{55}N_{18}O_{24}Eu$: Cu, 18.68; C, 331.75; H, 3.23; N, 14.82. Found: Cu, 19.6; C, 31.81; H, 3.25; N, 14.51. FAB+MS in methanol gave a molecular ion of $[M]^+$ 1273 *m/e* and $[M]^{2+}$ 606 *m/e*. The UV–vis spectrum in methanol has a maximum at 580 nm ($\varepsilon = 340$ cm^{-1} M^{-1}). The ^1H NMR spectrum had four resonances at 90.75, 37.14, 36.88, and 12.28 ppm in d^4 methanol. Magnetic moment at 300 K is 5.34 BM. The molecule is soluble and stable in lower alcohols and water. The compound will decompose in moderately (pH < 6) acidic solutions.

References

1. V. L. Pecoraro, A. J. Stemmler, B. R. Gibney, J. J. Bodwin, H. Wang, J. W. Kampf, and A. Barwinski, in *Prog. Inorg. Chem.* Vol. **45**, Chapter 2, K. Karlin (Ed.), Pergamon Press, 1996, p. 83.
2. M. S. Lah and V. L. Pecoraro, *Commun. Inorg. Chem.* **11**, 59 (1990).
3. V. L. Pecoraro, *Inorg. Chim. Acta* **155**, 171 (1989).
4. B. R. Gibney, A. J. Stemmler, S. Pilotek, J. W. Kampf, and V. L. Pecoraro, *Inorg. Chem.* **32**, 6008 (1993).

5. M. S. Lah and V. L. Pecoraro, *J. Am. Chem. Soc.* **111**, 7258, (1989).
6. M. S. Lah and V. L. Pecoraro, *Inorg. Chem.* **30**, 878 (1991).
7. B. R. Gibney, H. Wang, J. W. Kampf, and V. L. Pecoraro, *Inorg. Chem.* **35**, 6184 (1996).
8. D. P. Kessissiglou, J. W. Kampf, and V. L. Pecoraro, *Polyhedron* **13**, 1379–1391 (1994).
9. M. S. Lah, M. L. Kirk, W. Hatfield, and V. L. Pecoraro, *J. Chem. Soc., Chem. Commun.*, 1606 (1989).
10. A. J. Stemmler, J. W. Kampf, M. L. Kirk, B. H. Atasi, and V. L. Pecoraro, *Inorg. Chem.* **38**, 2807–2817 (1999).
11. B. R. Gibney, J. W. Kampf, D. P. Kessissiglou, and V. L. Pecoraro, *Inorg. Chem.* **33**, 4840 (1994).
12. M. S. Lah, B. R. Gibney, D. L. Tierney, J. E. Penner-Hahn, and V. L. Pecoraro, *J. Am. Chem. Soc.* **115**, 5857–5858 (1993).
13. A. J. Stemmler, J. W. Kampf, and V. L. Pecoraro, *Angew. Chem.* **35**, 2841 (1996).
14. A. J. Stemmler, J. W. Kampf, and V. L. Pecoraro, *Inorg. Chem.* **34**, 2271 (1995).
15. A. J. Stemmler, A. Barwinski, V. Young, Jr., and V. L. Pecoraro, *J. Am. Chem. Soc.* **118**, 11962 (1996).
16. J. Hase, J. Kobashi, K. Kawaguchi, and K. Sakamoto, *Chem. Pharm. Bull.* **19**, 363–368 (1971).
17. A. H. Blatt, (Ed.), *Org. Synth.* **2**, 67 (1943).

10. USE OF SALICYL- AND NAPHTHOYLHYDROXIMATE COMPLEXES IN PREPARATION OF MANGANESE AND COPPER 12-METALLACROWN-4 COMPLEXES: Mn(II)(ACETATE)$_2$[Mn(III)(SALICYLHYDROXIMATE)]$_4$ AND (TETRAMETHYLAMMONIUM)$_2$[Cu(II)5 (NAPHTHOYLHYDROXIMATE)$_4$]

Submitted by BRIAN R. GIBNEY and VINCENT L. PECORARO*
Checked by MINSHENG HUANG, D. COUCOUVANIS,* WYNDHAM B. BLANTON, and SCOTT W. GORDON-WYLIE†

The "metallacrown analogy" is a powerful synthetic methodology for the design and preparation of moderate-valence multinuclear complexes from hydroxamic acid and oxime ligands in straightforward high-yield syntheses.[1,2]

Using the salicylhydroxamic acid ligand, complexes with V(V),[3,4] Mn(III),[5–8] Fe(III),[9] Ni(II),[10] Cu(II)[11] and Ga(III)[12] in the ring positions and encapsulated Li(I), Na(I), K(I), Mn(II), Mg(II), Fe(III), Ni(II), Cu(II), and lanthanides[13] ions have been prepared in a variety of structural motifs, 9-MC-3,[3,4,9] 12-MC-4 (Fig. 1), 15-MC-5,[8] and metallacryptates.[12] Added complex stability can be obtained by anions such as acetate, sulfate, or chloride, which serve to bridge

*Department of Chemistry, The Willard H. Dow Chemical Laboratories, University of Michigan, Ann Arbor, MI 48109-1055.
†Department of Chemistry, Carnegie-Mellon University, Pittsburgh, PA 15213-2683.

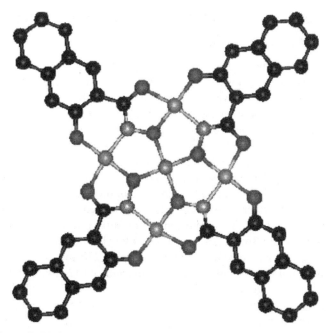

Figure 1. Molecular structure of the dianion, $Cu(II)[12-MC_{Cu(II)N(nha)}-4]$.

between the encapsulated and ring metals, resulting in metal salt selectivity rather than simple metal selectivity. Predictable metallacrown syntheses provide a plethora of coordination complexes by controlled variation in the metallacrown ring metal, encapsulated metal, ring size, bridging anion, and ligand architecture. This also leads to controllable methods for incorporating chirality into macrocycles.[13–15]

A. PREPARATION OF FREE-BASE HYDROXYLAMINE: KOH SOLUTION

Materials and General Procedures

Hydroxylamine hydrochloride (34.8 g, 500 mmol) is dissolved with rapid stirring in 185 mL of hot methanol. Potassium hydroxide (42.0 g, 750 mmol) is dissolved in 115 mL of room-temperature methanol. After cooling both solutions to ambient temperature the potassium hydroxide solution is added with stirring to the hydroxylamine hydrochloride (an exothermic process!). The resulting solution and precipitated potassium chloride are cooled to −20°C in a methanol ice bath before vacuum filtration. The clear colorless supernatant hydroxylamine: KOH solution is immediately used in the preparation of the hydroxamic acid ligands.

B. 3-HYDROXY-2-NAPHTHOHYDROXAMIC ACID (H₃NHA)

Materials and General Procedures

An amount of 3-hydroxy-2-naphthoic acid (18.8 g, 100 mmol) is refluxed in 500 mL of absolute ethanol with 5 mL concentrated sulfuric acid for 18 h. Periodic water removal is achieved via a Dean–Stark trap (8 × 15 mL). The resulting solution is cooled to room temperature and solvent is removed by flash evaporation. Ethyl-3-hydroxy-2-naphthoate crystallizes during this process. Recrystallization from methanol gives 14.0 g (64.8% yield) of ethyl 3-hydroxy-2-naphthoate. If crystals are not obtained by this procedure, one can purify the material by adding 300 mL of an aqueous Na_2CO_3 (pH 9–10) solution to the green-brown oil. The resulting aqueous solution is washed 4 times with 200 mL of diethyl ether. The diethylether is removed under reduced pressure to reveal an orange solid. This solid can be crystallized from hot methanol solutions, on cooling, to afford a tan, crystalline solid (13.6 g, 63 mmol, 63% yield). Free-base hydroxylamine:KOH solution (138 mL, 230 mmoles) is added to a 300-mL methanol solution of 3-hydroxy-2-naphthoate (24.5g, 115 mmol) and stirred for 5 h. The golden potassium salt of 3-hydroxy-2-naphthohydroxamic acid precipitates on standing in a fume hood. This salt is collected by gravity filtration, allowed to dry, and dissolved in warm 1.25 M acetic acid in methanol to crystallize. Yield: 18.7 g of H₃nha (46% yield based on ester). A higher yield of product (6.84 g, 80.6%) is achieved starting with 7.13 g of 3-hydroxy-2-naphthoate (33 mmol).2.

Properties

^{1}H-NMR (CD₃OD, 300 MHz) δ 7.12 (s, 1H), 7.21 (t, $J = 7.4$ Hz, 1H), 7.37 (t, $J = 7.4$ Hz, 1H), 7.58 (d, $J = 8.3$ Hz, 1H), 7.75 (d, $J = 8.3$ Hz, 1H), 8.29 (s, 1H). Mass spectroscopic molecular weight calculated for $C_{11}H_9N_1O_3$: 203.0582, observed 203.0580 *m/e*.

C. (TETRAETHYLAMMONIUM)₂[Cu(II)₅ (NAPHTHOYLHYDROXIMATE)₄]DIMETHYLFORMAMIDE {(Et₄N)₂[Cu(II)12-MC$_{Cu(II)N(nha)}$-4] · DMF}

$$4H_3nha + 5Cu(acetate)_2 \cdot H_2O + 8TMAOH \; TMA_2[Cu(II)12\text{-}MC_{Cu(II)N(nha)}\text{-}4]$$

An amount of $Cu(OAc)_2 \cdot H_2O$ (1.25 g, 6.25 mmol) and tetraethylammonium acetate $\cdot$ 4H₂O (2.60 g, 10 mmol) are dissolved in 50 mL of DMF. A 50-mL solution of DMF containing H₃nha (1.01 g, 5 mmol) is added. The solution immediately turns deep green on addition of the H₃nha and is allowed to stir for 2 h prior to gravity filtration. Vapor diffusion of diethyl ether to the filtrate affords 1.320 g (77%) of forest green crystalline rods. A second recrystallization gives the final product in 74.2% yield.

Properties

(Tetraethylammonium)2[Cu(II)12-MC$_{Cu(II)N(nha)}$-4] shows an electronic absorption at λ_{max} 616 nm ($e = 237$ M^{-1} cm^{-1}).

Anal. Calcd for Cu$_5$C$_{60}$H$_{64}$N$_6$O$_{12}$: Cu, 23.0; C, 52.2; H, 4.7; N, 6.1. Found: Cu, 22.4; C, 51.6; H, 5.1; N, 6.1. FAB-MS($-$) molecular ion 1118 *m/z* (88% of base).

D. Mn(II)(ACETATE)$_2$[Mn(III)(SALICYLHYDROXIMATE)]$_4$ {Mn(OAc)$_2$[12-MC$_{Mn(III)N(shi)}$-4]}

$$4H_3shi + 5Mn(OAc)_2 \cdot 4H_2O + 4Na(acetate)Mn(OAc)_2[12\text{-}MC_{Mn(III)N(shi)}\text{-}4]$$

A 50-mL solution of *N,N*-dimethylformamide containing Mn(OAc)$_2$·4H2O (1.53 g, 6.25 mmol) is added to a 50-mL solution of DMF containing salicylhydroxamic acid, H$_3$shi, (0.77 g, 5.0 mmol, Aldrich) and sodium acetate (0.41 g, 5.0 mmol). The solution immediately turns olive green and is gravity-filtered following 3 h of continuous stirring. Within a day, slow evaporation of solvent lead to 1.11 g (81% yield) of dark green crystals that were collected by gravity filtration. (The checkers reduced the volume of the solution to a third of its original volume to obtain the green crystalline product in 84% yield.)

Properties

Elemental analysis calculated for the metallacrown with six bound and one lattice DMF, Mn$_5$C$_{56}$H$_{78}$N$_{12}$O$_{24}$: Mn, 17.4; C, 42.6; H, 4.9; N, 10.6. Found: Mn, 17.8; C, 42.2; H, 4.5; N, 10.2. The ESI-MS($+$) molecular ion is observed because of loss of one of the two bridging acetate ions, (M-OAc)$^+$: 934 *m/e* (base).

References

1. V. L. Pecoraro, A. J. Stemmler, B. R. Gibney, J. J. Bodwin, H. Wang, J. W. Kampf, and A. Barwinski, in *Prog. Inorg. Chem.*, K. Karlin (Ed.), Pergamon Press, 1996, Vol. 45, Chapter 2, p. 83.
2. M. S. Lah and V. L. Pecoraro, *Commun. Inorg. Chem.* **11**, 59 (1990).
3. V. L. Pecoraro, *Inorg. Chim. Acta* **155**, 171 (1989).
4. B. R. Gibney, A. J. Stemmler, S. Pilotek, J. W. Kampf, and V. L. Pecoraro, *Inorg. Chem.* **32**, 6008 (1993).

5. M. S. Lah and V. L. Pecoraro, *J. Am. Chem. Soc.* **111**, 7258 (1989).
6. M. S. Lah and V. L. Pecoraro, *Inorg. Chem.* **30**, 878 (1991).
7. B. R. Gibney, H. Wang, J. W. Kampf, and V. L. Pecoraro, *Inorg. Chem.* **35**, 6184 (1996).
8. D. P. Kessissiglou, J. W. Kampf, and V. L. Pecoraro, *Polyhedron* **13**, 1379–1391 (1994).
9. M. S. Lah, M. L. Kirk, W. Hatfield, and V. L. Pecoraro, *J. Chem. Soc., Chem. Commun.*, 1606 (1989).
10. G. Psomas, A. J. Stemmler, C. Dendrinou-Samara, J. J. Bodwin, M. Schneider, M. Alexiou, J. W. Kampf, D. P. Kessissoglou, and V. L. Pecoraro, *Inorg. Chem.* (in press).
11. B. R. Gibney, J. W. Kampf, D. P. Kessissiglou, and V. L. Pecoraro, *Inorg. Chem.* **33**, 4840 (1994).
12. M. S. Lah, B. R. Gibney, D. L. Tierney, J. E. Penner-Hahn, and V. L. Pecoraro, *J. Am. Chem. Soc.* **115**, 5857–5858, (1993).
13. A. J. Stemmler, J. W. Kampf, and V. L. Pecoraro, *Angew. Chem., Int. Ed. Engl.* **35**, 2841 (1996).
14. J. A. Halfen, J. J. Bodwin, and V. L. Pecoraro, *Inorg. Chem.* **37**, 5416 (1998).
15. A. J. Stemmler, J. W. Kampf, M. L. Kirk, B. H. Atasi, and V. L. Pecoraro, *Inorg. Chem.* **38**, 2807–2817 (1999).

Chapter Two

USEFUL REAGENTS AND LIGANDS

11. HOMOLEPTIC TRANSITION METAL ACETONITRILE CATIONS WITH TETRAFLUOROBORATE OR TRIFLUOROMETHANESULFONATE ANIONS

Submitted by ROBERT A. HEINTZ, JENNIFER A. SMITH, PAUL S. SZALAY,
AMY WEISGERBER, and KIM R. DUNBAR[*]
Checked by KARSTEN BECK and DIMITRI COUCOUVANIS[†]

Transition metal cations solvated by weakly coordinating solvents are useful precursors for a variety of nonaqueous synthetic applications. The dissociation of acetonitrile ligands opens up vacant coordination sites for catalysis,[1,2] and allows transition metals to be introduced into extended arrays formed by condensation reactions with polydentate ligands.[3–5] Although salts of homoleptic acetonitrile metal cations with tetrafluoroborate anions have been known since the 1960s, there has been no detailed description of the syntheses and characterization of these compounds.[6] Likewise, there have been general descriptions of the syntheses of trifluoromethanesulfonate salts, but no specific outline for individual metals has been published.[7] In this report, the syntheses of $[M^{II}(CH_3CN)_x][BF_4]_2$ (M = Cr,Mn,Fe,Co,Ni,Cu) and $[M^{II}(CH_3CN)_x][SO_3CF_3]_2$ (M = Mn,Fe,Co,Ni) are described. Two different methods are used to prepare tetrafluoroborate salts: (1) the oxidation of metals with nitrosonium tetrafluoroborate, first described by Hathaway et al.;[8] and (2) the dehydration of aqueous tetrafluoroborate salts in a Soxhlet extractor with molecular sieves. The oxidation reaction is rapid, anhydrous, and recommended for smaller-scale preparations because of the relatively

[*]Department of Chemistry, Texas A&M University, College Station, TX 77843-3255.
[†]Department of Chemistry, University of Michigan, Ann Arbor, MI 48109-1055.

high cost of $NOBF_4$. The water dehydration method is slow, but is recommended for larger reaction scales and for those concerned with cost. Alternative water dehydration methods using triethylorthoformate or acetic anhydride were reported by Driessen and Reedijk.[9] The trifluoromethanesulfonate derivatives are prepared by dissolution of the anhydrous salts $M(SO_3CF_3)_2$ in CH_3CN.

Materials and General Procedures

Acetonitrile was dried by distillation over 3-Å molecular sieves that had been activated by heating in a column to 200°C under a stream of dry nitrogen gas. All other solvents used in the syntheses were dried with appropriate drying agents and freshly distilled before use. Manganese chips (99%), iron chips (99%), cobalt (99.5%), nickel powder (99.99%), and copper (99%) were purchased from Aldrich Chemical Company. The reagents nitrosonium tetra-fluoroborate, $NOBF_4$, and tetrafluoroboric acid, HBF_4 (48–50%), were obtained from commercial sources. $[M(H_2O)_x](BF_4)_2$ salts of Fe(II), Co(II), Ni(II), and Cu(II) were purchased from Aldrich, whereas $[Mn(H_2O)_x](BF_4)_2$ was prepared directly from the metal and aqueous tetrafluoroboric acid (HBF_4) (48–50%) under an inert atmosphere as reported for V(II).[10] The aqueous salts of the other metals can also be prepared by the same route. Tetracetatodichromium[11] and barium triflate were prepared as described in the literature.[12] All manipulations were carried out under a dry nitrogen atmosphere using standard Schlenk line techniques.

A. BISTETRAFLUOROBORATE–HEXAKISACETONITRILE METAL(II) $\{[M(CH_3CN)_6](BF_4)_2\}$ (M = Fe,Co,Ni)

This method, which involves preparation of acetonitrile cation salts of tetrafluoroborate from oxidation of metals with $NOBF_4$, is the one reported in a general manner by Hathaway and co-workers according to the following equation:

$$M + NOBF_4 + xCH_3CN \rightarrow [M(CH_3CN)_x][BF_4]_2 + NO$$

Procedure

■ **Caution.** *Nitrosonium tetrafluoroborate ($NOBF_4$) is corrosive and moisture sensitive and should be stored under nitrogen in a freezer.*

In a typical reaction, 0.50 g of $NOBF_4$ (4.32 mmol) and a molar excess of the solid, powdered metal are loaded into a 250-mL round-bottomed flask. Acetonitrile (40 mL) is added to the mixture with stirring and the flask is exposed to vacuum until the solution begins to bubble vigorously. The system is then back-filled with nitrogen and the process is repeated 3 additional times to remove the liberated NO gas. The flask is then isolated from the vacuum manifold and stirred for 12 h under reduced pressure. The solution volume is reduced under vacuum to ~10 mL (or until cloudy), transferred via cannula to a clean flask, treated with diethyl ether (50 mL) to complete the precipitation, and cooled in an ice bath to allow the solid to settle. The liquid is removed through a cannula, and the solid product is washed with copious amounts of diethyl ether and thoroughly dried under vacuum. Typical yields are 0.658 g (65%) for the white Fe product, 0.897 g (87%) for the pink Co solid, and 0.453 g (42%) for the blue Ni compound.

Slight modifications of this procedure are used to prepare $[Mn(CH_3CN)_4]$ $(BF_4)_2$ and $[Cu(CH_3CN)_4](BF_4)_2$.

B. BISTETRAFLUOROBORATE–TETRAKISACETONITRILE MANGANESE(II) $\{[Mn(CH_3CN)_4](BF_4)_2\}$

Procedure

Manganese chips (1.60 g, 29.1 mmol, excess) and $NOBF_4$ (1.64 g, 14.0 mmol) are added to a 250-mL round-bottomed flask under an inert atmosphere. Aceto-nitrile (40 mL) is added with stirring and the NO(g) is allowed to escape through a needle outlet placed in a septum. After stirring for 20 h, the reaction mixture is filtered and the colorless filtrate is concentrated under a dynamic vacuum to ~5 mL, after which time diethylether (40 mL) is added to produce a white precipitate. The supernatant is removed from the solid with a cannula and is discarded. The solid is washed with diethyl ether and dried under vacuum to yield a white crystalline product. A typical yield is 2.09 g (76%).

C. BISTETRAFLUOROBORATE–TETRAKISACETONITRILE COPPER(II) $\{[Cu(CH_3CN)_4](BF_4)_2\}$

Procedure

An amount of $NOBF_4$ (0.425 g, 3.64 mmol) and a slight molar excess of Cu pow-der are loaded into a round-bottomed flask no smaller than 250 mL in capacity. Ethyl acetate (30 mL) is added to the mixture with stirring, and the flask is exposed to a vacuum until the solution begins to bubble vigorously. The flask is then backfilled with inert gas and the vacuum is again briefly applied. This process is repeated 3 additional times to remove NO gas, after which time the flask is isolated from the vacuum manifold and stirred for 12 h under reduced pressure. The blue solution is reduced to 10 mL (or until saturated), transferred via cannula to a separate flask, and treated with acetonitrile (30 mL) to precipitate the desired pale blue product. The mixture is then cooled in an ice bath to settle out the product. The solid product is finally washed with copious amounts of diethylether, dried under vacuum, and stored in a drybox. The yield is 0.584 g (75%). Ethyl acetate is used in the first part of the procedure because direct oxidation of copper metal in acetonitrile leads to a mixture of cuprous [Cu(I)] and cupric [Cu(II)] products. Evidently, ethyl acetate is not a good ligand for stabilizing the cuprous ion in solution.

The remaining procedures presented in this section (syntheses D–H) involve preparation of acetonitrile-solvated cations from the water-solvated cations. The common formula for these compounds is as follows:

$$[M(H_2O)_x](BF_4)_2 + CH_3CN \xrightarrow[-H_2O]{\Delta,\, sieves} [M(CH_3CN)_x](BF_4)_2$$

D. $[M(CH_3CN)_x][BF_4]_2$ FROM $[M(H_2O)_x][BF_4]_2$ (M = Mn,Fe,Co,Ni,Cu)

Procedure

■   **Caution.**   *The fluoroboric acid aquo complex (HBF_4) is corrosive and a lachrymator. Triethyloxonium tetrafluoroborate, $[(Et_3O)(BF_4)]$ is toxic, corrosive, a neurologic hazard, a suspected carcinogen, and moisture-sensitive and should be stored under nitrogen in a freezer.*

This method involves the exchange of the H_2O ligand for CH_3CN and subsequent removal of the water from the acetonitrile solution. This method has been used with the Mn(II), Fe(II), Co(II), Ni(II), and Cu(II) solvated cations.

Amounts of $[M(H_2O)_x](BF_4)_2$ in the range of 3–5 g are transferred to a 250-mL Schlenk flask in a nitrogen-purged glovebag. The Schlenk flask is fitted with a Soxhlet extractor (3-cm-diameter 15-cm-high column) that has been loaded with 3-Å molecular sieves, and dry acetonitrile ($\sim$120 mL) is added through the top of the extractor to cover the sieves. It is important that sufficient acetonitrile be added to ensure that when the extractor is filled there is acetonitrile remaining in the bottom Schlenk flask. A condenser is added to the top of the extractor and the acetonitrile solution is refluxed and extracted through the sieves for 4–5 days. The mixture is then cooled and filtered to remove any insoluble materials. The solution is then concentrated under vacuum, diethyl ether is added to precipitate the product, and the solvent is removed with the use of a cannula. The solid is washed with copious amounts of diethyl ether and dried thoroughly under vacuum. Typical yields are 3.34 g (91%) for Mn, 3.73 g (85%) for Fe, 4.33 g (87%) for Co, 3.96 g (88%) for Ni, and 3.47 g (91%) for Cu.

If larger quantities of the water precursors are used, one must use multiple batches of freshly activated sieves during the dehydration process. If water is detected in the infrared spectrum of the product, then it is necessary to redissolve the solids in dry acetonitrile and repeat the Soxhlet extraction reflux procedure.

E. BISTETRAFLUOROBORATE–TETRAKISACETONITRILE CHROMIUM(II) $[Cr(CH_3CN)_4](BF_4)_2$ FROM CHROMOUS ACETATE $[Cr_2(O_2CCH_3)_4]$

Procedure

Because of its extreme sensitivity, it is recommended that the Cr complex be prepared by an alternative method to the two outlined above.[13] To a stirring suspension of 1 g of $Cr_2(O_2CCH_3)_4$ (2.94 mmol) in acetonitrile (50–60 mL) is added 13 mL (13 mmol, 4.4 equiv) of a 1 M CH_2Cl_2 solution of $(Et_3O)(BF_4)$. The reaction is allowed to stir overnight to give a blue solution and a small quantity of green precipitate. The blue solution is filtered to remove the solid, which is discarded. The volume of the filtrate is reduced under vacuum to approximately 10–15 mL and two portions of CH_2Cl_2 (20 mL) are added to precipitate the blue product. (Slow and careful addition of CH_2Cl_2 can produce crystals.) The solution is removed from the solid with a cannula and the product is washed with CH_2Cl_2 (20 mL). The solid is then dried under vacuum to give a light blue powder and transferred for storage into the drybox. A typical yield is 1.13 g (49%).

F. PREPARATION OF ACETONITRILE CATIONS WITH TRIFLUOROMETHANESULFONATE (TRIFLATE) ANIONS

In applications for which greater solubility is desired the trifluoromethanesulfonate (triflate) counterion is a better choice than tetrafluoroborate. Although $(CF_3SO_3)^-$ is more highly coordinating than $(BF_4)^-$, it is quite labile and displays high thermal stability. Triflate salts are used widely as anions for highly cationic clusters in order to increase their solubility and volatility for mass spectrometric analysis.[14–16]

Procedure

All glassware is pretreated with the commercially available reagent glassclad,[17] dried in an oven at 125°C for at least one hour, and cooled in a vacuum desiccator over drierite (100% CaSO$_4$). All compounds are handled in a dry anaerobic atmosphere.

G. ANHYDROUS METAL TRIFLATE COMPLEXES

$$[M^{II}(H_2O)_x][SO_3CF_3]_2 \xrightarrow[-H_2O]{\Delta} M^{II}(SO_3CF_3)_2 (M = Mn, Fe, Co, Ni)$$

The metal aqua triflate complexes are prepared as previously reported.[7,12] It was found that drying the solids by rotary evaporation is not sufficient to remove all of the water; therefore the samples should be heated for at least 12 h at ∼90°C under vacuum. Prolonged heating should be avoided, as decomposition of the sample occurs. Extended X-ray absorption fine structure (EXAFS) and vibrational spectral data of anhydrous metal triflates have already been reported and can be used to verify the identity of the products.[18] The colors of the anhydrous metal triflate complexes are white for Mn, white/off-white for Fe, pink for Co, and yellow for Ni.

H. METAL ACETONITRILE COMPLEX CATIONS
WITH TRIFLATE ANIONS

Procedure

Acetonitrile is refluxed over 3-Å molecular sieves and further dried by elution from an activated alumina column.

The anhydrous triflate complex (Mn (1.133 g, 3.21 mmol)–Fe (0.382 g, 1.08 mmol)–Co (0.257 g, 0.720 mmol)–Ni (0.731 g, 2.05 mmol) is added to a dry 250-mL round-bottomed Schlenk flask and treated with ∼150 mL of acetonitrile for 12 h. The volume of the solution is reduced to ∼5 mL under vacuum and the product is precipitated by the addition of diethyl ether (40 mL). The solution is removed via cannula, and the solid is dried under vacuum, redissolved in 20 mL acetonitrile, and finally filtered through a medium-porosity Schlenk frit to remove minor decomposition products carried over from the M(O$_3$SCF$_3$)$_2$ precursors. The solution is reduced to ∼5 mL under vacuum and precipitated by the addition of ∼40 mL of diethyl ether. The colors and typical yields are 0.917 g (55%) for the white Mn product, 0.326 g (50%) for a white/off-white Fe product, 0.255 g (59%) for a pink-peach Co solid, and 0.873 g (71%) for a blue Ni compound. The products are exceedingly moisture-sensitive.

General Properties of Metal Reagents

1. *Tetrafluoroborate Salts.* $[Cr(CH_3CN)_4][BF_4]_2$ is extremely air/moisture-sensitive. The Mn, Fe, Co, Ni, and Cu derivatives are moisture-sensitive. Accordingly, all of these compounds should be stored and handled only under an inert atmosphere. The salts are soluble in coordinating solvents such as methanol, ethanol, nitriles, acetone, nitromethane, and tetrahydrofuran, and are insoluble in essentially all other solvents. The infrared spectra display (CN bands in the 2200–2300 cm^{-1} region typical of coordinated acetonitrile. The electronic spectra exhibit the expected $d \to d$ transitions and the room-temperature (RT) effective magnetic moments are typical for high-spin $3d$ metal complexes. The spin-only values are exceeded in the cases of Co(II) and Fe(II) as expected as a result of to spin–orbit coupling. Infrared, magnetic susceptibility, and electronic spectral data are summarized in Table I.

2. *Triflate Salts.* The anhydrous triflate complexes are exceedingly moisture-sensitive; thus special care must be taken during transfer and storage. If the salts are exposed to moist ambient air, they will revert to the aqueous complexes in several hours. Infrared spectra of $[M(CH_3CN)_{4-6}](SO_3CF_3)_2$ salts measured as Nujol mulls exhibit two sharp CN vibrations of approximately equal intensity. The electronic spectra of the $[M(CH_3CN)_{4-6}](SO_3CF_3)_2$ compounds are similar

TABLE I. UV–Visible, Magnetic Susceptibility and Infrared Data for $[M(CH_3CN)_x][BF_4]_2$ [M = Cr, Mn, Cu ($x = 4$); Fe, Co, Ni ($x = 6$)]

M^{II}	Color	ν_{CN} (cm^{-1})	μ_{eff} (B.M.)	λ_{max} (nm)	ε (M^{-1} cm^{-1})
Crz	blue	2333 (s), 2305 (s)	3.84	643	10.3
Mn	white	2312 (m), 2284 (m)	5.81	536	0.3
				408	1.2
Fe	white	2310 (m), 2287 (m)	5.30	912	2.8
Co	pink	2316 (m), 2292 (m)	5.00	492	11.3
				476	11.0
Ni	blue	2316 (m), 2292 (m)	2.93	958	3.1
				582	2.5
				358	10.8
Cu	blue	2322 (m), 2300 (m)	1.83	744	10.8

- Infrared: Samples were prepared as nujol mulls placed between KBr plates.
- Magnetic susceptibility: Measurements were recorded on a Squid susceptometer at 298 K.
- UV-Visible: Concentrations from 30–24 mM in acetonitrile were used for Cr, Fe, Co, Ni, and Cu; 374 mM acetonitrile solutions were used for Mn; path length of 1 cm; background was solvent vs solvent.

TABLE II. UV-Visible, Magnetic Susceptibility and Infrared Data for
[M(CH₃CN)ₓ][SO₃CF₃]₂ [M = Mn (x = 4); Fe, Co, Ni (x = 6)]

M^{II}	Color	ν_{CN} (cm^{-1})	ν_{SO} (cm^{-1})	μ_{eff} (B.M.)	λ_{max} (nm)	ε (M^{-1}cm^{-1})
Mn	white	2311(s)	1043(s)	6.1	unresolvable	
		2281(s)				
Fe	white	2314(s)	1039(s)	5.8	904	4.0
		2285(s)				
Co	pink	2320(s)	1039(s)	4.0	413	4.7
		2293(s)			491	19.8
Ni	blue	2326(s)	1032(s)	3.3	499	7.0
		2299(s)			572	9.2
					947	4.2

- Infrared: Samples were prepared as nujol mulls placed between KBr plates.
- Magnetic susceptibility: Measurements were recorded on a Squid susceptometer at 298 K.
- UV-Visible: Concentrations from 15–42 mM in acetonitrile were used for Fe, Co, Ni, and Mn: path length of 1 cm; background was solvent vs solvent.

to the tetrafluoroborate salts and are in accord with divalent metal ions in weak to moderate ligand field environments.[19] The RT effective magnetic moments for the products are consistent with the spin-only values for isolated M(II) ions. Co(II) and Fe(II) have higher RT effective magnetic moments because of spin–orbit coupling. Magnetic susceptibility and infrared and electronic spectral data are summarized in Table II.

Sample Elemental Analyses. Calcd. for MnN₄C₈H₁₂B₂F₈: C, 24.46%; H, 3.08%; N, 14.27%. Found: C, 24.27%; H, 3.13%; N, 14.03%. Calc'd. for CuN₄C₈H₁₂B₂ F₈: C, 23.94%; H, 3.01%; N, 13.96%. Found: C, 23.77%; H, 2.96%; N, 13.85%.

References and Notes

1. A. Sen and T. Lai, *J. Am. Chem. Soc.* **103**, 4627 (1981).
2. A. Renzi, A. Panunzi, and A. J Vitagliano, *J. Chem. Soc., Chem. Commun.*, 471, (1976).
3. G. M. Finniss, E. Canadell, and K. R. Dunbar, *Angew. Chem., Int. Ed. Engl.* **35**, 2772 (1996).
4. H. Zhao, R. A. Heintz, R. D. Rogers, and K. R. Dunbar, *J. Am. Chem. Soc.* **118**, 12844 (1996).
5. W. E. Buschmann and J. S. Miller, *Inorg. Chem. Commun.* **1**, 174 (1998).
6. W. L. Groeneveld, J. J. Van Houte, and A. P. Zuur, *Recl. Trav. Chim. Pays-Bas*, 755 (1968).
7. N. E. Dixon, G. A. Lawrance, P. A. Lay, A. M. Sargeson, and H. Taube, *Inorg. Synth.* **24**, 243 (1986).
8. B. J. Hathaway, D. G. Holah, and A. E. Underhill, *J. Chem. Soc.*, 2444 (1964).
9. W. L. Driessen and J. Reedijk, *Inorg. Synth.* **29**, 111 (1992).

10. D. G. L. Holt, L. F. Larkworthy, D. C. Povey, G. W. Smith, and G. J. Leigh, *Inorg. Chim. Acta* **169**, 201 (1990).

11. F. Hein and S. Herzog, in *Handbook of Preparative Inorganic Chemistry*, Vol. 2, 2nd ed., G. Brauer (Ed.), Academic Press, London, 1965.

12. W. C. Kupferschmidt and R. B. Jordan, *Inorg. Chem.* **21**, 2089 (1982).

13. The formulation of this compound as containing four CH_3CN ligands has been verified by X-ray crystallography; see R. T. Henriques, E. Herdtweck, F. E. Kuhn, A. D. Lopes, J. Mink, and C. C. Ramao, *J. Chem. Soc., Dalton Trans.*, 1293 (1998).

14. B. Olenyuk, A. Fechtenkotter, and P. J Stang, *J. Chem. Soc., Dalton Trans.*, 1707 (1998).

15. B. Olenyuk and P. J., Stang, *Acc. Chem. Res.* **30**, 502 (1997).

16. G. A. Lawrance, *Chem. Rev.* **86**, 17 (1986).

17. M. A. Drezdzon and D. F. Shriver, *The Manipulation of Air-Sensitive Compounds*, 2nd ed., Wiley New York, 1986, p. 11 (this product is commercially available from United Chemical Technologies, Inc.).

18. K. Boumizane, M. H. Herzog-Cance, D. J. Jones, J. L. Pascal, J. Potier, and J. Roziere, *Polyhedron* **10**, 2757 (1991).

19. The Mn sample did not exhibit any observable transitions even at very high concentrations in a 1-cm-pathlength cell. Typically 5-cm-pathlength cells are required to measure the very weak transitions associated with Mn(II) ions.

12. SYNTHESIS OF $[M^{II}(NCMe)_6]^{2+}$ (M = V,Cr,Mn,Fe,Co,Ni) SALTS OF TETRA[3,5-BIS(TRIFLUOROMETHYL) PHENYL]BORATE

Submitted by WAYNE E. BUSCHMANN and JOEL S. MILLER*
Checked by KRISTIN BOWMAN-JAMES and CYNTHIA. N. MILLER[†]

Sources of metal ions that are soluble in nonaqueous solvents and possess very weakly or, ideally, noncoordinating redox-stable anions are essential for developing many areas of inorganic and materials chemistry. Anions such as $(CF_3SO_3)^-$ and $(BF_4)^-$ require moderately strong Lewis bases (i. e., MeCN, NH_3) to displace them from the coordination sphere of a metal ion.[1] Other nominally inert anions such as $(BPh_4)^-$ can react with a metal center.[1] The solubility of these salts is limited to more coordinating solvents such as MeCN and is generally much less in THF. Tetrakis[3,5-bis(trifluoromethyl)phenyl]borate $(TFPB)^{-\,2}$ avoids these limitations as it is essentially noncoordinating[1a] and redox-stable between ~ 1.6 and -1.8 V versus Ag/AgCl(aq).[3,4]

Its salts are extremely soluble (even in diethyl ether) and can be prepared with many different cations, including Na^+,[2] Ag^+,[5] Tl^+,[6] [tetraalkylammonium]$^+$,[3]

*Department of Chemistry, University of Utah, Salt Lake City, UT 84112-0850.
[†] Department of Chemistry, University of Kansas, Lawrence, KS 66045.

and ferrocenium.[7] The complex $[M^{II}(NCMe)_6](TFPB)_2$ (M = V,Cr,Mn, Fe,Co,Ni), because of the its facile ability to lose MeCN, provides metal cation sources that are soluble in MeCN, THF, and Et_2O.[8] Preparation of $[M^{II}(NCMe)_6]$-$(TFPB)_2$ (M = Mn,Fe,Co,Ni) is achieved by anion exchange between the metal dihalide and Ag(TFPB). A nonaqueous preparation of the Ag(TFPB) is included that avoids the tedious light-sensitive extraction and drying that existing aqueous routes require.[5] When M = Cr and V, the silver salts back-react (oxidize) the divalent metal ions, thereby requiring a different strategy based on nonaqueous V^{II} and Cr^{II} precursors. Reliable and convenient nonaqueous routes to $[V^{II}(NC\text{-}Me)_6](BPh_4)_2$ are also presented. In another contribution to this volume the preparation of $[Cr^{II}(NCMe)_4](BF_4)_2$ is reported.[9]

Materials and General Procedures

All manipulations are performed under nitrogen or argon using standard Schlenk techniques or in a Vacuum Atmospheres inert-atmosphere DriLab enclosure. Dichloromethane is dried and distilled under N_2 from CaH_2. Acetonitrile is dried and twice distilled under N_2 from CaH_2. Diethylether and tetrahydrofuran are dried and distilled under N_2 from sodium benzophenone ketyl radical. $Ag(CF_3SO_3)$, $[(Ph_3P)_2N]Cl$ (Aldrich), $AgNO_3$ (Alfa), $MnCl_2$ (Allied Chemical, anhydrous, 99%), $NiBr_2$, $CoCl_2$, and $FeBr_2$ (Mallinckrodt, anhydrous, 99%) are used as received. $VCl_3(MeCN)_3$ is prepared by the literature procedure.[11] $[(Ph_3P)_2N][TFPB]$ is obtained by mixing together dichloromethane solutions containing stoichiometric amounts of $[(Ph_3P)_2N]Cl$ and K(TFPB), filtering off the KCl byproduct, and removing the dichloromethane under reduced pressure leaving a white solid.[*]

The preparation of $[V(NCMe)_6][BPh_4]_2$ is based on a published procedure[10] with only slight modifications as follows. A solution of $VCl_3(MeCN)_3$ (0.749 g, 2.67 mmol) in 85 mL of MeCN is added to a 125-mL Erlenmeyer flask containing $NaBPh_4$ (2.79 g, 8.14 mmol). This green solution is stirred at room temperature for 2 h, concentrated under reduced pressure to ~ 60 mL, and left at room temperature for 3 weeks to give a mixture of blue-green prisms of $[V(NCMe)_6](BPh_4)_2$ and colorless NaCl. This mixture is filtered and the solid (~ 2.3 g) stirred in 60 mL of MeCN, the undissolved NaCl (~ 0.130 g) filtered off, and the blue solution cooled to $-40°C$ to recrystallize 2.05 g (82% yield) of blue-green prisms in 2 days. IR (Nujol): 2310(m), 2281(m) cm^{-1}.

---

[*] It should be noted that the checkers obtained $\sim 10\%$ lower yields for the $[M^{II}(NCMe)_6][TFPB]_2$ (M = V,Cr,Mn,Fe,Co,Ni) complexes.

A. POTASSIUM TETRA-3,5-BIS(TRIFLUOROMETHYL) PHENYLBORATE, (K{B[C$_6$H$_3$(CF$_3$)$_2$]$_4$})

$$Mg + C_6H_3I(CF_3)_2 \rightarrow MgI[C_6H_3(CF_3)_2]$$
$$8MgI[C_6H_3(CF_3)_2] + 2BF_3 \rightarrow Mg\{B[C_6H_3(CF_3)_2]_4\}_2 + 3MgF_2 + 4MgI_2$$
$$K_2CO_3 + Mg\{B[C_6H_3(CF_3)_2]_4\}_2 \rightarrow K\{B[C_6H_3(CF_3)_2]_4\} + MgCO_3$$

A flame-dried 200-mL round-bottomed flask equipped with a reflux condenser, addition funnel, stir bar, and Mg turnings (0.824 g, 33.9 mmol) is evacuated and backfilled with N$_2$ 3 times and charged with 10 mL Et$_2$O. A 20-mL Et$_2$O solution of 3,5-(trifluoromethyl)iodobenzene (3.84 g, 11.3 mmol) is added dropwise over ~30 min with stirring and the reaction mixture turns green-brown. After stirring for 1 h a 5-mL Et$_2$O solution of BF$_3$·Et$_2$O (0.320 g, 2.26 mmol) is added drop-wise over ~30 min with stirring. The reaction mixture is refluxed overnight,* after which, it is poured portionwise into a 500-mL Erlenmeyer flask containing a stirring solution of K$_2$CO$_3$ (25 g in 100 mL H$_2$O) to generate the potassium salt. The MgCO$_3$ precipitate is filtered off and washed 4 times with 25 mL of Et$_2$O. The Et$_2$O layer is removed and the aqueous layer saturated with KCl and, in a 250-mL separatory funnel, extracted 4 times with 25 mL of Et$_2$O. The combined Et$_2$O layers are concentrated to dryness under reduced pressure, leaving an orange residue. The residue (~2 g) is redissolved in 20 mL CH$_2$Cl$_2$/THF (1 : 1) and layered with 20-mL hexane portions to recrystallize the product in 4 days.† The product is recrystallized a second time in the same manner to give colorless prisms (1.77 g, 87%) that become opaque when dried in vacuo due to solvent loss. This compound is stable at room temperature indefinitely. (dp: >300°C). IR (Nujol): 1615(w), 1360(m), 1285(s), 1143(s), 1084(m). NMR: ^{1}H (CD$_2$Cl$_2$): 7.74 (m, o, 8H), 7.58 (m, p, 4H) ppm. ^{19}F (CD$_2$Cl$_2$): 99.77 (s) ppm relative C$_6$F$_6$.‡

Anal. Calcd. for BC$_{38}$F$_{24}$H$_{21}$K: C, 42.61, H, 1.34. Found: C, 42.98; H, 1.52.

---

* Yields can be increased (from 48.4 to 71.8%) when the reaction is allowed to reflux for 24–30 h with a dry-ice condenser according to a checker of this synthesis.

† Layering the ether solution with hexane each day for 3 days at freezer temperatures was a preferred method of crystallization for a checker of this synthesis.

‡ This value was observed by a checker at-134.233 ppm relative to CFCl$_3$.

B. SILVER TETRA-3,5-BIS(TRIFLUOROMETHYL) PHENYLBORATE (Ag{B[C₆H₃(CF₃)₂]₄})

B. SILVER TETRA-3,5-BIS(TRIFLUOROMETHYL) PHENYLBORATE $(Ag\{B[C_6H_3(CF_3)_2]_4\})$

$$AgNO_3 + K\{B[C_6H_3(CF_3)_2]_4\} \rightarrow Ag\{B[C_6H_3(CF_3)_2]_4\} + KNO_3$$

$Ag\{B[C_6H_3(CF_3)_2]_4\}$ is prepared by adding a dry 7-mL MeCN solution of $Ag(NO_3)$ (1.086 mmol, 0.1846 g) to a 10-mL MeCN solution of $K\{B[C_6H_3-(CF_3)_2]_4\}$ (1.109 mmol, 1.000 g) in a 50-mL Erlenmeyer flask. After stirring for 10 min, 10 mL of Et_2O is added to help precipitate KNO_3, which is removed by filtration. The filtrate is concentrated under reduced pressure to dryness (without heating above $\sim 80°C$). The white residue is dissolved in 25 mL of THF while minimizing light exposure and the remaining KNO_3 removed by filtration. The filtrate is layered with 40 mL of hexane and kept at $-30°C$ to crystallize the product. Colorless prisms (0.834 g, 79% yield) that become slightly cloudy when dried in vacuo are recovered after 7 days. This compound is stable for several months when stored cold in darkness. IR (Nujol): 1609(m), 1357(s), 1281(vs), 1144(vs), 1124(vs) cm^{-1}. TGA: 87.8% weight loss between 122°C (15°C/min) and 250°C leaving a white residue; calculated 86.9% weight loss for AgF as residue. NMR: ^1H (CD_2Cl_2): 7.77 (*m*, *o*, 8H), 7.62 (*m*, *p*, 4H) ppm. ^{19}F (CD_2Cl_2): 99.77(s) ppm relative C_6F_6.

Anal. Calcd. for the 0.28 hexane solvate $AgBC_{33.68}H_{14.24}F_{24}$: C, 40.65; H, 1.61. Found: C, 40.55; H, 1.84.

C. HEXAKIS(ACETONITRILE)MANGANESE(II)BIS-TETRA-3,5-BIS(TRIFLUOROMETHYL)PHENYLBORATE ([Mnᴵᴵ(NCMe)₆]{B[C₆H₃(CF₃)₂]₄}₂)

C. HEXAKIS(ACETONITRILE)MANGANESE(II)BIS-TETRA-3,5-BIS(TRIFLUOROMETHYL)PHENYLBORATE $([Mn^{II}(NCMe)_6]\{B[C_6H_3(CF_3)_2]_4\}_2)$

$$MnCl_2 + 2Ag\{B[C_6H_3(CF_3)_2]_4\} + 6MeCN \rightarrow$$
$$[Mn^{II}(NCMe)_6]\{B[C_6H_3(CF_3)_2]_4\}_2 + 2AgCl$$

A dry 15-mL MeCN solution of $Ag\{B[C_6H_3(CF_3)_2]_4\}$ (1.332 mmol, 1.293 g) is added to $MnCl_2$ (0.6662 mmol, 0.0838 g) in a 25-mL Erlenmeyer flask, and the mixture is stirred at room temperature in darkness for 12 h. The silver chloride is filtered off, leaving a pale pink solution. This is concentrated under reduced pressure to ~ 6 mL and cooled to $-40°C$ for 24 h to crystallize colorless needles that are filtered cold (>80% yield) and stored at $-40°C$. IR (Nujol): 2315(m), 2289(m) cm^{-1}. TGA (thermogravimetric analysis) weight loss between room temperature and 450°C (15°C/min): 84.2%.

Anal. Calcd. for B$_2$C$_{76}$F$_{48}$H$_{42}$MnN$_6$: C, 45.02; H, 2.09; N, 4.14. Found: C, 45.29; H, 2.00; N, 4.57.

D. HEXAKIS(ACETONITRILE)IRON(II)BIS-TETRA-3,5-BIS(TRIFLUOROMETHYL)PHENYLBORATE ([Fe^{II}(NCMe)$_6$]{B[C$_6$H$_3$(CF$_3$)$_2$]$_4$}$_2$)

$$FeBr_2 + 2Ag\{B[C_6H_3(CF_3)_2]_4\} + 6MeCN \rightarrow$$
$$[Fe^{II}(NCMe)_6]\{B[C_6H_3(CF_3)_2]_4\}_2 + 2AgBr$$

A dry 15-mL MeCN solution of Ag{B[C$_6$H$_3$(CF$_3$)$_2$]$_4$} (1.332 mmol, 1.293 g) is added to FeBr$_2$ (0.6662 mmol, 0.1437 g) in a 25-mL Erlenmeyer flask and the mixture is stirred at room temperature in darkness for 12 h. The silver bromide is filtered off, leaving a pale yellow solution. This is concentrated under reduced pressure to ~6 mL and cooled to −40°C for 24 h to crystallize colorless needles that are filtered cold (>80% yield) and stored at −40°C. IR (Nujol): 2318(m), 2291(m) cm^{-1}. TGA weight loss between room temperature and 450°C (15°C/min): 84.4%.

Anal. Calcd. B$_2$C$_{76}$F$_{48}$FeH$_{42}$N$_6$: C, 45.00; H, 2.09; N, 4.14. Found: C, 43.86; H, 1.88; N, 4.29.

E. HEXAKIS(ACETONITRILE)COBALT(II)BIS-TETRA-3,5-BIS(TRIFLUOROMETHYL)PHENYLBORATE ([Co^{II}(NCMe)$_6$]{B[C$_6$H$_3$(CF$_3$)$_2$]$_4$})

$$CoCl_2 + 2Ag\{B[C_6H_3(CF_3)_2]_4\} + 6MeCN \rightarrow$$
$$[Co^{II}(NCMe)_6]\{B[C_6H_3(CF_3)_2]_4\}_2 + 2AgCl$$

A dry 15-mL MeCN solution of Ag{B[C$_6$H$_3$(CF$_3$)$_2$]$_4$} (1.332 mmol, 1.293 g) is added to CoCl$_2$ (0.6662 mmol, 0.0865 g) in a 25-mL Erlenmeyer flask and the mixture is stirred at room temperature in darkness for 12 h. The silver chloride is filtered off, leaving a pink solution. This is concentrated under reduced pressure to ~6 mL and cooled to −40°C for 24 h to crystallize pale pink needles that are filtered cold (>80% yield) and stored at −40°C. IR (Nujol): 2321(m), 2295(m) cm^{-1}. TGA weight loss between room temperature and 450°C (15°C/min): 85.5%.

Anal. Calcd. for $B_2C_{76}CoF_{48}H_{42}N_6$: C, 44.92; H, 2.08; N, 4.14. Found: C, 44.09; H, 2.04; N, 4.17.

F. HEXAKIS(ACETONITRILE)NICKEL(II)BIS-TETRA-3,5-BIS(TRIFLUOROMETHYL)PHENYLBORATE ([Ni^{II}(NCMe)$_6$]{B[C$_6$H$_3$(CF$_3$)$_2$]$_4$})

$$NiBr_2 + 2Ag\{B[C_6H_3(CF_3)_2]_4\} + 6MeCN \rightarrow$$
$$[Ni^{II}(NCMe)_6]\{B[C_6H_3(CF_3)_2]_4\}_2 + 2AgBr$$

A dry 15-mL MeCN solution of $Ag\{B[C_6H_3(CF_3)_2]_4\}$ (1.332 mmol, 1.293 g) is added to $NiBr_2$ (0.6662 mmol, 0.456 g) in a 25-mL Erlenmeyer flask and the mixture is stirred at room temperature in darkness for 12 h. The silver bromide is filtered off, leaving a sky-blue solution. This is concentrated under reduced pressure to ~ 6 mL and cooled to $-40°C$ for 24 h to crystallize blue-purple nee-dles of the product that are filtered cold ($>80\%$ yield) and stored at $-40°C$. IR (Nujol): 2326(m), 2299(m) cm^{-1}. TGA weight loss between room temperature and 450°C (15°C/min): 95.1%.

Anal. Calcd. for $B_2C_{76}F_{48}H_{42}N_6Ni$: C, 44.94; H, 2.08; N, 4.14. Found: C, 44.87; H, 2.14; N, 4.89.

G. HEXAKIS(ACETONITRILE)CHROMIUM(II)BIS-TETRA-3,5-BIS(TRIFLUOROMETHYL)PHENYLBORATE ([Cr^{II}(NCMe)$_6$]{B[C$_6$H$_3$(CF$_3$)$_2$]$_4$}$_2$)

$$[Cr(NCMe)_4](BF_4)_2 + 2K\{B[C_6H_3(CF_3)_2]_4\} + 2MeCN \rightarrow$$
$$[Cr^{II}(NCMe)_6]\{B[C_6H_3(CF_3)_2]_4\}_2 + 2K[BF_4]$$

A dry 5-mL THF solution of $K\{B[C_6H_3(CF_3)_2]_4\}$ (0.9880 mmol, 0.8913 g) is added to a 6-mL THF/MeCN (5 : 1) solution/suspension of $[Cr(NCMe)_4](BF_4)_2$ (0.4940 mmol, 0.1926 g) in a 25-mL Erlenmeyer flask. The remaining $[Cr(NCMe)_4](BF_4)_2$ solid dissolves on addition, and a white precipitate of $K[BF_4]$ forms. The reaction mixture is stirred for 4 h at room temperature and then filtered. The filtrate is concentrated under reduced pressure to dryness and the sky-blue solid (~ 1 g) dissolved in 7 mL MeCN to recrystallize the product at $-40°C$. Recovered 0.742 g (74% yield) of sky-blue needles after 24 h that are

filtered cold and stored at $-40°C$. IR (Nujol): 2324(m), 2297(m), 2277(w) cm^{-1}.
TGA weight loss between room temperature and $450°C$ ($15°C$/min): 85.1%.

Anal. Calcd. for B$_2$C$_{76}$CrF$_{48}$H$_{42}$N$_6$: C, 45.08; H, 2.09; N, 4.15. Found: C, 45.40; H, 2.28; N, 5.59.

H. HEXAKIS(ACETONITRILE)VANADIUM(II)BIS-TETRA-3,5-BIS(TRIFLUOROMETHYL)PHENYLBORATE ([V^{II}(NCMe)$_6$]{B[C$_6$H$_3$(CF$_3$)$_2$]$_4$}$_2$)

$$[V^{II}(NCMe)_6][BPh_4]_2 + 2[(Ph_3P)_2N]\{B[C_6H_3(CF_3)_2]_4\} \rightarrow$$
$$[V^{II}(NCMe)_6]\{B[C_6H_3(CF_3)_2]_4\}_2 + 2[(Ph_3P)_2N][BPh_4]$$

A dry 20-mL MeCN/CH$_2$Cl$_2$ (1 : 1) solution of [(Ph$_3$P)$_2$N]{B[C$_6$H$_3$(CF$_3$)$_2$]$_4$} (1.315 mmol, 1.843 g) is added to a stirring 40-mL MeCN solution/suspension of [V(NCMe)$_6$][BPh$_4$]$_2$ (0.6413 mmol, 0.6001 g) in a 125-mL Erlenmeyer flask. The remaining solid [V(NCMe)$_6$][BPh$_4$]$_2$ dissolves rapidly on addition of the [(Ph$_3$P)$_2$N]{B[C$_6$H$_3$(CF$_3$)$_2$]$_4$} solution and [(Ph$_3$P)$_2$N][BPh$_4$] begins to crystallize. After 20 min the solution is concentrated under reduced pressure to ~15 mL and filtered to remove [(Ph$_3$P)$_2$N][BPh$_4$]. The blue solution is cooled to $-40°C$ to crystallize aqua-blue needles mixed with colorless prisms of [(Ph$_3$P)$_2$N][BPh$_4$] in 24 h. The mixture of crystals is isolated by filtration and the [V^{II}(NCMe)$_6$]{B[C$_6$H$_3$(CF$_3$)$_2$]$_4$}$_2$ is dissolved and washed away from the [(Ph$_3$P)$_2$N][BPh$_4$] with 30 mL of Et$_2$O. The Et$_2$O is removed under reduced pressure and the remaining solid (~1 g) is recrystallized from 7 mL of MeCN at $-40°C$. Recovered 0.888 g (68% yield) aqua-blue needles in 24 h are filtered cold and stored at $-40°C$. IR (Nujol): 2320(m), 2291(m) cm^{-1}. TGA weight loss between room temperature and $450°C$ ($15°C$/min): 74.3%.

Anal. Calcd. for B$_2$C$_{76}$F$_{48}$H$_{42}$N$_6$V: C, 45.11; H, 2.09; N, 4.15. Found: C, 44.49; H, 2.06; N, 4.25.

General Properties

The [M^{II}(NCMe)$_6$]{B[C$_6$H$_3$(CF$_3$)$_2$]$_4$}$_2$ (M = V,Cr,Mn,Fe,Co,Ni) complexes are moderately moisture-sensitive crystalline solids and the V^{II} and Cr^{II} complexes are O$_2$-sensitive. Significant solvent loss from the solid complexes occurs at room temperature within hours, but the complexes are stable for months when stored cold. All the complexes are soluble in coordinating organic solvents (e. g., acetonitrile, THF, diethyl ether). IR spectra, TGA, and magnetic susceptibility (Table I) can be used to identify the complexes in the solid state while UV–vis spectroscopy (Table I) best shows the purity of the metal center. The

TABLE I. Magnetic and UV–Visible Data for $[M^{II}(NCMe)_6]\{B[C_6H_3(CF_3)_2]_4\}_2$ (M = V,Cr,Mn,Fe,Co,Ni)

M^{II}	$\mu_{\text{eff}},^a$ m_B	λ_{max}, cm^{-1}	$\varepsilon,^b$ M^{-1} cm^{-1}
V	3.92(1)	16,300	46.3
		22,500	46.3
Cr	5.09(3)	9,000	1.8
		15,500	12.7
Mn	5.98(2)	15,100	∼0.2
		19,100	∼0.3
		23,700	∼0.5
		26,900	∼0.6
Fe	5.85(1)	10,900	9.2
Co	5.38(1)	9,200	5.2
		20,300	14.3
		21,200	14.7
Ni	3.31(1)	10,400	7.1
		17,200	5.8
		27,500	8.0

^{a} Each entry is an average of four measurements at 300 K. Numbers in parentheses are estimated standard deviations of the measurements in the least significant digits.
^{b} Concentrations were 1–2.5 mM in MeCN; pathlength 5 cm; background was solvent versus solvent; $(TFPB)^-$ cutoff above 33,000 cm^{-1}.

IR spectra show two ν_{CN} bands of approximately equal intensity. In TGA experiments decomposition with loss of MeCN and metal abstraction of fluoride from the anion occurs above 130°C with the production of other decomposition side-products, leaving dark residues at 450°C. Obtaining reasonable elemental analyses is problematic because of the ease of solvent loss at room temperature and combustion problems. Nonetheless, the elemental analyses of $[M^{II}(NCMe)_6]$-$(TFPB)_2$ (M = V,Cr,Mn,Fe,Co,Ni) are consistent with the proposed formulation, but insensitive to the metal ion. The room temperature effective moments $[M^{II}(NCMe)_6](TFPB)_2$ are in the range typically found for other octahedral divalent salts of these metal ions, with values greater than the spin-only values observed for the Cr^{II}, Fe^{II}, Co^{II}, and Ni^{II}. UV–vis spectra of the complexes in acetonitrile show the expected $d \rightarrow d$ transitions for these octahedral divalent metal centers.

References and Notes

1. (a) S. H. Strauss, *Chem. Rev.* **93**, 927 (1993); (b) D. H. Geske, *J. Phys. Chem.* **63**, 1062 (1959). (c) D. H. Geske, *J. Phys. Chem.* **66**, 1743 (1962).

2. H. Nishida, N. Takada, and M. Yoshimura, *Bull. Chem. Soc. Jpn.* **57**, 2600 (1984).

3. M. G. Hill, W. M. Lamanna, and K. R. Mann, *Inorg. Chem.* **30**, 4687 (1991).

4. In comparison, the $E_{1/2}$ for $[BPh_4]^-$ was reported to be ~ 0.5 V versus Ag/AgNO$_3$ (MeCN).[1b]

5. (a) Y. Hayashi, J. J. Rohde, and E. J. Corey, *J. Am. Chem. Soc.* **118**, 5502 (1996); (b) J. H. Golden, P. F. Mutolo, E. B. Lobkovsky, and F. J. DiSalvo, *Inorg. Chem.* **33**, 5374 (1994); (c) J. Powell, A. Lough, and T. Saeed, *J. Chem. Soc., Dalton Trans.*, 4137 (1997).

6. R. P. Hughes, D. C. Lindner, A. L. Rheingold, and G. P. A. Yap, *Inorg. Chem.* **36**, 1726 (1997).

7. F. Calderazzo, G. Pampaloni, L. Rocchi, and U. Englert, *Organomet.* **13**, 2592 (1994).

8. W. E. Buschmann and J. S. Miller, *Chem. Eur. J.* **4**, 1731 (1998).

9. K. R. Dunbar, *Inorg. Synth.*, 33 (2000).

10. S. J. Anderson, F. J. Wells, G. Wilkinson, B. Hussain, and M. B. Hursthouse, *Polyhedron* **7**, 2615 (1988).

11. A. T. Casey, R. J. H. Clark, R. S. Nyholm, and D. E. Scaife, *Inorg. Synth.* **13**, 165 (1972).

13. TRANSITION METAL *p*-TOLUENESULFONATES

Submitted by STEPHEN M. HOLMES,* SCOTT G. MCKINLEY,* and GREGORY S. GIROLAMI*
Checked by PAUL S. SZALAY and KIM R. DUNBAR[†]

Binary salts of transition metal cations with weakly coordinating anions are extremely useful starting materials. Perchlorate salts have long been known for many transition metal cations, but these salts are potentially explosive in the presence of organic ligands and solvents. For this reason, researchers have increasingly turned to binary salts with nonoxidizing anions such as hexafluorophosphate, tetrafluoroborate, tetraarylborate, and trifluoromethanesulfonate (triflate). Some drawbacks may be associated with the use of these anions. The PF_6^- and BF_4^- anions have a tendency to hydrolyze in aqueous solution to generate HF,[1] whereas tetraarylborate anions (including fluorinated analogs) have relatively reactive B–C bonds, and the aryl rings can coordinate in a pi fashion to metal centers.[2-4] The triflates[5,6] are moderately expensive to prepare and occasionally difficult to crystallize.

In the present contribution, we describe the preparation of binary transition metal *p*-toluenesulfonates (tosylates), which have been known[7] since the 1870s but have been little used as synthetic starting materials. These salts are relatively inexpensive to prepare, can be isolated in high crystalline yields, and are soluble in many polar solvents. Tosylate salts are typically less soluble than the corresponding triflates, but more soluble than corresponding halide salts.

*School of Chemical Sciences, University of Illinois at Urbana—Champaign, 600 South Mathews Ave., Urbana, IL 61801.

[†] Department of Chemistry, Texas A&M University, College Station, TX 77843–3255.

Transition metal tosylates have been described for Ti[III],[8] Cr[II],[9] Cr[III],[10,11] Mn[II],[11,12] Fe[II,III],[11,13] Co[II],[11,14-16] Ni[II],[11,14-17] Cu[II],[11,15-18] and Ru[II,III],[19] as well as for Sc[III],[20] Y[III],[20] and the lanthanides Sm[III], Gd[III], Dy[III], Ho[III], Er[III], and Yb[III].[20] Alkali metal tosylates have also been briefly described for Li, Na, K, and Cs.[21] These compounds are typically prepared by addition of *p*-toluenesulfonic acid to a metal carbonate, hydroxide, or carboxylate, or by addition of silver tosylate to a metal chloride.

In the present contribution, we describe the preparations of several binary transition metal tosylates directly from the metal and *p*-toluenesulfonic acid under an inert atmosphere. This method is easy to carry out, and affords products that are completely free of contaminating counterions. The presence of excess metal provides a reducing environment, so that divalent products are obtained for all first-row transition metals except for Ti and V, which form trivalent products under these reaction conditions. The Cr[II] salt may be converted to Cr[III](OTs)$_3$ by air oxidation in the presence of excess *p*-toluenesulfonic acid. Tosylate salts of Ti[III], V[III], Fe[III], and Cs can also be prepared by treatment of the corresponding metal chloride with *p*-toluenesulfonic acid, and a vanadyl salt has been prepared by similar treatment of vanadyl acetylacetonate.

Methods for preparing anhydrous tosylate salts are also described. Analytical and infrared data for the anhydrous tosylate salts are presented in Tables I and II. The IR spectra of the hydrated salts are essentially identical to those of the anhydrous salts, except extra bands due to water are present near 3150–3500 and 1650–1700 cm^{-1}.

Procedure

Except where noted, all procedures were carried out under a dry argon atmosphere using Schlenk and cannula techniques. VO(acac)$_2$ was prepared by

TABLE I. Analytical Data for the Anhydrous Tosylate Salts[a]

Compound	Color	C	H	M
Ti(OTs)$_3$	Pale green	44.9 (44.5)	3.78 (3.79)	8.53 (9.13)
V(OTs)$_3$	Pale green	44.6 (42.4)	3.76 (4.18)	9.02 (9.51)
VO(OTs)$_2$	Gray	41.0 (40.9)	3.45 (3.47)	12.5 (12.4)
Cr(OTs)$_2$	Pale green	42.6 (41.8)	3.59 (3.99)	13.2 (12.0)
Mn(OTs)$_2$	White	42.3 (42.1)	3.56 (3.50)	13.8 (14.0)
Fe(OTs)$_2$	Buff	42.2 (41.9)	3.55 (3.38)	14.0 (15.4)
Fe(OTs)$_3$	Orange	44.3 (41.8)	3.72 (3.57)	9.81 (9.57)
Co(OTs)$_2$	Lavender	41.9 (41.1)	3.52 (3.64)	14.7 (14.5)
Ni(OTs)$_2$	Yellow	41.9 (41.7)	3.53 (3.85)	14.6 (14.3)

[a] Calculated (found).

TABLE II. Infrared Data for the Hydrated Tosylate Salts[a]

Compound	ν(OH) (vs, br)	ν(CH) (W)	Overtone (W)	δ(OH) (s)	ν_a(SO$_2$) (vs)	ν_s(SO$_2$) (vs)	δ(CH) (vs)	δ(CH) (vs)	δ(CH) (vs)	ν(SO) (vs)	δ(CS)? (vs)
[Ti(OH$_2$)$_4$][OTs]$_3$	3329	3067 3050	1924	1653	1254 1153	1109	1033	1009	818	684	567
[V(OH$_2$)$_6$[OTs]$_3$	3399 3289	3040 3028	1914	1676 1650	1190	1127	1040	1013	814	682	566
[VO(OH$_2$)$_5$][OTs]$_2$	3425 3188		1918	1697 1654	1196	1127	1038	1011	811	685	569
[Cr(OH$_2$)$_4$][OTs]$_2$	3484 3355	3064 3032	1912	1666 1649	1234 1205 1183	1124	1040	1014	812	682	567
[Cr(OH$_2$)$_6$][OTs]$_3$	3155		1927	1662	1211 1155	1127	1037	1014	823	687	563
[Mn(OH$_2$)$_4$] [OTs]$_2$·H$_2$O	3426 3268	3040 3028	1915	1669 1644	1192	1128	1041	1013	814	684	568
[Fe(OH$_2$)$_6$][OTs]$_2$	3396	3066 3041 3027	1914	1669 1646	1192	1128	1041	1013	814	684	569
[Co(OH$_2$)$_6$][OTs]$_2$	3408	3066 3028	1914	1670 1646	1189	1128	1040	1013	814	684	568
[Ni(OH$_2$)$_6$][OTs]$_2$	3407		1915	1669 1650	1191	1128	1040	1013	814	683	566

[a] Frequencies in cm^{-1}.

a published method.[22] Ti (Cerac), V (Cerac), Cr (Cerac), Mn (Aesar), Fe (Baker), Co (Fisher), Ni (Allied Chemical), TiCl$_3$ (Cerac), VCl$_3$ (Aldrich), FeCl$_3$ (Cerac), CsCl (Cerac), and *p*-toluenesulfonic acid hydrate (Acros, Aldrich) were used as received. Anhydrous *p*-toluenesulfonic acid was prepared by heating the hydrate to 160°C under vacuum for 1 h. Solvents were dried over CaH$_2$ (acetonitrile), Mg turnings (methanol), or Na/benzophenone (diethylether) under nitrogen. Deionized water was sparged with argon before use.

■ **Caution.** *Solid p-toluenesulfonic acid and its solutions are corrosive.*

A. TETRAAQUOTITANIUM(III)TRIS-*p*-TOLUENESULFONATE

$$\text{Ti} + 3\text{HOTs} \cdot \text{H}_2\text{O} + \text{H}_2\text{O} \rightarrow [\text{Ti(OH}_2)_4](\text{OTs})_3 + \frac{3}{2}\text{H}_2$$

To $\sim$325-mesh titanium powder (1.19 g, 24.8 mmol) in a 250-mL, round-bottomed flask equipped with a reflux condenser is added *p*-toluenesulfonic acid monohydrate (18.96 g, 99.7 mmol) and water (40 mL). The mixture is heated to reflux for 5 h; after this time, some metal powder is still present. The resulting purple solution is filtered while hot through Celite, and the filter cake is washed with water (2 × 50 mL). The purple filtrate is concentrated to 25 mL by vacuum distillation, and then allowed to cool to room temperature. After 2 h, the purple needles that deposit from solution are collected by filtration and dried overnight under vacuum at room temperature. Subsequent crops are obtained by concentrating the mother liquor by vacuum distillation and then cooling the solution to room temperature. Yield: 11.34 g (72.0%).

Anal. Calcd. for $C_{21}H_{29}O_{13}S_3Ti$: C, 39.8; H, 4.62; S, 15.2. Found: C, 39.5; H, 5.13; S, 15.8.

Properties

Pale purple $[Ti(OH_2)_4](OTs)_3$ can be stored indefinitely under an inert atmosphere at room temperature. It is soluble in tetrahydrofuran, methanol, and water, but is insoluble in acetonitrile.

B. TITANIUM(III)TRIS-*p*-TOLUENESULFONATE

$$TiCl_3 + 3HOTs \rightarrow Ti(OTs)_3 + 3HCl$$

To $TiCl_3$ (1.77 g, 11.5 mmol) in a 250-mL, round-bottomed flask equipped with a reflux condenser is added anhydrous *p*-toluenesulfonic acid (6.20 g, 36.0 mmol). The mixture is heated for 15 min in a bath kept at 160°C, after which time HCl evolution ceases. The mixture is dried for 20 min under vacuum in a bath kept at 160°C, and then cooled to room temperature. The resulting solid is treated with a mixture of acetonitrile (25 mL) and methanol (2 mL) at reflux for 20 min. The resulting purple solution is cooled to room temperature, and a pale green solid precipitates. The solid is collected by filtration, washed with Et_2O (2 × 20 mL), and dried overnight under vacuum at room temperature. Additional material can be obtained from the filtrate by concentrating it to 10 mL, adding Et_2O (5 mL), and cooling to −20°C. Yield: 4.12 g (67.1%).

Anal. Calcd for $C_{21}H_{21}O_9S_3Ti$: C, 44.9; H, 3.78; Ti, 8.53. Found: C, 44.5; H, 3.79; Ti, 9.13.

Properties

Pale green Ti(OTs)$_3$ can be stored indefinitely under an inert atmosphere at room temperature. It is soluble in methanol and water, sparingly soluble in tetrahydrofuran, and insoluble in acetonitrile.

C. HEXAAQUOVANADIUM(III)TRIS-*p*-TOLUENESULFONATE

$$V + 3HOTs \cdot H_2O + 3H_2O \rightarrow [V(OH_2)_6][OTs]_3 + \frac{3}{2}H_2$$

To ~325-mesh vanadium powder (3.87 g, 76.0 mmol) in a 250-mL round-bottomed flask equipped with a reflux condenser is added *p*-toluenesulfonic acid monohydrate (43.3 g, 228 mmol) and water (20 mL). The mixture is heated to reflux for 16 h. A deep purple solution is obtained, and some metal powder is still present. The mixture is filtered while hot through Celite and the purple filtrate concentrated to 10 mL by vacuum distillation. The solution is cooled to room temperature, and large pale purple needles are deposited. The crystals are collected by filtration and dried overnight under vacuum at room temperature. Additional crops can be obtained from the filtrate by concentrating it to 5 mL by vacuum distillation, and then cooling the resulting solution to room temperature. Yield: 23.0 g (45.1%).

Anal. Calcd. for C$_{21}$H$_{33}$O$_{15}$S$_3$V: C, 37.5; H, 4.96; S, 14.3. Found: C, 37.6; H, 5.17; S, 14.2.

Properties

Pale purple [V(OH$_2$)$_6$](OTs)$_3$ can be stored indefinitely under an inert atmosphere at room temperature. It is soluble in methanol and acidified (pH = 2) water. It is sparingly soluble in tetrahydrofuran, and insoluble in acetonitrile. In water at pH 7, the salt is unstable, as shown by the immediate formation of a brown color.

D. VANADIUM(III)TRIS-*p*-TOLUENESULFONATE

$$VCl_3 + 3HOTs \rightarrow V(OTs)_3 + 3HCl$$

Anhydrous VCl$_3$ (2.20 g, 13.98 mmol) and anhydrous *p*-toluenesulfonic acid (7.19 g, 41.75 mmol) are combined in a 250-mL, round-bottomed flask equipped with a reflux condenser, and the solid mixture is heated for 10 h in a bath kept at 160°C. The mixture is allowed to cool and the resulting pale green solid is washed with Et_2O (30 mL) to remove any remaining acid. The solid is extracted with a boiling 3 : 1 mixture of acetonitrile and methanol (100 mL). The extract is filtered while still hot, and then was cooled to $-20°C$ overnight. The apple-green plates that form are collected and dried for 16 h under vacuum at room temperature. The filtrate is evaporated to dryness under vacuum at room temperature, and the resulting solid extracted with a refluxing mixture of 10 : 1 acetonitrile and methanol (10 mL). Cooling the hot extract to room temperature affords additional green material, which is dried as described above. Yield: 7.20 g (91.6%).

Anal. Calcd. for $C_{21}H_{21}O_9S_3V$: C, 44.6; H, 3.76; N, 0; Cl, 0; V, 9.02. Found: C, 42.4; H, 4.18; N, 0.15; Cl, <0.15; V, 9.51.

Properties

Pale green $V(OTs)_3$ can be stored indefinitely under an inert atmosphere at room temperature. It is soluble in methanol and acidified (pH = 2) water. It is sparingly soluble in tetrahydrofuran and insoluble in acetonitrile. Anhydrous $V(OTs)_3$ can also be prepared by dehydration of $[V(OH_2)_6](OTs)_3$ at 160°C under vacuum (see below).

E. PENTAQUOOXOVANADIUM(IV)BIS-*p*-TOLUENESULFONATE HEMIHYDRATE

$$VO(acac)_2 + 2HOTs \cdot H_2O + \frac{7}{2}H_2O \rightarrow$$

$$[VO(OH_2)_5](OTs)_2 \cdot 0.5H_2O + 2Hacac$$

Vanadyl acetylacetonate (5.03 g, 19.0 mmol) and *p*-toluenesulfonic acid monohydrate (7.23 g, 38.0 mmol) are combined as solids in a 250-mL, round-bottomed flask equipped with a reflux condenser. The solids are heated under vacuum for 1 h in a bath kept at 160°C. The mixture is cooled to room temperature, and the resulting green solid extracted with hot H_2O (10 mL). The extract is filtered, and then most of the water is removed by vacuum distillation; a green foam and some viscous green syrup remain. This residue is treated with acetonitrile (20 mL), and the resulting mixture is heated to reflux for 20 min. A large quantity of blue-green plates form on cooling. The crystals are isolated, washed with acetonitrile (20 mL) and Et_2O (20 mL), and dried overnight under vacuum at room temperature. Yield: 8.97 g (94.7%).

Anal. Calcd. for $C_{14}H_{25}O_{12.5}S_2V$: C, 33.0; H, 4.97; S, 12.6; V, 10.0. Found: C, 32.9; H, 5.01; S, 12.6; V, 10.0.

Properties

Pale green $[VO(OH_2)_5](OTs)_2 \cdot 0.5H_2O$ can be stored indefinitely at room temperature. It is soluble in methanol and water, but is insoluble in acetonitrile.

F. TETRAAQUOCHROMIUM(II)BIS-*p*-TOLUENESULFONATE

$$Cr + 2HOTs \cdot H_2O + 2H_2O \rightarrow [Cr(OH_2)_4](OTs)_2 + H_2$$

To $-100/+200$-mesh chromium granules (6.0 g, 115 mmol) in a 500-mL, round-bottomed flask equipped with a reflux condenser is added *p*-toluene sulfonic acid monohydrate (43.26 g, 227 mmol) and water (90 mL). The mixture is heated to reflux for 24 h. A deep blue solution is obtained, and some excess metal powder remains. The hot mixture is filtered through Celite, and the filtrate is allowed to cool to room temperature. The blue needles that form are collected by filtration and dried overnight under vacuum at room temperature. A second crop of crystals is obtained by concentrating the filtrate to ~ 20 mL by vacuum distillation and cooling the resulting solution to room temperature. Yield: 45.0 g (84.8%).

Anal. Calcd. for $C_{14}H_{22}O_{10}S_2Cr$: C, 36.0; H, 4.76; S, 13.8; Cr, 11.2. Found: C, 34.5; H, 4.77; S, 13.8; Cr, 11.2.

Properties

Pale blue $[Cr(OH_2)_4](OTs)_2$ can be stored indefinitely under an inert atmosphere at room temperature, but slowly dehydrates to the dihydrate over several months. It is soluble in methanol and water, but is insoluble in acetonitrile.

G. HEXAQUOCHROMIUM(III)TRIS-*p*-TOLUENESULFONATE

$$Cr + 3HOTs \cdot H_2O + \frac{1}{4}O_2 + \frac{5}{2}H_2O \rightarrow [Cr(OH_2)_6](OTs)_3 + H_2$$

To $-100/+200$-mesh chromium granules (1.00 g, 19.2 mmol) in a 250-mL, round-bottomed flask equipped with a reflux condenser is added *p*-toluenesulfonic acid monohydrate (11.28 g, 59.3 mmol), and water (40 mL). The mixture is heated to reflux for 3 h, leaving little unreacted metal. The blue mixture is filtered while hot through Celite and the filtrate stirred in air overnight. The dark blue-green solution is concentrated to 25 mL by vacuum distillation. Dry acetonitrile (100 mL) is added, and pale green microcrystals are deposited. The crystals are isolated by filtration, washed with acetonitrile (30 mL) and diethyl ether (60 mL), and dried under vacuum overnight at 25°C. Yield: 11.5 g (88.5%).

Anal. Calcd. for $C_{21}H_{33}O_{15}S_3Cr$: C, 35.1; H, 5.34; N, 0.0; S, 13.4; Cr, 7.24. Found: C, 35.5; H, 5.37; N, 0.0; S, 13.2; Cr, 6.58.

Properties

Pale green $[Cr(OH_2)_6](OTs)_3$ can be stored indefinitely at room temperature. It is soluble in methanol and water, but is insoluble in acetonitrile.

H. TETRAQUOMANGANESE(II)BIS-*p*-TOLUENESULFONATE HYDRATE

$$Mn + 2HOTs \cdot H_2O + 3H_2O \rightarrow [Mn(OH_2)_4][OTs]_2 \cdot H_2O + H_2$$

To manganese powder (6.28 g, 114 mmol) in a 250-mL, round-bottomed flask equipped with a reflux condenser is added *p*-toluenesulfonic acid monohydrate (46.52 g, 245 mmol) and water (150 mL). The mixture is heated to reflux for 1 h. The resulting pale pink solution is filtered through Celite while hot, and the filtrate allowed to cool to room temperature. The deposited white crystals are collected by filtration and dried overnight under vacuum at room temperature. A second crop of white crystals can be obtained by concentrating the mother liquor. Yield: 45.3 g (81.3%).

Anal. Calcd. for $C_{14}H_{24}O_{11}S_2Mn$: C, 34.5; H, 4.97; S, 13.2; Mn, 11.3. Found: C, 34.8; H, 4.29; S, 12.1; Mn, 10.8.

Properties

White $[Mn(OH_2)_4](OTs)_2 \cdot H_2O$ can be stored indefinitely at room temperature under an inert atmosphere. It is soluble in methanol and water, but is insoluble in acetonitrile and diethylether.

I. HEXAQUOIRON(II)BIS-*p*-TOLUENESULFONATE

$$Fe + 2HOTs \cdot H_2O + 4H_2O \rightarrow [Fe(OH_2)_6][OTs]_2 + H_2$$

To iron powder (4.07 g, 72.9 mmol) in a 250-mL, round-bottomed flask equipped with a reflux condenser is added *p*-toluenesulfonic acid monohydrate (13.45 g, 70.7 mmol) and water (20 mL). The mixture is heated to reflux for 5 h. The pale green solution is filtered while hot through Celite, and the filter cake is washed with water (3 × 10 mL). The filtrate is allowed to cool to room temperature, and pale green needles form within 2 h. The crystals are collected by filtration and dried overnight under vacuum at room temperature. A second crop of crystal can be obtained by concentrating the mother liquor. Yield: 15.85 g (88.5%).

Anal. Calcd. for $C_{14}H_{26}O_{12}S_2Fe$: C, 33.2; H, 5.19; Fe, 11.0. Found: C, 33.7; H, 5.16; Fe, 11.5.

Properties

Pale green $[Fe(OH_2)_6](OTs)_2$ can be stored indefinitely at room temperature under an inert atmosphere. It is soluble in methanol and water, but is insoluble in acetonitrile and diethylether.

J. IRON(III)TRIS-*p*-TOLUENESULFONATE

$$FeCl3 + 3HOTs \cdot H_2O \rightarrow Fe(OTs)_3 + 3HCl + 3H_2O$$

A mixture of $FeCl_3$ (3.18 g, 19.6 mmol) and *p*-toluenesulfonic acid monohydrate (11.20 g, 58.9 mmol) in a 250-mL, round-bottomed flask equipped with a reflux condenser is heated under vacuum for 1 h in a bath kept at 160°C. The resulting orange solid is washed with diethyl ether (30 mL). The solid is treated with methanol (30 mL), and the mixture heated to reflux for 20 min. The resulting solution is filtered and the filtrate taken to dryness under vacuum. A 3 : 1 mixture of acetonitrile/methanol (30 mL) is added and the mixture heated to reflux for 20 min. The resulting suspension is allowed to cool to room temperature and the orange precipitate is collected by filtration. Additional product can be obtained from the filtrate by concentrating it to 5 mL by vacuum distillation, and adding acetonitrile (10 mL). The orange solids are combined, washed with diethylether (20 mL), and dried for 30 min under vacuum at 160°C. Yield: 10.6 g (95.0%).

Anal. Calcd. for $C_{21}H_{21}O_9S_3Fe$: C, 44.3; H, 3.72; N, 0; S, 16.9; Fe, 9.81. Found: C, 41.8; H, 3.57; N, 0; S, 16.3; Fe, 9.57.

Properties

Orange Fe(OTs)$_3$ can be stored indefinitely at room temperature under an inert atmosphere. It is soluble in methanol and water, and sparingly soluble in tetrahydrofuran. It is insoluble in acetonitrile, diethyl ether, and hydrocarbons.

K. HEXAQUOCOBALT(II)BIS-*p*-TOLUENESULFONATE

$$Co + 2HOTs \cdot H_2O + 4H_2O \rightarrow [Co(OH_2)_6][OTs]_2 + H_2$$

To cobalt powder (4.23 g, 71.8 mmol) in a 250-mL, round-bottomed flask equipped with a reflux condenser is added *p*-toluenesulfonic acid monohydrate (23.26 g, 122 mmol) and water (20 mL). The mixture is heated to reflux for 5 h. The resulting reddish-pink solution is filtered while hot through Celite, and the filter cake is washed with water (3 × 10 mL). The filtrate is cooled to room temperature, and pale orange blocks form after 2 h. The crystals are collected by filtration and dried overnight under vacuum at room temperature. A second crop of crystals are obtained by concentration and cooling of the mother liquor. Yield: 16.75 g (53.8%).

Anal. Calcd. for C$_{14}$H$_{26}$O$_{12}$S$_2$Co: C, 32.4; H, 5.26; Co, 11.4. Found: C, 32.5; H, 5.05; Co, 10.0.

Properties

Pale orange [Co(OH$_2$)$_6$](OTs)$_2$ can be stored indefinitely at room temperature under an inert atmosphere. It is soluble in methanol and water.

L. HEXAQUONICKEL(II)BIS-*p*-TOLUENESULFONATE

$$Ni + 2HOTs \cdot H_2O + 4H_2O \rightarrow [Ni(OH_2)_6](OTs)_2 + H_2$$

To nickel powder (5.0 g, 85.2 mmol) in a 250-mL, round-bottomed flask equipped with a reflux condenser is added *p*-toluenesulfonic acid monohydrate (32.23 g, 169 mmol) and water (20 mL). The mixture is heated to reflux for 15 h; after this time, some metal powder remains. The green mixture is filtered while hot through Celite and the filter cake washed with water (3 × 10 mL). The filtrate is cooled to room temperature, and large green needles deposit on standing. The crystals are collected by filtration and dried overnight under vacuum at room temperature. A second crop of crystals is obtained by concentrating and cooling the mother liquor. Yield: 22.61 g (52.4%).

Anal. Calcd. for $C_{14}H_{26}O_{12}S_2Ni$: C, 33.0; H, 4.96; Ni, 11.5. Found: C, 33.2; H, 5.17; Ni, 11.5.

Properties

Pale green $[Ni(OH_2)_6](OTs)_2$ can be stored indefinitely at room temperature under an inert atmosphere. It is soluble in methanol and water, but is insoluble in acetonitrile and diethylether.

M. CESIUM-*p*-TOLUENESULFONATE

$$CsCl + HOTs \cdot H_2O \rightarrow CsOTs + HCl + H_2O$$

Solid CsCl (5.83 g, 34.6 mmol) and *p*-toluenesulfonic acid monohydrate (6.60 g, 34.7 mmol) are combined in a 250-mL, round-bottomed flask equipped with a reflux condenser. The mixture is heated for 15 min in a bath kept at 160°C. The resulting yellow slurry solidifies when cooled to room temperature. The solid is treated with methanol (60 mL) and acetonitrile (5 mL), and the mixture is heated to reflux. The solution is filtered while hot, and the pale yellow filtrate is cooled to room temperature. The resulting white plates are isolated by filtration and dried under vacuum at room temperature overnight. A second crop of crystals is obtained by concentrating the mother liquor to 30 mL and adding diethylether (30 mL). Yield: 8.01 g (76.1%).

Anal. Calcd. for $C_7H_7O_3SCs$: C, 27.6; H, 2.32; Cl, 0. Found: C, 27.2; H, 2.56; Cl, <0.15.

Properties

White Cs(OTs) can be stored indefinitely at room temperature under an inert atmosphere. It is soluble in methanol and water, but is insoluble in acetonitrile and diethylether.

N. PREPARATION OF ANHYDROUS TOSYLATE SALTS

Anhydrous tosylate salts can be prepared by heating the corresponding hydrated salts under vacuum to 160°C for 1 h. Yields are nearly quantitative. Infrared data for these compounds are shown in Table III.

TABLE III. Infrared Data for the Anhydrous Tosylate Salts[a]

Compound	ν(CH) (w)	Overtone (w)	ν_a(SO$_2$) (vs)	ν_s(SO$_2$) (vs)	δ(CH) (vs)	δ(CH) (vs)	δ(CH) (vs)	ν(SO) (vs)	δ(CS)? (vs)
Ti(OTs)$_3$	3094	1919	1298	1149	1040	1010	819	683	564
	3062	—	—	1120	—	—	—	—	—
V(OTs)$_3$	3060	1923	1291	1130	1073	1016	820	685	571
	3040	—	—	—	—	—	—	—	—
VO(OTs)$_2$	3066	1921	1282	1129	1063	1015	814	683	561
	3031	—	1267	—	—	—	—	—	—
Cr(OTs)$_2$	3060	1917	1235	1161	1060	1015	817	688	575
	3039	—	—	—	—	—	—	—	—
Mn(OTs)$_2$	3060	1920	1200	1144	1065	1017	815	690	574
	3037	—	—	—	—	—	—	—	—
Fe(OTs)$_2$	3061	1920	1195	1143	1065	1018	815	688	576
	3039	—	—	—	—	—	—	—	—
Fe(OTs)$_3$	3062	1920	1298	1125	1040	1010	814	687	562
	3039	—	1287	1116	—	—	—	—	—
Co(OTs)$_2$	3061	1920	1194	1142	1065	1018	819	686	576
	3039	—	—	—	—	—	—	—	—
Ni(OTs)$_2$	3063	1921	1201	1142	1067	1018	816	686	577
	3041	—	—	—	—	—	—	—	—
Cs(OTs)	3045	1940	1223	1128	1037	1015	823	682	564
			1195						

[a] Frequencies in cm^{-1}.

Properties

The anhydrous tosylate salts are soluble in methanol and water, but are insoluble in acetonitrile and diethylether.

References

1. T. R. Musgrave and T. S. Lin, *J. Coord. Chem.* **2**, 323–324 (1973).
2. M. Aresta, E. Quaranta, and I Tommasi, *New J. Chem.* **21**, 595–613 (1997).
3. W. V. Konze, B. L. Scott, and G. J. Kubas, *Chem. Commun.*, 1807–1808 (1999).
4. J. Powell, A. Lough, and T. Saeed, *J. Chem. Soc, Dalton Trans.*, 4137–4138 (1977).
5. N. E. Dixon, G. A. Lawrance, P. A. Lay, A. M Sargeson, and H. Taube, *Inorg. Synth.* **24**, 243–250 (1986).
6. G. A. Lawrance, *Chem. Rev.* **86**, 17–33 (1986).
7. P. Claesson and K. Wallin, *Chem. Ber.* **12**, 1848–1854 (1879).
8. M. A. S. Aquino, W. Clegg, Q. Liu, and A. G. Sykes, *Acta Crystallogr.* **C51**, 560–562 (1995).
9. W. Duczmal, *Roczniki Chem.* **51**, 369–371 (1977).
10. G. Jantsch and K. Meckenstock, *Monatsh. Chem.* **52**, 169–184 (1929).
11. V. P. Kapila, B. Kapila, and S. Kumar, *Ind. J. Chem.* **30A**, 908–911 (1991).

12. S. Kumar, S. K. Gupta, and S. K. Sharma, *Thermochim. Acta* **71**, 193–197 (1983).
13. J. S. Haynes, J. R. Sams, and R. C. Thompson, *Can. J. Chem.* **59**, 669–678 (1981).
14. A. L. Arduini, M. Garnett, R. C. Thompson, and T. C. T. Wong, *Can. J. Chem.* **53**, 3812–3819 (1975).
15. M. Bombín, A. Guerrero, M. A. Martinez-Zaporta, and A. Ramirez, *Thermochim. Acta* **146**, 341–252 (1989).
16. T. Nortia and S. Laitinen, *Suomen Kemistilehti* **B41**, 136–141 (1968).
17. W. G. Wright, *J. Chem. Soc.*, 263–266 (1942).
18. C. Couldwell, K. Prout, D. Robey, R. Taylor, and F. J. C. Rossotti, *Acta Crystallogr.* **B34**, 1491–1499 (1978).
19. P. Bernhard, H.-B. Bürgi, J. Hauser, H. Lehmann, and A. Ludi, *Inorg. Chem.* **21**, 3936–3941 (1982).
20. Y. Ohki, Y. Suzuki, T. Takeuchi, and A Ouchi, *Bull Chem. Soc. Jpn.* **61**, 393–405 (1988).
21. K. O. Debevc, C. Pohar, and V. J. Vlachy, *Soln. Chem.* **25**, 787–795 (1996).
22. B. E. Bryant and W. C. Fernelius, *Inorg. Synth.* **5**, 113–116 (1957).

14. SALTS OF BINUCLEAR IRON(II) AND IRON(III) DIANIONS [$(Fe_2Cl_6)^{2-}$ AND $(Fe_2OCl_6)^{2-}$]

Submitted by KIM R. DUNBAR,[*] JOHN J. LONGRIDGE,[†]
JEREMY M. RAWSON,[†] and JUI-SUI SUN[*]
Checked by KARL S. HAGEN and BAO DO[‡]

Salts of the oxo-bridged dianion, $(Fe_2OCl_6)^{2-}$, particularly its tetra alkyl ammonium derivatives $(R_4N)_2(Fe_2OCl_6)$, have proved to be convenient starting materials for a series of polynuclear iron complexes.[1] The most common procedure utilized for the syntheses of these dinuclear complexes is that described by Armstrong and Lippard,[2] whose yields are significantly greater than those reported in alternative syntheses.[3]

We now report convenient synthetic routes to salts of this binuclear ferric dianion, $(Fe_2OCl_6)^{2-}$, and the related binuclear ferrous dianion $(Fe_2Cl_6)^2$. Previously we found[4–6] that reaction of ferric chloride with phosphines such as tris(2,4,6-trimethoxyphenyl)phosphine (tmpp) unexpectedly yielded the previously unkown $(Fe_2Cl_6)^{2-}$ dianion as the salt $(Htmpp)_2(Fe_2Cl_6)$. Benzene solutions of $(Htmpp)_2(Fe_2Cl_6)$ reacted[7] with molecular oxygen to yield the phosphine-oxide complex, $FeCl_3$ (tmppO), while oxidation in protic solvents such as MeOH and EtOH formed[8] $(Htmpp)_2(FeCl_4)$ and $(Htmpp)(FeCl_4)$ as intermediates, with $FeCl_3.(tmppO)$ formed only on prolonged oxygenation. We have found that simple salts such as tetraethyl ammonium [$(Et_4N)^+$], tetraphenylarsonium

[*] Department of Chemistry, Texas A & M university, College Station TX 77843-3255 48824.
[†] Department of Chemistry, University of Cambridge, Lensfield Road, Camgridge, CB2 1EW, UK.
[‡] Department of Chemistry, Emory University, 1515 Pierce Dr., Atlanta, GA 30322.

[(AsPh$_4$)$^+$], tetraphenylphosphonium [(PPh$_4$)$^+$], and bis(triphenylphosphine imminium) [(PPN)$^+$] of (Fe$_2$Cl$_6$)$^{2-}$ can be conveniently prepared from anhydrous iron(II) chloride and the corresponding chloride salt, such as (Et$_4$N)Cl, in acetone.[9] In contrast to (Htmpp)$_2$(Fe$_2$Cl$_6$), aerial oxidation of these salts proceeds smoothly to form the oxo-bridged binuclear ferric dianion, (Fe$_2$OCl$_6$)$^{2-}$. Aerial oxidation of ferrous salts was previously proved to be a convenient route to oxo-bridged ferric complexes; for instance, reaction of FeCl$_2$·4H$_2$O with NaOAc yields[10] the trimetallic salt [Fe$_3$O(OAc)$_6$]Cl.3H$_2$O.

Herein we report convenient high-yield routes to both the binuclear ferrous and ferric salts, (PPh$_4$)$_2$(Fe$_2$Cl$_6$) and (PPh$_4$)$_2$(Fe$_2$OCl$_6$). These procedures can be readily adapted to other quaternary group 15 cations (R$_4$N$^+$, R$_4$P$^+$, R$_4$As$^+$ and also the iminium cation, PPN$^+$). We also report a simple two-step preparation of the (Fe$_2$OCl$_6$)$^{2-}$ dianion, which can be carried out on a large scale on the open bench.

The starting materials used in the following syntheses were obtained from commercial sources and used without further purification.

A. DI(TETRAPHENYLPHOSPHONIUM)HEXACHLORO-DIFERRATE(II) [(Fe$_2$Cl$_6$)(PPh$_4$)$_2$]

$$PPh_4Cl + FeCl_2 \rightarrow (Fe_2Cl_6)(PPh_4)_2$$

Procedure

A solution of (PPh$_4$)Cl (0.887 g, 2.367 mmol) in degassed acetone (15 mL) is added to a suspension of FeCl$_2$ (0.300 g, 2.367 mmol) in degassed acetone (15 mL). The resultant solution is stirred for 12 h to yield a pale yellow precipitate under a yellow solution. The solution is filtered through Celite and the precipitate washed with acetone (20 mL). The combined filtrates are concentrated and Et$_2$O is added slowly to yield an off-white solid that is dried in vacuo. Yield: 0.854 g, 72%.

Anal. Calcd. for Fe$_2$Cl$_6$P$_2$C$_{48}$H$_{40}$ C, 57.5; H, 4.0; Cl, 21.1. Found: C, 57.1; H, 3.9; Cl, 21.0%.

Properties

The IR spectrum of (PPh$_4$)$_2$(Fe$_2$Cl$_6$) shows a medium–strong ν_{FeCl} band at 340 cm^{-1}. The magnetic properties of these salts reveal an interesting dependence on the counterion that is manifested in terms of differences of the Fe–Cl–Fe bridge angles. Full details[9] of the temperature-dependent magnetic susceptibility measurement of Mössbauer spectra will be the subject of a separate report, although these findings support the existence of ferromagnetic coupling in some of the salts and weak or no coupling between Fe(II) centers in other analoges. For $(PPh_4)_2(Fe_2Cl_6)$, an effective magnetic moment of $7.19\mu_B$ at 300 K was observed (using an applied field of 0.1 T).

B. DI(TETRAPHENYLPHOSPHONIUM)(μ-OXO)BIS (TRICHLOROFERRATE(III) [$(PPh_4)_2(Fe_2OCl_6)$]

Procedure

$$(PPh_4)_2(Fe_2Cl_6) + O_2 \rightarrow (PPh_4)_2(Fe_2OCl_6) + \frac{1}{2}O_2$$

A solution of $(PPh_4)_2(Fe_2Cl_6)$ (0.119 g, 0.119 mmol) in acetone (20 mL) is purged with O_2 gas at $-15°C$ for 5 min, which results in a color change from yellow to orange after I min. The reaction solution is stirred under an O_2 atmosphere for a further 30 min. The solvent is reduced to a small volume and Et_2O is added to precipitate a yellow solid that is collected by filtration, washed with Et_2O, and dried in vacuo. Yield 0.097 g, 80%.

Anal. Calcd. for $Fe_2Cl_6ON_2C16H_{14}$: C, 32.0; H, 6.7; Cl, 35.4. Found, C, 32.5; H, 7.0; Cl, 35.7%.

Properties

The yellow solid $(PPh_4)_2(Fe_2OCl_6)$ is air-stable in the solid state. The IR spectrum shows broad, strong absorptions at 362 and 874 cm^{-1} that have been assigned as ν_{FeCl} and $\nu_{Fe-O-Fe}$, respectively.

C. DI(TETRAETHYLAMMONIUM)(μ-OXO)BIS (TRICHLOROFERRATE(III) $(Et_4N)_2(Fe_2OCl_6)$

Procedure

$$(Et_4N)Cl \cdot H_2O + FeCl_2 \cdot 4H_2O \rightarrow (Et_4N)_2(Fe_2Cl_6) \cdot 4H_2O + H_2O$$

$$(Et_4N)_2(Fe_2Cl_6) + O_2 \rightarrow (Et_4N)_2(Fe_2OCl_6) + \frac{1}{2}O_2$$

The preparation of this compound is typically carried out with a slight molar excess of $FeCl_2 \cdot 4H_2O$ as an excess of $(Et_4N)Cl$ has been found[2] to retard crystallization of the product. It is not necessary to degas solvents for this experiment.

A solution of $(Et_4N)Cl \cdot H_2O$ (1.53 g, 8.33 mmol) in MeOH (10 mL) is added to solution of $FeCl_2 \cdot 4H_2O$ (1.66 g, 8.33 mmoL) in H_2O (10 mL). The solution is stirred for 10 min and the solvent removed in vacuo, yielding a pale yellow solid of composition $(Et_4N)_2(Fe_2Cl_6) \cdot 4H_2O$ in quantitative yield. A freshly prepared sample of solid (2.5 g, 3.94 mmol) is dissolved in MeCN (40 mL), and the solution turns an immediate red-brown color. The solution is stirred for 30 min, and filtered to remove a small quantity of insoluble matter, and the filtrate is concentrated to a red-brown oil on a rotary evaporator. Crystals of $(Et_4N)_2(Fe_2OCl_6)$ are recovered from the oily residue using a procedure similar to that reported previously.[9] The oil is redissolved in MeCN (4 mL) and after addition of THF (30 mL), the solution is cooled to $-20°C$. After one day the first crop of red-brown crystals are collected by filtration and washed with THF (2 × 10 mL). The combined filtrate and washings are again cooled to $-20°C$ for 24 h to yield a second crop of material, and the procedure is repeated to yield a third crop. The three crops are separately ground to a golden-brown powder and dried in vacuo to remove lattice solvent. The yields for the three crops after drying are 1.365 g (58%), 0.303 g (13%) and 0.094 g (4%), respectively, providing a total yield of 1.762 g (75%).

Anal. Calcd. for $C_{16}H_{48}N_2Fe_2Cl_6O_4$: C, 29.3; H, 7.4; N, 4.3. Found: C, 29.9; H, 7.1; N, 4.3%.

Anal. Calcd. for $C_{16}H_{40}N_2Fe_2OCl_6$: C, 32.0; H, 6.7; N. 4.7. Found: C, 31.9; H, 6.7; N, 4.5%.

Properties

Samples of the pale yellow intermediate, $(Et_4N)_2(Fe_2Cl_6) \cdot 4H_2O$, slowly undergo decomposition on exposure to the atmosphere, turning orange, and it is preferable to use a freshly obtained sample for the second step of the reaction presented above. Older samples of this material yield a greater proportion of insoluble material on dissolution in MeCN, thereby reducing the overall yield of $(Et_4N)_2$ (Fe_2OCl_6).

The red-brown crystalline needles of $(Et_4N)_2(Fe_2OCl_6)$ lose solvent from the lattice on standing in the air, yielding a golden-brown powder, but show no signs of further decomposition. Th IR spectrum (Nujol mull) exhibits a $\nu_{Fe-O-Fe}$ band at 855 cm^{-1}. The UV–vis spectrum (MeCN) shows absorption maxima at 244, 292, 319(sh), and 376(sh) nm. Solutions of $(Et_4N)_2(FeOCl_6)$ in MeCN decompose on standing for several hours at room temperature.[2]

Acknowledgment

We would like to thank the University of Cambridge, the EPSRC, and the Newton Trust for financial support (JJL) KRD acknowledges the Centre for Fundamental Materials Research at Michigan State University for SQUID susceptibility measurements and NSF for partial funding of this research.

References

1. (a) W. H. Armstrong, M. E. Roth, and S. J. Lippard, *J. Am. Chem. Soc.* **109**, 6318 (1987); (b) P. N. Turnowski, W. H. Turnowski, W. H. Armstrong, S. Liu, S. N. Brown, and S. J. Lippard, *Inorg. Chem.* **33**, 636 (1994); (c) S. Parsons, G. A. Solan, and R. E. P. Winpenny, *J. Chem. Soc., Chem. Commun.*, 1967 (1995); (d) C. Benelli, S. Parsons, G. A. Solan, and R. E. P. Winpenny, *Angew. Chem., Int. Ed., Engl.* **35**, 1825 (1996).
2. W. H. Armstrong and S. J. Lippard, *Inorg. Chem.* **24**, 981–982 (1985).
3. G. Haselhorst, K. Wieghardt, S. Keller, and B. Schrader, *Inorg. Chem.* **32**, 520 (1993), and references cited therein.
4. A. Quillevere, Ph.D. dissertation, Michigan State University, 1992.
5. K. R. Dunbar and A. Quillevere, *Angew. Chem., Int. Ed. Engl.* **32**, 293 (1993).
6. K. R. Dunbar and J.-S. Sun, *Mol. Cryst Liq. Cryst: Proc. IVth Int. Conf. Molecule-Based Materials*, 1995, Vol. 274, pp. 57–62.
7. K. R. Dunbar, S. C. Haefner, and A. Quillevere, *Pollyhedron* **9**, 1965 (1990).
8. K. R. Dunbar and A. Quillevere, *Polyhedron* **12**, 807 (1993).
9. (a) J.-S. Sun, Ph.D. dissertation, Michigan State University, 1994; (b) manuscript in preparation.
10. L. Meesuk, U. A. Jayasooriya, and R. D. Cannon, *J. Am. Chem. Soc.* **109**, 2009 (1987).

15. TETRAETHYLAMMONIUM-TETRATHIOPERRHENATE
Et₄N(ReS₄)

Submitted by JONATHAN T. GOODMAN and THOMAS B. RAUCHFUSS[*]
Checked by MADELI CASTRUITA, RAQUEL TERROBA,
and JONATHAN M. McCONNACHIE[†]

Known since the work of Berzelius, the tetrathiometallates, anions of the formula MS_4^{n-}, represent one of the fundamentally important classes of soluble metal sulfides.[1] Of these anions, ReS_4^- is unique in displaying high reactivity toward unsaturated organic compounds.[2-5] ReS_4^- has also been shown to form complexes with a variety of metal ions.[6] Tetrathioperrhenate is reduced at -1.58 V

[*]Department of Chemistry, University of Illinois at Urbana—Champaign, Urbana, IL 61801.
[†]Exxon Research and Engineering Co., Annandale, NJ 0880.

versus $Fc^{0/+}$ to form a persistent d^1 species.[7] The anion has been characterized by [185/187] Re NMR spectroscopy,[8] XANES (X-ray absorption near-edge spectroscopy),[9] X-ray crystallography,[10] and electronic structure studies.[9] Salts of ReS_4^- have been used as precursors to binary rhenium sulfides such as ReS_2 and Re_2S_7, which are active hydrogenation and dehydrogenation catalysts.[11,12] Rhenium sulfides are also known to prevent coke formation on platinum reforming catalysts.[13] The synthetic method reported by Müller et al.[8] and workers at Exxon[14] has proved reliable. The synthesis involves the action of a polysulfide solution on ReO_4^-. In contrast, MoS_4^{2-} and WS_4^{2-} are synthesized from the corresponding oxides using solutions of SH^-.[15] The present procedure can be modified to produce other rhenium sulfides, including $ReO(S_4)_2^-$ and $ReS(S_4)_2^-$.[16]

Procedures

All manipulations were carried out in air. Reagent-grade solvents and distilled water were used throughout. Any water-soluble perrhenate salt should be a viable replacement for $NaReO_4$, such as $KReO_4$ and $NH_4(ReO_4)$.

■ **Caution.** *Because of the toxic and corrosive nature of many of the compounds used in this preparation, it is recommended that all procedures be performed in a fume hood and that nitrile gloves be warn at all times. Hydrogen sulfide is highly toxic* $(LD_{50\ mice} + 673)$.

A. STOCK AMMONIUM SULFIDE SOLUTION (20%)

Such solutions can be obtained commercially with no loss in yield.

A 250-mL Erlenmeyer flask is charged with 66 mL of a 30% aqueous NH_3 and 134 mL H_2O. This solution is saturated with H_2S (**Caution**—*toxic!*) by passing the gas vigorously through the solution. The solution is initially warmed to $\sim 50°C$. After ~ 1.5 h the solution is cooled to room temperature, at which time the H_2S flow is discontinued and the flask is stoppered. The yellow solution should be stored in a screwcap bottle (no metal) in a well-ventilated area. This solution darkens on prolonged storage but remains active.

B. TETRAETHYLAMMONIUM-TETRASULFIDORHENATE [Et₄N(ReS₄)]

$$NaReO_4 + 2S_x^{2-} + Et_4NBr \rightarrow Et_4NReS_4 + 2S_{x-2}O_2^{2-} + NaBr$$

A 150-mL Schlenk flask is charged with 100 mL of the stock ammonium sulfide solution followed by 3.75 g (14.62 mmol) of S_8. The mixture is stirred vigorously for 10 min to produce a dark orange homogeneous solution. To this polysulfide solution is then added, in order, 11.5 g (55 mmol) of Et_4NBr and 8.0 g (29.3 mmol) of $NaReO_4$. These salts dissolve rapidly. After 18 h the dark precipitate is collected on a glass-fritted funnel. The red-violet solid is washed with H_2O (300 mL), EtOH (100 mL), MeOH (300 mL), and Et_2O (200 mL) and air-dried. Yield: 12.7 g (98%) of dark violet powder. This material is sufficiently pure for most purposes.

The product can be recrystallized by first dissolving 2.02 g in 400 mL of MeCN. The dark violet solution is filtered and concentrated to ~ 200 mL. The concentrate is diluted with 200 mL of toluene, and the solution is again concentrated to a volume of 200 mL. Black-green crystals or a purple powder is collected and washed with toluene (50 mL) and Et_2O (50 mL). Both forms analyze to be $(Et_4N)ReS_4$. The solid is dried in air. Recovery: 1.72 g (85%).

Properties

Solid $Et_4N(ReS_4)$ decomposes in air over a period of weeks to a dark brown material that can be removed by washing with small amounts of MeCN. $Et_4N(ReS_4)$ is soluble in polar organic solvents, including MeCN, DMF, DMSO, and CH_2Cl_2. The IR spectrum of $Et_4N(ReS_4)$ displays a strong $\nu_{Re=S}$ at 484 cm^{-1}. The UV–vis spectrum (MeCN solution) consists of bands at 232 (27,400), 312 (23,400), 348 (2150), and 508 nm (12,800). An additional absorption can be observed at ~ 582 nm in very concentrated solutions.[17] $Et_4N(ReS_4)$ can be efficiently converted to $Ph_4P(ReS_4)$ by treating an MeCN solution of Et_4N^+ salt with Ph_4PBr followed by concentration of the solution. Often it is useful to repeat this operation to ensure complete metathesis, as can be assayed by ^1H NMR analysis. The PPh_4^+ salt, which exhibits lower solubility than the Et_4N^+ derivative, is useful in growing single crystals suitable for X-ray diffraction.

References

1. A. Müller, E. Diemann, R. Jostes, and H. Bögge, *Angew. Chem. Int. Ed. Engl.* **20**, 934 (1981).
2. J. T. Goodman, S. Inomata, and T. B. Rauchfuss, *J. Am. Chem. Soc.* **118**, 11674 (1996).
3. J. T. Goodman and T. B. Rauchfuss, *Angew. Chem. Int. Ed. Engl.* **36**, 2083 (1997).
4. J. T. Goodman and T. B. Rauchfuss, *Inorg. Chem.* **37**, 5040 (1998).
5. J. T. Goodman and T. B. Rauchfuss, *J. Am. Chem. Soc.* **121**, 5017 (1999).
6. M. A. Massa, T. B. Rauchfuss, and S. R. Wilson, *Inorg. Chem.* **30**, 4667 (1991).
7. R. Schäfer, W. Kaim, M. Moscherosch, and M. Krejčik, *J. Chem. Soc. Chem. Commun.*, 834 (1992).
8. A. Müller, E. Krickemeyer, H. Bögge, M. Peak, and D. Rehder, *Chimia* **40**, 50 (1986).

9. A. Müller, V. Wittneben, E. Diemann, J. Hormes, and U. Kuetgens, *Chem. Phys. Lett.* **225**, 359 (1994).
10. Y. Do, E. D. Simhon, and R. H. Holm, *Inorg. Chem.* **24**, 4635 (1985).
11. L. Y. Chiang, J. W. Swirczewski, R. R. Chianelli, and E. I. Stiefel, *Catal. Lett.* **1**, 177 (1985).
12. L. Y. Chiang, J. W. Swirczewski, R. Kastrup, C. S. Hsu, and R. B. Upasani, *J. Am. Chem. Soc.* **113**, 6574 (1991).
13. C. L. Pieck, P. Marecot, and J. Barbier, *Appl. Catal. A* **145**, 323 (1996).
14. U.S. Patent 4,997,962 (Exxon). 1990. *Chem. Abstr.* 1990, **113**, 214804
15. W.-H. Pan, M. E. Leonowicz, and E. I. Stiefel, *Inorg. Chem.* **22**, 672 (1983).
16. A. Müller, E. Krickemeyer, and H. Bögge, *Z. Arnorg. Allg. Chem.* **554**, 61 (1987).
17. R. Schäfer, W. Kaim, and J. Fiedler, *Inorg. Chem.* **32**, 3199 (1993).

16. LARGE-SCALE SYNTHESIS OF METHYLTRIOXORHENIUM (CH_3ReO_3)

Submitted by WOLFGANG A. HERRMANN and ROLAND M. KRATZER[*]
Checked by JAMES H. ESPENSON and WEIDONG WANG[†]

Methyltrioxorhenium(VII) (MTO), CH_3ReO_3, has been shown to be an organometallic compound with a broad variety of catalytic properties.[1-3] Two synthetic pathways have been established:[4,5] the direct alkylation of dirhenium heptoxide (Re_2O_7) with tetramethyltin yielding 50% of unreactive trimethylstannyl perrhenate, and the alkylation with tributylmethyltin in the presence of trifluoroacetic anhydride. In the latter case, the formation of the rhenium-containing byproduct can be avoided. The disadvantage of both methods is that the Re_2O_7 is extremely moisture-sensitive.

From earlier results,[6] we now report in detail on the large-scale synthesis of MTO starting from silver perrhenate. The reaction involves the activation of the perrhenate with 2 equiv of TMSCl followed by alkylation with tetramethyltin:

$$AgReO_4 + 2Me_3SiCl + SnMe_4 \rightarrow MeReO_3 + AgCl + (Me_3Si)_2O + ClSnMe_3$$

Procedure

Because of the easy scaleup, readily available starting materials, and the avoidance of air and moisture exclusion, this is the method of choice in

[*]Anorganisch-chemisches Institut der Technischen Universität München, Lichtenbergstrasse 4, D-85747 Garching, Germany.
[†]Department of Chemistry, Iowa State University Ames, IA 50011.

organorhenium(VII)oxide synthesis. Herein the synthesis is described on a 10-g scale. The same results were obtained on a 100-g scale.

Rhenium powder (Starck), Ag[NO₃] (Degussa), Me₃SiCl (TMSCl, 98%, Aldrich), SnMe₄ (99 + %, Aldrich), and CH₃CN (99%, Aldrich) were used as received. An alternative access to SnMe₄ is recycling of the reaction byproduct Me₃SnCl. In this case Me₃SnCl was methylated using MeMgCl (23% in THF, Chemetall).

■ **Caution.** *Tetramethyltin and trimethyltin chloride are very toxic and volatile. All transformations are therefore to be performed under a hood.*

A. SILVER TETRAOXORHENATE [Ag(ReO₄)]

To a slowly stirred suspension of 10.0 g (0.054 mol) Re powder in 20 mL of water in a 500-mL Erlenmeyer flask is added 100 mL of H_2O_2 (35%) under ice cooling over the course of 4 h. After stirring for 30 min at room temperature, the suspension is heated to 80°C for 3 h. The mixture is filtered from small amounts of insoluble materials, yielding a clear solution. Adding 1.1 equiv Ag(NO₃) (10.0 g, 0.059 mol) immediately results in the formation of Ag(ReO₄) as a white precipitate. The product is filtered and washed with diethylether (3 × 25 mL) to remove water and hydrogen peroxide. Drying under vacuum yields 97% of Ag(ReO₄) (18.7 g, 0.052 mol).

B. METHYLTRIOXORHENIUM (CH₃ReO₃)

First, 14.0 g Ag(ReO₄) (0.039 mol) is dissolved in 150 mL CH₃CN using a 250-mL flask. After the addition of 10.8 mL of Me₃SiCl (0.085 mol) a white precipitate is formed (AgCl) and the solution becomes orange. To this suspension 6.0 mL of SnMe₄ (0.043 mol) is added and the mixture stirred for 24 h.

The suspension is then filtered into a sublimation Schlenk tube (0.75 L volume). The solution is evaporated under reduced pressure (20 mbar). Because of the high volatility of the tin compounds it is highly recommended that a trap cooled with liquid nitrogen be installed between the Schlenk tube and the pump throughout the workup procedure. After removal of the solvent an appropriate water-cooled sublimation finger is inserted into the Schlenk tube. Thus, the resulting solid is sublimed at ambient temperature under reduced pressure (20 mbar), yielding Me₃SnCl. Because of tiny amounts of overalkylated rhenium compounds, the sublimate has a slightly orange color. After removal of Me₃SnCl the vacuum is adjusted to high vacuum (0.01 mbar), subliming MTO as a yellow-white crystalline solid. Alternatively, MTO can be sublimed from the residue under a pressure of 0.1 mbar, elevating the temperature to 40–60°C.[7] Sometimes the MTO fraction still contains a certain amount of the toxic tin compound, causing a pungent smell. These traces can be efficiently removed by storing the compound in air on a filter paper in the hood for 1–3 h. The purity is checked by elemental analysis.

Anal. Calcd. C, 4.82%; H, 1.21%. Found: C, 4.80%; H, 1.27%. Melting point 108°C. Yield: 7.8 g CH_3ReO_3 (0.031 mol, 80%).[8]

Acknowledgment

This work was supported by the Bayerische Forschungsverbund Katalyse. Re powder was a generous gift of Starck, $Ag(NO_3)$ of Degussa AG.

References and Notes

1. W. A. Herrmann and F. E. Kühn, *Acc. Chem. Res.* **30**, 169 (1997).
2. W. A. Herrmann, *J. Organomet. Chem.* **500**, 149 (1995).
3. W. A. Herrmann, R. M. Kratzer, H. Ding, H. Glass, and W. R. Thiel, *J. Organomet. Chem.* **555**, 293 (1998).
4. W. A. Herrmann, J. G. Kuchler, J. K. Felixberger, E. Herdtweck, and W. Wagner, *Angew. Chem.* **100**, 420 (1988); *Angew. Chem., Int. Ed. Engl.* **27**, 394 (1988).
5. W. A. Herrmann, F. E. Kühn, R. W. Fischer, W. R. Thiel, and C. C. Romão, *Inorg. Chem.* **31**, 4431 (1992).
6. W. A. Herrmann, R. M. Kratzer, and R. W. Fischer, *Angew. Chem.* **109**, 2767 (1997); *Angew. Chem., Int. Ed. Engl.* **36**, 2652 (1997).
7. The checkers comment that high-purity MTO could also be obtained by recrystallization from CH_2Cl_2/hexane (73% yield).
8. To our knowledge, a change in color from white to light gray during storage of MTO does not affect the purity or the activity of the compound in known applications. Nevertheless, color changes can be hampered by excluding exposure to light and by storing the compound under nitrogen.

17. 4,5-DIAMINOCATECHOL: A USEFUL BUILDING BLOCK IN SYNTHESIS OF MULTIMETALLIC COMPLEXES

Submitted by DELL T. ROSA, ROBERT A. REYNOLDS III,
STEVEN M. MALINAK, and DIMITRI COUCOUVANIS*
Checked by MD. MESER ALI and FREDERICK M. MacDONNELL[†]

4,5-Diaminocatechol was first synthesized in 1947,[1] but the product was extremely sensitive to air and decomposed readily, making isolation and further work

* Department of Chemistry, The University of Michigan, Ann Arbor, MI 48109.
† Department of Chemistry and Biochemistry, University of Texas at Arlington, Arlington, TX 76019.

with this material challenging. The synthesis of this compound, reported here, is the result of our interest in obtaining it in a convenient manner and using it in the synthesis of salphen complexes functionalized with a catechol site. These ditopic ligands could then be used as building blocks for multimetallic complexes by taking advantage of both the tetradentate and bidentate metal binding sites (see Fig. 1).

The synthesis for 4,5-diaminocatechol reported herein represents an improvement over the only known synthesis because the molecule is stabilized by quaternizing one of the amines. The synthesis of 4,5-diaminocatechol is readily accomplished in three steps. Starting with commercially available 1,2-dimethoxy benzene (veratrole), the first step involves nitration to give 4,5-dinitroveratrole in high yield. Reduction of the nitro groups is performed with hydrazine monohydrate to give 4,5-diaminoveratrole as a white solid. Demethylating 4,5-diaminoveratrole with BBr$_3$ is accomplished following a procedure similar to that reported for some mono- and disubstituted aryl ethers, except that methanol was used to quench the unreacted BBr$_3$ in place of H$_2$O.[2] This quenching method produces HBr in excess, which protonates one of the amines, leading to isolation of 4,5-diaminocatechol hydrobromide.

Catechol-functionalized salphen molecules can be readily synthesized from the reaction of 2 equiv of the desired salicylaldehyde derivative with 1 equiv of the 4,5-diaminocatechol after the latter is neutralized with 1 equiv of base, either CH$_3$O$^-$ or pyridine. Pyridine is particularly useful in the procedure since it will deprotonate the quaternized amine but not the catechol. Representative procedures for the syntheses of three different salphen–catechol ligands are provided in the submitted syntheses.

In addition to providing the synthesis of 4,5-diaminocatechol hydrobromide and three salphen–catechol ligands, we also have included representative syntheses of two metallated complexes to show the general utility of these new

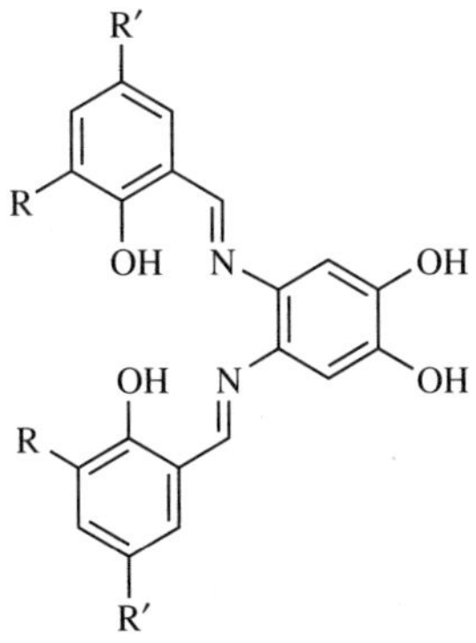

Figure 1

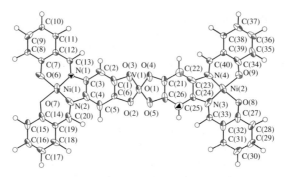

Figure 2

multifunctional ligands. One of these complexes, $(Bu_4N)_2\{VO[Ni^{II}\text{-}(H_4salphen)\text{-}(O)_2]_2\}$, is shown in Fig. 2. By reacting the metallated salphen–catechols in the appropriate ratios with other metal ions, a variety of di-[3] tri-,[4] tetra-,[4,5] and multi-nuclear[6] complexes have been obtained.

General Considerations

All manipulations were performed under an inert atmosphere using standard glovebox and Schlenk techniques, unless otherwise noted. Solvents were distilled under N_2 from the appropriate drying agents (THF and diethyl ether from sodium/benzophenone; CH_2Cl_2, pyridine from CaH_2; methanol from $MgSO_4$) and/or stored over 3-Å molecular sieves (acetone, absolute ethanol, and distilled pyridine) and thoroughly deoxygenated with N_2 prior to use. Abbreviations used: $[M^{II}\text{-}(R_2R_2{}'\ salphen)\text{-}(OH)_2]$ $[(N,N'\text{-bis(3-R-5-R'-salicylidene})\text{-}4,5\text{-dihydroxy-}$phenylenediamino($M^{II}$)] (see Fig. 1); pyH^+ (pyridinium).

■ **Caution.** *Hot concentrated nitric acid is extremely corrosive and dangerous. Complete body protection (gloves and appropriate clothing) are necessary and eye protection is imperative.*

A. 1,2-DIMETHOXY-4,5-DINITROBENZENE (4,5-DINITROVERATROLE)

$$(C_6H_4(CH_3O)_2 + xsHNO_3 \rightarrow (NO_2)_2(C_6H_4)(CH_3O)_2$$

A slightly modified version of a previously published procedure[7] is employed in the synthesis of 4,5-dinitroveratrole. Concentrated HNO_3 (100 mL) and H_2O (10 mL) are combined in a 500-mL flask equipped with a N_2 inlet, a condenser, and a dropping funnel and cooled to 0°C. 1,2-dimethoxybenzene (veratrole) (40 mL, 0.31 mol) is added dropwise with vigorous stirring over approximately 1 h to produce a yellow slurry. The slurry is then warmed to 60°C to liberate NO_2 as a brown gas. Once gas evolution ceases (3–4 h), the solution is poured into 1 L of ice water and the yellow solid isolated by suction filtration. The solid is then washed thoroughly with a saturated solution of $NaHCO_3$ followed by H_2O. Once dry, the product is recrystallized from 900 mL of hot ethanol. A highly crystalline yellow solid is then isolated by filtration, and after drying the yield is 65 g (80%).

B. 1,2-DIMETHOXY-4,5-DIAMINOBENZENE (4,5-DIAMINOVERATROLE)

■   **Caution.**   *Hydrazine is highly toxic. It also is a strong reducing agent and can oxidize violently. Appropriate safety must be exercised.*

$$(NO_2)_2(C_6H_4)(CH_3O)_2 + N_2H_4 \cdot H_2O \xrightarrow{10\% \ Pd/C} (NH_2)_2(C_6H_4)(CH_3O)_2$$

A slightly modified version of a previously published procedure was employed in the synthesis of 4,5-diaminoveratrole.[7] An amount of 4,5-dinitroveratrole (15 g, 66 mmol) and 10% Pd/C (0.6 g) is added to a 1-L round-bottomed flask equipped with a condenser and a N_2 inlet. The material is dissolved in 350 mL of fresh absolute ethanol, and then hydrazine monohydrate (33 mL, 0.66 mol) is added slowly. A vigorous exothermic reaction occurs within the first 30 min of reaction time. The reaction is then set to reflux. The color of the reaction solution proceeds from orange through yellow to colorless within 5 h. The slurry is then filtered through Celite to remove Pd/C and the colorless to pale yellow filtrate reduced to dryness under vacuum to yield a white to pale yellow microcrystalline solid. Traces of yellow color can be removed by washing the product with 3×10 mL of isopropyl alcohol followed by 3×20 mL of diethyl ether. Yield: 9.8 g (88%). ^{1}H NMR (DMSO-d_6, δ, ppm): $-OCH_3$, 3.57 (s, 6H); $-NH_2$, 4.07 (broad s, 4H); aromatic H, 6.26 (s, 2H).

C. 1,2-DIHYDROXY-4,5-DIAMINOBENZENEHYDROBROMIDE

$$(NH_2)_2(C_6H_4)(CH_3O)_2 + BBr_3 + CH_3OH \rightarrow$$
$$(NH_2)_2(C_6H_4)(OH)_2 \cdot HBr \ (unbalanced)$$

4,5-Diaminoveratrole (8.0 g, 48 mmol) is dissolved in 100 mL of CH_2Cl_2 in a flask equipped with a condenser and a N_2 inlet. BBr_3 (100 mL of a 1.0 M

CH_2Cl_2 solution, 100 mmol) is added slowly via cannula to the stirred solution. The initial reaction is mildly exothermic, with a gray precipitate forming. The slurry is stirred for 16 h. Then, to the mixture, which contains an excess of BBr_3, is added 200 mL of CH_3OH, producing a purple solution. The solvent is then removed under vacuum. The methanol treatment is repeated twice. The final crude purple residue is then recrystallized from methanol/ether (1 : 9) under N_2 to yield a purple crystalline solid that is isolated by filtration and washed with diethylether. Yield: 9.4 g (90%).

Anal. Calcd. for $C_6H_9N_2O_2Br$: C, 32.60; H, 4.10; N, 12.67. Found: C, 32.22; H, 4.05; N, 12.58. NMR (DMSO-d_6; δ, ppm): $-NH_2$, 8.70 (broad s, 4H); Aromatic *H*: 6.81 (s, 2H).

D. [*N,N'*-BIS(3,3'-H-5,5'-H'-SALICYLIDENE)-4,5-DIHYDROXYPHENYLENEDIAMINE] [H$_2$-(H$_4$SALPHEN)-(OH)$_2$]

$$(NH_2)_2C_6H_2(OH)_2 \cdot HBr + 2C_6H_4(CHO)(OH) \rightarrow$$
$$[H_{2-}(salphen)\text{-}(OH)_2] + HBr$$

4,5-Diaminocatechol hydrobromide (4.4 g, 20 mmol) and $Na(OCH_3) \cdot CH_3OH$ (1.7 g, 20 mmol) are dissolved in 100 mL of methanol to produce a yellow solution. Salicylaldehyde (4.3 mL, 40 mmol) is then added in one portion to produce an orange solution. Stirring for 16 h results in precipitation of an orange microcrystalline solid, which is isolated by filtration and washed with small portions of ethanol. Yield: 5.6 g (80%).

Anal. Calculated for $C_{20}H_{16}N_2O_4$: C, 68.97; H, 4.60; N, 8.05. Found: C, 68.09; H, 4.85; N, 7.84. NMR (DMSO-d_6; δ, ppm): aromatic *H*: 6.92 (s, 2H), 6.93 (s, 2H), 6.95 (s, 2H), 6.98 (s, 2H), 7.38 (t, 2H), 7.67 (d, 2H); N=C–*H*, 8.99 (s, 2H). Mid-IR (cm^{-1}, KBr disk): $\nu_{C=N}$, 1621(vs). Electronic spectrum [nm (ε M^{-1} cm^{-1}), CH_3OH solution]: 365(sh), 338 (20,900), 268 (23,000), 234(s).

E. [(*N,N'*-BIS(3,3'-*t*-BUTYL-5,5'-*t*-BUTYLSALICYLIDENE) 4,5-DIHYDROXYPHENYLENEDIAMINE]ETHERATE {[H$_2$-('BUTYL$_4$SALPHEN)(OH)$_2$]·ETHER}

$$(NH_2)_2C_6H_2(OH)_2 \cdot HBr + 2(t - C_4H_9)_2C_6H_3(CHO)(OH) + C_5H_5N \rightarrow$$
$$[H_{2-}(t\text{-}C_4H_9)_4(salphen)\text{-}(OH)_2] + pyHBr$$

4,5-Diaminocatechol hydrobromide (2.0 g, 9.0 mmol) is dissolved in 200 mL of methanol. A slight excess of pyridine (1.0 mL) is added to neutralize the HBr. 3,5-Di-*tert*-butylsalicylaldehyde (4.24 g, 18.0 mmol) is then added in one portion. The solution is refluxed overnight, turning a deep yellow color within the first few hours. The solvent is then removed in vacuo, and the residue is taken up in diethyl ether. The resulting slurry is then filtered in air over a pad of Celite to remove pyHBr. The filtrate is again taken to dryness, and the resulting yellow orange powder is dissolved in a minimum amount of warm hexanes. This latter solution is placed in the freezer and, after overnight standing, a yellow crystalline solid precipitates. The solid is obtained by filtration and dried under vacuum. Yield: 4–5 g (70–85%). Elemental analysis and the ^{1}H NMR spectrum support the formulation as a monoetherate.

Anal. Calcd. for $C_{40}H_{56}N_2O_5$: C, 74.25; H, 9.05; N, 4.33. Found: C, 73.77; H, 9.14; N, 4.29. NMR (acetone-d_6, δ, ppm): CH_3CH_2- from ether, 1.90 (t, 3H); *t*-butyl groups, 1.28 and 1.30 (singlets, 36H); CH_3CH_2- from ether, 3.45 (q, 2H); ring proton on catechol, 6.88 (s, 2H); ring protons on phenol, 7.23 and 7.43 (*meta*-coupled doublets, 4H); N=C–*H*, 8.62 (s, 2H). Mid-IR (cm^{-1}, KBr disk): $\nu_{C=N}$, 1661(vs). Electronic spectrum [nm (ε M^{-1} cm^{-1}), CH_3CN solution]: 386(sh), 356 (24,000), 301(sh), 279 (27,000).

F. ((*N*,N′-BIS(3-ETHOXY-5-H-SALICYLIDENE)-4,5-DIHYDROXYPHENYLENEDIAMINE(PYRIDINIUM BROMIDE) {[H₂(EtO₂H₂SALPHEN)(OH)₂]·pyHBr}

$$(NH_2)_2C_6H_2(OH)_2 \cdot HBr + 2(C_2H_5O)C_6H_3(CHO)(OH) + C_5H_5N \rightarrow$$
$$[H_{2-}(EtO_2H_2salphen)(OH)_2] \cdot pyHBr$$

4,5-Diaminocatechol hydrobromide (2.0 g, 9.0 mmol) was dissolved in 20 mL of ethanol. A slight excess of pyridine was added to neutralize the HBr. In a separate flask, 3-ethoxysalicylaldehyde (18 mmol) was dissolved in 20 mL of ethanol. The solutions were then combined with stirring. The solution developed an intense orange color immediately. A bright orange powder began to precipitate within 20 min. Stirring was continued for 2 h, after which time the reaction was filtered to remove the orange solid, which was subsequently washed well with diethylether and dried under vacuum. Elemental analysis and the ^{1}H NMR spectrum both support the formulation as a coprecipitation of the ligand with one equivalent of pyridinium bromide. Reaction yield was 70%.

Anal. Calcd. for $C_{29}H_{30}N_3O_6Br$: C, 58.38; H, 5.08; N, 7.05. Found: C, 58.08; H, 5.29; N, 6.97. NMR (DMSO-d_6, δ, ppm): CH_3CH_2- from ethoxy, 1.34 (t, 6H); CH_3CH_2- from ethoxy, 4.05 (q, 4H); ring proton on catechol, 7.00 (s, 2H); ring protons on phenol, 7.23 (d, 2H), 7.08 (d, 2H), 6.90 (doublet of doublets, 2H); N=C–*H*, 8.79 (s, 2H). Pyridinium appears at 7.96 (triplet), 8.46 (triplet), 8.88 (doublet). Mid-IR (cm^{-1}, KBr disk): ($\nu_{\text{N–H}}$ from pyH^{+}), 3096(m); $\nu_{\text{C=N}}$, 1615(vs). Electronic spectrum [nm (ε, M^{-1} cm^{-1}), CH$_3$CN solution]: 449(sh), 380(sh), 350 (22,000), 326(sh), 278(24,000), 263(sh), 258(sh).

G. [(*N*,*N*′-BIS(3,3′-H-5,5′-H′-SALICYLIDENE)-4,5-DIHYDROXYPHENYLENEDIAMINO] NICKEL(II) {[Ni^{II}-(H$_4$SALPHEN)(OH)$_2$]}

[H$_2$-(H$_4$salphen)(OH)$_2$]·pyHBr + Ni(OAc)$_2$·4H$_2$O →

$\qquad$ [Ni^{II}-(H$_4$salphen)-(OH)$_2$] + Py + HOAc (unbalanced)

[H$_2$-(H$_4$salphen)(OH)$_2$] (2.81 g, 5.00 mmol) is added to 100 mL of methanol. The slurry is warmed until the ligand dissolves. Ni(OAc)$_2$·4H$_2$O (1.25 g, 5.00 mol) is then dissolved in 40 mL of methanol and this solution is added to the ligand solution in one portion. An orange powder precipitates immediately. The solid is isolated by filtration and washed with diethyl ether. Yield = 1.6 g (80%).

Anal. Calcd. for C$_{20}$H$_{14}$N$_2$O$_4$Ni (MM = 405): C, 59.31; H, 3.48; N, 6.92; Ni, 14.5. Found: C, 59.57; H, 3.81; N, 6.91; Ni, 13.9. MS (EI 70 eV): 404. NMR (DMSO-d_6; δ, ppm). Aromatic H: 7.54 (d, 2H), 7.45 (s, 2H), 7.26 (t, 2H), 6.83 (d, 2H), 6.62 (t, 2H); N=C–*H*, 8.44 (s, 2H). Mid-IR (cm^{-1}, KBr disk): $\nu_{\text{C=N}}$, 1616(vs). Electronic spectrum [nm (ε, M^{-1} cm^{-1}), DMF solution]: 480(sh), 399(sh), 379 (26,200), 364(sh), 318 (13,500).

H. BIS-TETRABUTYLAMMONIUM-BIS[(*N*,*N*′-BIS(3,3′-H-5,5′-H′-SALICYLIDENE)CATECHOLATODIAMINO]NICKEL(II)-OXOVANADIUM(IV) (Bu$_4$N)$_2${VO[Ni^{II}-(H$_4$SALPHEN)(O)$_2$]$_2$}

2[Ni^{II}-(H$_4$salphen)(OH)$_2$] + 4Bu$_4$N(OH) + VO(acac)$_2$ →

$\qquad$ (Bu$_4$N)$_2$(VO){[Ni^{II}-(H$_4$salphen)-(O)$_2$]$_2$}

[Ni^{II}-(H$_4$salphen)(OH)$_2$] (0.50 g, 1.2 mmol) is dissolved in 40 mL of DMF with stirring. Then 1.0 M Bu$_4$N(OH) (2.5 mL, 2.5 mmol) is added by syringe and the solution stirred for 30 min. VO(acac)$_2$ (0.15 g, 0.62 mmol) is dissolved separately in 10 mL of in methanol and then added in one portion to the ligand solution to produce a deep red solution. The reaction is stirred for 20 min and filtered and then a layer of diethyl ether (160 mL) is added and allowed to diffuse slowly into the reaction filtrate. Dark red crystals are obtained by filtration and washed well with diethylether. Yield: 0.66 g (75%).

Anal. Calcd. for $C_{72}H_{96}N_6O_9Ni_2V \cdot DMF$ (MM = 1430): C, 62.95; H, 7.25; N, 6.85; Ni, 8.20; V, 3.56. Found: C, 62.79; H, 7.23; N, 6.35; Ni, 7.48; V, 3.63. MS (FAB-): Calculated m/e (M-DMF) = 1358; found, 1358. Mid-IR (cm^{-1}, KBr disk): $\nu_{C=O}$, DMF), 1669(m); $\nu_{C=N}$, 1603(vs). Electronic spectrum [nm $(7\varepsilon$, M^{-1} cm^{-1}), CH$_3$OH solution]: 466(sh), 436 (62,700), 416(sh), 375 (40,100), 340 (37,200), 326(sh), 297 (25,100), 251 (77,300). Magnetic moment, $\mu_{eff}^{corr.}$: 2.09 BM (300 K), 1.84 BM (4 K).

Properties

4,5-Diaminocatechol hydrobromide is a purple microcrystalline solid that is stable to air for approximately 3–5 days if kept dry. On deprotonation of the quaternized amine, solutions of 4,5-diaminocatechol are pale orange and are extremely air-sensitive but react with a variety of salicylaldehyde derivatives to produce catechol-functionalized salphen derivatives. These bifunctional ligands are yellow to orange in color and have been used in the synthesis of a variety of multimetallic complexes.[3–6]

References and Notes

1. J. Ehrlich and M. T. Bogert, *J. Org. Chem.* **12**, 522 (1947). [The checkers recommend using concentrated HNO₃ (90 mL) to H₂O (10 mL) and the same conditions listed in the synthesis to obtain complete nitration to dinitroveratrol.]
2. (a) J. M. Lansinger and R. C. Ronald, *Synth. Commun.* **9**, 341 (1979); (b) J. B. Press, *Synth. Commun.* **9**, 407 (1979).
3. T. F. Baumann, personal communication.
4. D. T. Rosa, Ph.D. thesis, The University of Michigan, 1998.
5. (a) S. M. Malinak, D. T. Rosa, and D. Coucouvanis, *Inorg. Chem.* **37**, 1175 (1998); (b) S. M. Malinak, Ph.D. thesis, The University of Michigan, 1998.
6. R. A. Reynolds, III, Ph.D. thesis, The University of Michigan, 1998.
7. J. L. Sessler, T. D. Mody, R. Ramasamy, and A. D. Sherry, *New J. Chem.* **16**, 541 (1994).

18. A CONVENIENT SYNTHESIS OF 1,5,9-TRITHIACYCLODODECANE ($S_3C_9H_{18}$)

Submitted by RICHARD D. ADAMS* and JOSEPH L. PERRIN*
Checked by ANDREA B. MITCHELL and GREGORY J. GRANT†

Polythioether macrocycles or "thiacrowns," as they have been called, have attracted attention because of their ability to serve as good ligands for the transition

* Department of Chemistry and Biochemistry, University of South Carolina, Columbia, SC 29208.
† Department of Chemistry, The University of Tennessee at Chattanooga, Chattanooga, TN 37403-2598.

metals.[1] Historically, these ligands have been difficult to synthesize, although more recent approaches that utilize the base character of cesium carbonate have led to significant improvements in the yields of some of these reactions.[2] We have discovered new procedures that provide certain thiacrowns catalytically in relatively good yields from simple thietane precursors.[3]

In this synthesis, our procedure for the preparation of 1,5,9-trithiacyclododecane 12S3 from the strained cyclic thioether known as thietane ($SCH_2CH_2CH_2$) is described [see Eqs. (1) and (2)].

$$3\ SC_3H_6 \rightarrow S_3C_9H_{18} \tag{1}$$

General Procedure

Thietane (trimethylene sulfide) can be purchased from Aldrich Chemical Company and used without further purification. The thietane complex $Re_2(CO)_9$ ($SCH_2CH_2CH_2$), which is used as the catalyst, is obtained in good yield from the reaction of $Re_2(CO)_9(NCMe)$ with thietane by the published procedure:[3a]

$$\tag{2}$$

A. 1,5,9-TRITHIACYCLODODECANE

A dry 25-mL greaseless round-bottomed flask is equipped with a Teflon-coated magnetic stirring bar, a greaseless condenser, and a nitrogen-inlet adapter on the condenser. A Variac-controlled heating mantle is placed under the flask. A 15.0 mg (0.0215 mmol) amount of crystalline $Re_2(CO)_9(SCH_2CH_2CH_2)$ and 5.0 g (67.7 mmol) of thietane are added to the flask and the reaction mixture is briefly evacuated and refilled with nitrogen 3 times to remove traces of oxygen. The reaction apparatus is then wrapped completely in aluminum foil to minimize possible effects of light on the reaction. The solution is then heated to the reflux temperature of thietane (94°C) for 72 h. *Note*: Vigorous heating can lead to pyrolytic polymerization and loss of product. An ^1H NMR spectrum of the reaction mixture taken at the end of the reaction period shows a 49% conversion to 12S3 and a collection of low-molecular-weight polymers in a 0.7/1 ratio by weight (TON = 213 for 12S3). The excess thietane is removed in vacuo, and the resulting residue weighs 2.43 g. The residue is then thoroughly extracted with five

25-mL portions of hot hexane, filtered through cotton, and evaporated to yield 0.914 g of 12S3, 38% yield based on the amount of consumed thietane. This product is readily purified by recrystallization by dissolving in minimum amount of warm hexane and then cooling to $-15°C$ for overnight. The principal sideproduct(s) are a range of low-molecular-weight polymers (MW = 8700), ^{1}H NMR, $d = 2.60$ (t, 12H), and 1.83 (q, 6H) in CDCl₃ that have poor solubility in hexane solvent and remain in the residue after the extraction step.[4] It should be noted that this synthesis can be performed similarly using $Re_2(CO)_9(NCMe)$ as the catalyst, but the yield of 12S3 is lower.[3a]

Properties

Compound 12S3 is a white crystalline solid at 25°C (melting point 97–98°C). It is readily soluble in dichloromethane, hexane, and benzene, and is stable in air for long periods of time. Its structure has been determined crystallographically.[5] Its ^{1}H NMR spectrum shows resonances at $\delta = 2.67$ (t, 12H) and 1.87 (q, 6H) in CDCl₃. The above-mentioned procedure can be used to prepare a variety of substituted derivatives of 12S3, but the yields are lower and the purification procedures are more extensive.[6] This procedure has also been used to prepare some polyselenaether macrocycles.[7]

References

1. (a) A. J. Blake, M. Schröder, *Adv. Inorg. Chem. Radiochem.* **35**, 1 (1990); (b) S. R. Cooper and S. C. Rawle, *Struct. Bond.* **72**, 1 (1990); (c) S. R. Cooper, in *Crown Compounds: Towards Future Applications*, S. R. Cooper (Ed.), VCH, New York, 1992, Chapter 15; (d) G. Reid and M. Schröder, *Chem. Soc. Rev.* **19**, 239 (1990).

2. R. M. Kellogg, in *Crown Compounds: Towards Future Applications*; S. R. Cooper (Ed.), VCH, New York, 1992, Chapter 14.

3. (a) R. D. Adams, J. E. Cortopassi, and S. B. Falloon, *Organometallics* **14**, 1748 (1995); (b) R. D. Adams and S. B. Falloon, *J. Am. Chem. Soc.* **116**, 10540 (1994); (c) R. D. Adams and S. B. Falloon, *Organometallics* **14**, 4594 (1995); (d) R. D. Adams, S. B. Falloon, J. L. Perrin, J. A. Queisser, and J. H. Yamamoto, *Chem. Ber.* **129**, 313 (1995); (e) R. D. Adams, and S. B. Falloon, *Chem. Rev.* **95**, 2587 (1995); (f) R. D. Adams, in *Catalysis by Di- and Polynuclear Metal Complexes*, R. D. Adams and F. A. Cotton (Eds.), Wiley-VCH, New York, 1998, Chapter 8.

4. R. D. Adams and S. B. Falloon, *Organometallics* **16**, 3866 (1997).

5. S. C. Rawle, G. A. Admans, and S. R. Cooper, *J. Chem. Soc. Dalton Trans.*, 93 (1988).

6. (a) R. D. Adams, J. A. Queisser, and J. H. Yamamoto, *Organometallics* **15**, 2489 (1996); (b) R. D. Adams, J. L. Perrin, J. A. Queisser, and J. B. Wolfe, *Organometallics* **16**, 2612 (1997).

7. R. D. Adams, K. T. McBride, and R. D. Rodgers, *Organometallics* **16**, 3895 (1997).

Chapter Three

SOLID-STATE MATERIALS AND CLUSTERS

19. SYNTHESIS OF QUATERNARY SELENOPHOSPHATES USING MOLTEN SALT FLUXES: $Rb_8Hg_4(Se_2)_2(PSe_4)_4$, K_4In_2 $(PSe_5)_2(P_2Se_6)$, $Rb_4Ti_2(P_2Se_7)(P_2Se_9)_2$, $Rb_4U_4(Se)_2(Se_2)_4(PSe_4)_4$

Submitted by KONSTANTINOS CHONDROUDIS
and MERCOURI G. KANATZIDIS[*]
Checked by JOE KOLIS[†]

The application of salt fluxes in the synthesis of new solid-state compounds has witnessed significant development since 1990. Particularly important has been the development of the molten polychalcogenide flux method in the exploratory synthesis of complex chalcogenides.[1] This relatively new approach to chalcogenides has simplified access to low and intermediate temperatures (160–600°C) and has contributed to the discovery of some very interesting materials. In many cases the compounds stabilized under polychalcogenide flux conditions are only kinetically stable and cannot be synthesized at higher temperatures; however, thermodynamic influences are not entirely avoided by this approach. Lower temperatures also make possible the use of molecular assemblies as building blocks for incorporation into solid-state structures.

The polychalcophosphate fluxes are formed by simple in situ fusion of $A_2Q/$ P_2Q_5/Q (Q = S,Se,Te) in the same manner as was done for the A_2Q_x fluxes. Although the fluxes are conceptually $A_xP_yQ_z$, a more accurate description is probably that of $(P_yQ_z)^{n-}$ species solubilized in excess polychalcogenide flux,

[*] Department of Chemistry and Center for Fundamental Materials Research, Michigan State University, East Lansing, MI 48824.
[†] Department of Chemistry, Clemson University, Clemson, SC, 29634-1905.

which also continues to serve as the oxidant to dissolve metallic elements into the flux. The inclusion of P into the fluxes renders them somewhat more basic than their all-chalcogenide counterparts (i.e., A_2Q_x), but the melting points of $A_xP_yQ_z$ are also in the intermediate temperature range (minimum melting points: 300–400°C), making possible the synthesis of metastable phases. In addition, the $[P_yQ_z]^{n-}$ species act as effective mineralizers, promoting crystal growth. Notice that these compositions always have excess chalcogen, which is important in two ways: (1) it helps to lower the melting point of the flux, and (2) it serves as an electron acceptor when the metal is dissolved in the flux. The chemical properties of these melts can be controlled by the ratios of their constituent elements. The relatively good solubility properties of $A_x[P_yQ_z]$ salts in water and organic solvents allow for easy isolation of products. Because $[P_xQ_y]^{n-}$ containing solids are still few, and because, from a synthetic chemist's point of view, well-defined discrete anionic fragments are potential building blocks for new compounds, methodologies that facilitate the production and usage of such "synthons" are particularly attractive. Thus, we developed a general methodology by which new selenophosphate (or chalcophosphate) compounds can be consistently obtained. The high number of different $[P_yQ_z]^{n-}$ ligands along with their impressive bonding versatility creates many possibilities for new structure types and compositions, most of which were unanticipated.[2-5] In this article we provide examples of detailed syntheses of compounds prepared by the alkali metal *polychalcophosphate* flux method, namely, the molecular $Rb_8Hg_4(Se_2)_2$ $(PSe_4)_4$,[4d] the one-dimensional (1D) $K_4In_2(PSe_5)_2(P_2Se_6)$,[3h] the 2D Rb_4Ti_2 (P_2Se_7) $(P_2Se_9)_2$,[4b] and the 3D $Rb_4U_4(Se)_2(Se_2)_4(PSe_4)_4$.[5b]

A. DIPHOSPHORUSPENTASELENIDE (P_2Se_5)

$$2P + 5Se \rightarrow P_2Se_5$$

Procedure

A glass with the nominal composition P_2Se_5 is used as a starting material for the synthesis of the quaternary compounds.

First, 1.356 g (43.78 mmol) of P powder (black form) and 8.644 g (109.47 mmol) of Se powder (black form) are weighed and mixed thoroughly (no grinding necessary) and then transferred to a Pyrex tube (15 mL in volume) inside a nitrogen-filled glovebox. The Pyrex tube is then flame-sealed under vacuum at a pressure of $\sim 10^{-4}$ Torr. The reaction tube is placed in a programmable furnace, and the reaction temperature is programmed as follows (where $T_1 =$ starting temperature; $T_2 =$ ending temperature):

Step	T_1 (°C)	T_2 (°C)	Time (h)
1	50	460	12
2	460	460	48
3	460	50	12

To isolate the product, the tube is opened with a glass cutter and the material is mechanically removed from it and ground into powder using a mortar and pestle inside a nitrogen-filled glovebox. The air-sensitive black powder is stored in a vial with a screwcap and can be used in the same manner as for the following syntheses. This is P_2Se_5 in composition only. It is probably a mixture of various binary P/Se compounds, most of them amorphous. The actual phase composition is not important since it is used only as starting material.

■ **Caution.** *Reactions in sealed tubes are always dangerous because of possible explosions. Eye protection is highly recommended, and the sealed tubes should be opened in a well-ventilated hood. Selenium and selenium compounds are toxic.*

B. OCTARUBIDIUM-TETRAMERCURY-BISDISELENIDO-TETRAKISTETRASELENOPHOSPHATE [$Rb_8Hg_4(Se_2)_2(PSe_4)_4$]

$$4HgSe + 2P_2Se_5 + 8Rb_2Se + 20Se \rightarrow Rb_8Hg_4P_4Se_{20} + \text{``}Rb_8Se_{22}\text{''}$$

■ **Caution.** *Traces of mercury might be present in the solution, and all waste must be discarded in a manner appropriate to avoid heavy-metal contamination*

Procedure

First, 0.168 g (0.60 mmol) of HgSe powder (any chemical supplier), 0.137 g (0.30 mmol) of P_2Se_5, 0.300 g (1.20 mmol) of Rb_2Se (see Ref. 3), and 0.237 g (3.00 mmol) of Se powder are weighed and mixed thoroughly and then transferred to a Pyrex tube (6 mL in volume) inside a nitrogen-filled glovebox. The Pyrex tube is then flame-sealed under vacuum at a pressure of $\sim 10^{-4}$ Torr. The reaction tube is placed in a programmable furnace, and the reaction temperature is programmed as follows:

Step	T_1 (°C)	T_2 (°C)	Time (h)
1	50	495	24
2	495	495	110
3	495	300	130
4	300	250	24
5	250	50	24

To isolate the product, the tube is opened with a glass cutter and the contents are placed into a 250-mL flask under N_2. Excess Rb_xSe_y is removed by adding 150 mL of degassed dimethylformamide (DMF) with occasional stirring, resulting in a dark green-black solution dissolving the flux. After the remaining solid settles completely, the solution is slowly decanted

The washings with degassed DMF are repeated until the solution remains colorless, indicating total removal of the flux. Then the product is washed with ether (> 200 mL or as needed to dry compound), dried, and put into a small vial with a screwcap (5 mL). Tri-*n*-butylphosphine (~ 3 mL) is added to remove elemental selenium.

■ **Caution.** *Tri-n-butylphosphine is toxic and should be handled in a fume hood, wearing gloves.*

After occasional stirring for ~ 20 min the phosphine is removed by decanting and the product is washed with copious amounts of acetone and then with ether. Small yellow needles of $Rb_8Hg_4(Se_2)_2(PSe_4)_4$ are obtained ($\sim 78\%$ yield based on Hg).

Properties

Observed X-ray powder pattern spacings (Å Cu K_a) ($I/I_{max}\%$ in parentheses): 12.477(100), 6.687(88), 4.410(5), 3.951(3), 3.694(17), 3.618(20), 3.465(33), 3.340(8), 3.175(53), 3.122(31), 3.029(90), 2.987(52), 2.794(69), 2.693(14), 2.665(8), 2.577(3), 2.499(20), 2.473(18), 2.387(11), 2.321(19), 2.300(18), 2.209(2), 2.141(10), 2.096(10), 2.054(14), 1.990(15), 1.952(17), 1.855(7), 1.807(5), 1.785(25), 1.777(10), 1.715(26). Fourier transform–infrared (FTIR) spectrum (cm^{-1}, CsI pellet): 461(vs), 434(s), 417(vs), 176(m), 159(w).

C. TETRAPOTASSIUM-DIINDIUM-
BISPHOSPHORUSPENTASELENIDE-
DIPHOSPHORUSHEXASELENIDE [$K_4In_2(PSe_5)_2(P_2Se_6)$]

$$2In + 4P_2Se_5 + 4K_2Se + 20Se \rightarrow K_4In_2P_4Se_{16} + \text{``}K_4P_4Se_{28}\text{''}$$

Procedure

First, 0.034 g (0.30 mmol) of In powder (any chemical supplier), 0.274 g (0.60 mmol) of P_2Se_5, 0.094 g (0.60 mmol) of K_2Se (see Ref. 3), and 0.237 g (3.00 mmol) of Se powder are weighed and mixed thoroughly and then transferred to a Pyrex tube (6 mL in volume) inside a nitrogen-filled glovebox. The Pyrex tube is then flame-sealed under vacuum at a pressure of $\sim 10^{-4}$ Torr. The reaction tube is placed in a programmable furnace, and the reaction temperature is programmed as follows:

Step	T_1 (°C)	T_2 (°C)	Time (h)
1	50	480	24
2	480	480	72
3	480	175	92
4	175	50	2

To isolate the product, the tube is opened with a glass cutter and the contents are placed into a 250-mL flask under N_2. Excess $K_xP_ySe_z$ is removed by adding 150 mL of degassed dimethylformamide (DMF) with occasional stirring, resulting in a dark green-black solution dissolving the flux. After the remaining solid settles completely, the solution is slowly decanted. The washings with degassed DMF are repeated until the solution remains colorless, indicating total removal of the flux. Then the product is washed with ether, dried, and put into a small vial with a screwcap (5 mL). Tri-*n*-butylphosphine (~ 3 mL) is added to remove elemental selenium.

■ **Caution.** *Tri-n-butylphosphine is toxic and should be handled in a fume hood wearing gloves.*

After occasional stirring for ~ 20 min, the phosphine is removed and the product is washed with copious amounts of acetone and then with ether. Brown–brick red needles of $K_4In_2(PSe_5)_2(P_2Se_6)$ are obtained ($\sim 77\%$ yield based on In).

Properties

Observed X-ray powder pattern spacings (Å Cu K_a) ($I/I_{max}\%$ in parentheses): 8.286(100), 6.027(63), 4.055(5), 3.770(25), 3.641(9), 3.542(3), 3.477(4), 3.246(14), 3.217(37), 3.197(8), 3.139(8), 3.096(10), 3.080(6), 2.952(7), 2.895(84), 2.880(32), 2.783(19), 2.755(9), 2.720(7), 2.699(8), 2.657(15), 2.603(9), 2.580(17), 2.524(4), 2.450(6), 2.358(7), 2.325(4), 2.254(5), 2.200(4), 2.134(4), 2.105(5), 2.073(8), 2.028(6), 1.995(15), 1.967(6), 1.919(4), 1.883(22), 1.859(7), 1.819(10), 1.797(4), 1.777(5), 1.734(6), 1.694(7). Fourier transform–infrared (FTIR) spectrum (cm^{-1}, CsI pellet): 486(s), 469(s), 451(s), 427(vs), 304(m), 225(w), 184(m), 150(w), 135(w).

D. TETRARUBIDIUM-DITITANIUM-
DIPHOSPHORUSHEPTASELENIDE-
BISDIPHOSPHORUSNONASELENIDE [Rb$_4$Ti$_2$(P$_2$Se$_7$)(P$_2$Se$_9$)$_2$]

$$2Ti + 6P_2Se_5 + 4Rb_2Se + 20Se \rightarrow Rb_4Ti_2P_6Se_{25} + \text{``}Rb_4P_6Se_{29}\text{''}$$

Procedure

First, 0.014 g (0.30 mmol) of Ti powder (any chemical supplier), 0.411 g (0.90 mmol) of P$_2$Se$_5$, 0.150 g (0.60 mmol) of Rb$_2$Se, and 0.237 g (3.00 mmol) of Se powder are weighed and mixed thoroughly and then transferred to a Pyrex tube (6 mL in volume) inside a nitrogen-filled glovebox. The Pyrex tube is then flame-sealed under vacuum at a pressure of $\sim 10^{-4}$ Torr. The reaction tube is placed in a programmable furnace, and the reaction temperature is programmed as follows:

Step	T_1 (°C)	T_2 (°C)	Time (h)
1	50	490	24
2	490	490	72
3	490	175	92
4	175	50	2

To isolate the product, the tube is opened with a glass cutter and the contents are placed into a 250-mL flask under N$_2$. Excess Rb$_x$P$_y$Se$_z$ is removed by adding 150 mL of degassed dimethylformamide (DMF) with occasional stirring, resulting in a dark green-black solution dissolving the flux. After the remaining solid settles completely, the solution is slowly decanted. The washings with degassed DMF are repeated until the solution remains colorless, indicating total removal of the flux. Then the product is washed with ether, dried, and put into a small vial with a screwcap (5 mL). Tri-*n*-butylphosphine (2 $\times$ $\sim$2 mL) is added to remove elemental selenium.

■   **Caution.** *Tri-n-butylphosphine is toxic and should be handled in a fume hood wearing gloves.*

After occasional stirring for $\sim$20 min the phosphine is removed and the product is washed with copious amounts of acetone and then with ether. Black needles of Rb$_4$Ti$_2$(P$_2$Se$_7$)(P$_2$Se$_9$)$_2$ are obtained ($\sim$80% yield based on Ti).

Properties

Observed X-ray powder pattern spacings (Å Cu K_a) (I/I_{max}% in parentheses): 12.869(30), 9.327(24), 8.887(32), 8.260(11), 8.023(10), 6.368(30), 5.735(53),

5.218(24), 5.096(23), 4.885(11), 4.554(11), 4.053(15), 4.011(26), 3.650(16),
3.577(15), 3.537(18), 3.489(14), 3.440(38), 3.371(65), 3.361(26), 3.321(26),
3.266(16), 3.236(29), 3.208(41), 3.185(17), 3.156(18), 3.108(41), 3.085(17),
3.052(20), 2.961(32), 2.934(47), 2.925(21), 2.905(64), 2.884(49), 2.867(20),
2.857(100), 2.824(33), 2.774(32), 2.761(33), 2.702(12), 2.656(14), 2.635(54),
2.592(24), 2.547(19), 2.466(14), 2.443(14), 2.432(15), 2.414(10), 2.392(21),
2.336(14), 2.299(9), 2.250(10), 2.204(10), 2.185(17), 2.145(12), 2.024(14),
2.000(12), 1.986(10), 1.914(17), 1.891(18), 1.859(25), 1.838(14). Fourier
transform–infrared (FTIR) spectrum (cm^{-1}, CsI pellet): 503(w), 491(vs),
447(m), 422(w), 390(s), 351(s), 262(vs), 233(s), 155(w).

E. TETRARUBIDIUM-TETRAURANIUM-
BISSELENIDOTETRAKISDISELENIDO-
TETRAKISTETRASELENOPHOSPHATE [Rb$_4$U$_4$(Se)$_2$(Se$_2$)$_4$(PSe$_4$)$_4$]

$$4U + 2P_2Se_5 + 4Rb_2Se + 20Se \rightarrow Rb_4U_4P_4Se_{26} + \text{``}Rb_4Se_8\text{''}$$

■ **Caution.** *Traces of uranium might be present in the solution, and all waste must be discarded in a manner appropriate for radioactive metal contamination.*

Procedure

First, 0.143 g (0.60 mmol) of U powder (Alpha, AESAR), 0.137 g (0.30 mmol) of P$_2$Se$_5$, 0.150 g (0.60 mmol) of Rb$_2$Se, and 0.237 g (3.00 mmol) of Se powder are weighed and mixed thoroughly and then transferred to a Pyrex tube (6 mL in volume) inside a nitrogen-filled glovebox. The Pyrex tube is then flame-sealed under vacuum at a pressure of $\sim 10^{-4}$ Torr. The reaction tube is placed in a programmable furnace, and the reaction temperature is programmed as follows:

Step	T_1 (°C)	T_2 (°C)	Time (h)
1	50	515	24
2	515	515	100
3	515	300	150
4	300	50	48

To isolate the product, the tube is opened with a glass cutter and the contents are placed into a 250-mL flask under N$_2$. Excess Rb$_x$Se$_z$ is removed by adding 150 mL of degassed dimethylformamide (DMF) with occasional stirring, resulting in a dark green-black solution dissolving the flux. After the remaining solid settles completely, the solution is slowly decanted.

The washings with degassed DMF are repeated until the solution remains colorless, indicating total removal of the flux. Then the product is washed with ether, dried, and put into a small vial with a screwcap (5 mL). Tri-*n*-butylphosphine (2 × 2 mL) is added to remove elemental selenium.

■ **Caution.** *Tri-n-butylphosphine is toxic and should be handled in a fume hood, wearing gloves.*

After occasional stirring for ∼20 min the phosphine is removed and the product is washed with copious amounts of acetone and then with ether. Black rod-shaped crystals of $Rb_4U_4(Se)_2(Se_2)_4(PSe_4)_4$ are obtained (∼81% yield based on U).

Properties

Observed X-ray powder pattern spacings (Å Cu K_a) ($I/I_{max}\%$ in parentheses): 9.294(100), 6.245(7), 6.021(28), 5.698(17), 5.373(40), 5.199(11), 4.936(11), 4.587(8), 4.404(5), 4.183(4), 3.877(2), 3.599(18), 3.567(7), 3.492(4), 3.319(9), 3.197(15), 3.092(90), 3.080(33), 2.998(32), 2.858(21), 2.773(20), 2.675(37), 2.600(3), 2.484(5), 2.443(22), 2.365(6), 2.336(29), 2.327(22), 2.246(22), 2.217(13), 2.198(11), 2.183(17), 2.111(14), 2.048(6), 2.015(10), 1.923(11), 1.852(13). Fourier transform–infrared (FTIR) spectrum (cm^{-1}, CsI pellet): 453(s), 444(s), 416(s), 278(sh), 270(sh).

Semiquantitative Microprobe Analyses. Elemental analyses were performed using a JEOL JSM-6400V scanning electron microscope (SEM) equipped with a TN 5500 EDS detector. Data acquisition was performed with an accelerating voltage of 20 kV and a thirty second accumulation time. The compositions found with this technique are very close to the actual compositions determined by single-crystal crystallographic analysis. In order to obtain good analyses (to within ±3 atomic weight), averaging of the results from more than four or five crystals may be required.

References

1. M. G. Kanatzidis and A. C. Sutorik, *Progr. Inorg. Chem.* **43**, 151–265 (1995).
2. M. G. Kanatzidis, *Curr. Opin. Solid State Mater. Sci.* **2**, 139–149 (1997).
3. (a) T. J. McCarthy and M. G. Kanatzidis, *Chem. Mater.* **5**, 1061–1063 (1993); (b) T. J. McCarthy and M. G. Kanatzidis, *J. Chem. Soc., Chem. Commun.* 1089–1090 (1994); (c) T. J. McCarthy, T. Hogan, C. R. Kannewurf, and M. G. Kanatzidis, *Chem. Mater.* **6**, 1072–1079 (1994); (d) T. J. McCarthy and M. G. Kanatzidis, *J. Alloys Comp.* **236**, 70–85 (1996); (e) K. Chondroudis, and M. G. Kanatzidis, Materials Research Society, Fall 1996 Meeting, Boston, MA; (f) K. Chondroudis, T. J. McCarthy, and M. G. Kanatzidis, *Inorg. Chem.* **35**, 840–844 (1996); (g) K. Chondroudis and M. G. Kanatzidis,

J. Chem. Soc., Chem. Commun. 1371–1372 (1996); (h) K. Chondroudis and M. G. Kanatzidis, *Inorg. Chem.* (in press).

4. (a) T. J. McCarthy and M. G. Kanatzidis, *Inorg. Chem.* **34**, 1257–1267 (1995); (b) K. Chondroudis and M. G. Kanatzidis, *Inorg. Chem.* **34**, 5401–5402 (1995); (c) K. Chondroudis, and T. J. McCarthy, and M. G. Kanatzidis, *Inorg. Chem.* **35**, 3451–3452 (1996); (d) K. Chondroudis and M. G. Kanatzidis, *J. Chem. Soc., Chem. Commun.* 401–402 (1997); (e) K. Chondroudis and M. G. Kanatzidis, *Angew. Chem.* **36**, 1324–1326 (1997); (f) K. Chondroudis, J. A. Hanko and M. G. Kanatzidis, *Inorg. Chem.* **34**, 2623–2632 (1997).

5. (a) K. Chondroudis and M. G. Kanatzidis, *C. R. Acad. Sci. Paris, Ser. B* **322**, 887–894 (1996); (b) K. Chondroudis and M. G. Kanatzidis, *J. Am. Chem. Soc.* **119**, 2574–2575 (1997).

20. HYDROTHERMAL SYNTHESIS OF SULFOSALTS: Ag_3AsS_3, Ag_3SbS_3, Ag_7SbS_6, $Cu_{12}Sb_4S_{13}$, $CuInS_2$, $FeIn_2S_4$

Submitted by MICHAEL B. KORZENSKI[*] and JOSEPH W. KOLIS[*]
Checked by SHANNON BROWN and MERCOURI G. KANATZIDIS[†]

Hydrothermal crystallization is defined as the use of a high-temperature/high pressure aqueous medium to dissolve and subsequently crystallize materials that would normally be insoluble at ambient conditions. It is usually employed in cases when certain compounds are difficult or impossible to produce by other synthetic methods.[1] For example, if a phase is unstable at higher temperatures, if it melts incongruently, or if starting materials are very volatile, crystallization from the melt or via ceramic methods may not be satisfactory. Metal sulfides are particularly demanding synthetic targets because the volatility of sulfur and its tendency to oxidize requires that syntheses be performed in sealed tubes. However, the high vapor pressure of sulfur during the heating cycle creates problems. Thus high-temperature synthesis of metal sulfides often requires long reaction times with low yields.[2] There are, however, a large number of metal sulfide minerals grown hydrothermally as single crystals in the earth's lithosphere.[3] This technique can be transferred to the benchtop to grow high-quality crystals of scientifically and technologically important metal sulfosalts.[4]

General Procedures

We have found that the most convenient general method of sulfosalt crystal growth is a variation of the Rabenau method, whereby the reactants and hydro-

[*] Department of Chemistry, Clemson University, Clemson, SC, 29634-1905.
[†] Department of Chemistry, Michigan State University, East Lansing, MI, 48824.

thermal fluid are sealed in quartz ampoules at an appropriate level of loading.[5] The pressure is generated internally by the expansion of the fluid on heating. The quartz is inert to most fluids except those that contain an appreciable amount of OH^- or F^- ions, as they attack the silica. However, oxidizing, halide, or acidic solutions are easily handled in the quartz ampoules. The ampoules are placed in autoclaves that are counterpressured to prevent the ampoules from bursting as the temperature is raised. The ampoules are not easily imploded as as long as the pressure within the autoclave equals or exceeds the pressure generated within the ampoule by the expanding fluid; if this is the case, the ampoule will remain intact. Two methods of counterpressuring are presented here.

The simplest autoclaves can be assembled from components purchased from High Pressure Equipment Company (1222 Linden Avenue, Erie, PA 16505) for minimal cost. The autoclave can be prepared from a so-called medium-pressure nipple, typically 8 in. long and $\frac{9}{16}$ in. o.d. and $\frac{5}{16}$ in. i.d.. One end can be closed off with a simple cap, while the other can be closed with a cap or connector containing a safety head with a rupture disk. These components are all standard and can be purchased for less than $200. Such an autoclave is shown in Fig. 1, and is capable of containing two quartz ampoules (8 cm in length; 7 mm o.d.) at a time. An autoclave with these dimensions has cone and thread fittings made of 316 stainless steel, and has an internal volume of 10.1 mL and pressure rating of 20,000 psi (lb/in.^{2}) at 25°C. Larger sized autoclaves can be assembled using nipples and fittings as large as 1in. o.d. and $\frac{9}{16}$: i.d. if desired. They can contain more quartz ampoules but at a somewhat greater cost. Parr autoclaves (model 4740) like the type shown in Fig. 2 can be purchased from Parr Instrument Company (211 53rd. Street, Chicago, IL 61265) for around $1000 and can hold up to eight quartz ampoules at a time. This autoclave, which is rated to 8500 psi, is equipped with a rupture disk designed to vent the closed system in case of buildup of internal pressure. The gauge block can be purchased from Parr or easily assembled from components for somewhat less cost. It should be noted that both types of pressure vessel are efficiently designed such that they can be easily reused many times without damage to the treads or seals. The initial purchase price is no higher than any specialty glassware, and the vessels are far more durable.

■ **Caution.** *Closed vessels at high temperatures and pressures are inherently energetic systems. They obviously require careful and conscientious attention, and safety is always a central issue. However, the autoclaves are carefully designed for the purpose of containing the specified pressures and temperatures. It is our experience that they are of uniformly high quality, and if used properly, present no problems. In addition, each design presented here contains a safety rupture disk designed to blow out and release the pressure should it inadvertently exceed the rated level of the particular autoclave for any reason. Nevertheless, to*

HIGH PRESSURE CELL
316 STAINLESS STEEL
RATED TO 20,000 p.s.i.
10.1 ml. TOTAL VOLUME
STANDARD CONE AND THREAD FITTINGS

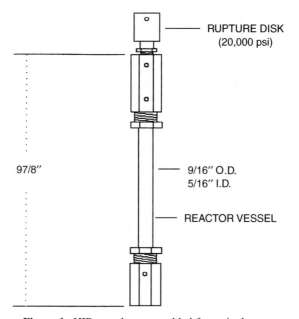

Figure 1. HIP autoclave assembled from single components.

*minimize risk, it is recommended that the ovens be located in an isolated room
with minimum traffic and few, if any, other uses.*

A. TRISILVER ARSENIC TRISULFIDE (PROUSTITE; Ag₃AsS₃)

$$12Ag + As_4S_4 \rightarrow 4Ag_3AsS_3$$

Procedure

All reagents were of analytical grade (Strem Chemical Company) and used
without further purification. Silver powder (0.0545 g, 0.505 mmol), arsenic(II)
sulfide (0.0155 g, 0.0724 mmol), and ammonium sulfide (0.100 mL, 40 wt%)
are added to a quartz ampoule (7 mm o.d., 5 mm i.d.) approximately 9 cm
long inside a glovebox filled with an argon atmosphere. The quartz ampoule

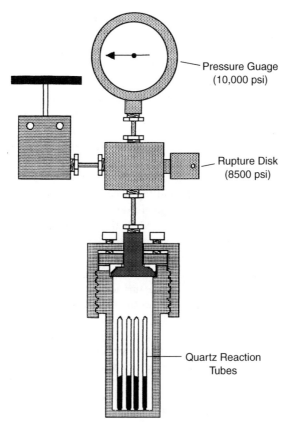

Figure 2. Parr autoclave filled with quartz reaction tubes.

was previously flame-sealed at one end. A rubber septum is fitted over the open end of the ampoule before leaving the inert atmosphere of the glovebox. The ampoule is then filled with distilled water by syringe until approximately 35–40% full with the combined water and starting materials. The ampoule is fitted to a vacuum line using an Ace-Thread tube adapter (model 5027, no.7). A Dewar filled with liquid N_2 is placed on a lab jack and slowly raised until the water in the bottom tip of the ampoule begins to freeze. The stopcock to the vacuum line is then slowly opened to allow for the evacuation of the ampoule. The Dewar is gradually raised in a slow manner until all the water is frozen.

■ **Caution.** *This step must be done very slowly to avoid any bumping of the solution inside the ampoule.*

Once the solution is completely frozen, the ampoule is sealed at a position two-thirds higher than the water level using an oxygen torch. A blue cone flame approximately $\frac{1}{4}$ in. long is used to soften the quartz in a triangular three-point fashion. This step is also done slowly to avoid oversoftening, which leads to concave portions in the sealed end of the ampoule. These areas tend to be thinner and more prone to breakage at the maximum reaction temperature and pressure. Once the three sides become fused together, the flame is directed at each of the three corners for about 5–10 sec while the tube is pulled downward with a pair of stainless-steel tweezers. At this point the ampoule is about 50% filled with solution. Such a fill level will generate about 3000 psi pressure at maximum reaction temperature. Once the sealed end is cool to the touch, the ampoule is placed in a beaker filled with cold tap water to thaw it out. The ampoule is then placed inside a 316 stainless-steel cone and thread fitted HIP autoclave with screw-on endcaps. Water is added to the autoclave to a 70% fill mark. This will provide the counterpressure to prevent the quartz ampoules from bursting during heating. The endcaps are then screwed on and tightened with an adjustable wrench. The autoclave is placed in a tube furnace set at 375°C for 3–5 days. After removal from the furnace, it is allowed to cool to room temperature. The autoclave is vented inside a hood and opened to remove the quartz tubes.

■   **Caution.**   *Safety goggles and gloves should be worn throughout this step in case any residual pressure builds up during the reaction.*

The ampoule is frozen once again. Using thick gloves or thickly folded paper towels, the ampoules are scored in the middle using a glass cutter and opened and placed in a Schlenk flask to thaw.

■   **Caution.**   *These reactions have the ability to produce poisonous gases such as H_2S as byproducts, so opening of ampoules and filtering procedures should be done in a well-ventilated hood.*

Once thawed, the solid products are filtered using the vacuum filter flask and washed several times with acetone and distilled water in approximately 20-mL portions each. The product is allowed to dry on the filter paper under vacuum for 20 min or until completely dry (78.1% yield).

Properties

Ag_3AsS_3 forms as transparent orange polyhedron-shaped crystals that are stable in air. In the presence of concentrated KOH solutions crystals of proustite immediately tarnish black.[6] Observed X-ray powder pattern spacings [Å Cu K_α) (I/I_{max}% in parentheses]: 5.402(16), 3.697(6), 3.276(96), 3.185(77), 3.119(51),

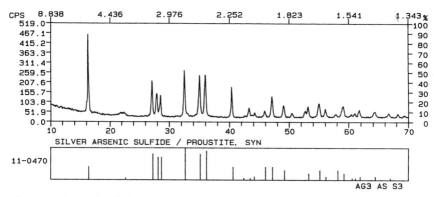

Figure 3. A powder diffraction pattern of proustite (Ag_3AsS_3) (top, observed; bottom, JCPDS), typical of the results obtained via these syntheses.

2.744(100), 2.703(7), 2.554(88), 2.488(93), 2.229(29), 2.118(5), 2.085(19), 2.075(6), 2.045(7), 1.974(9), 1.926(32), 1.853(17), 1.803(7), 1.733(13), 1.712(14), 1.668(21), 1.639(8), 1.636(8), 1.561(12), 1.517(6), 1.500(7). (JCPDS card no. 11-0470) (see Fig. 3).

In cases where the Parr 4740 autoclave is used as the pressure containment device, multiple quartz reaction ampoules can be used in each run. These autoclaves have the advantage that up to eight quartz ampoules can be run at once, while the autoclave is of an appropriate size to fit in a normal tube furnace. In this case, pressurization is most conveniently accomplished by attaching the autoclave directly to a new cylinder of inert gas such as N_2 or argon. Most commercial gas suppliers supply gases in cylinders that are pressurized to 2200–2400 psi. A control valve and gauge on the cylinder is not necessary since the Parr autoclave can easily contain these pressures. Direct inlet adapters are inexpensive and can be purchased from standard compressed-gas dealers. Swage-type and NPT connectors are suitable for these connections, but any tubing must be medium-walled copper or steel.

■ **Caution.** *Reinforced plastic tubing or thin-walled metal, such as flexible steel tubing, is not sufficient to contain these pressures.*

On heating to 375°C, the pressure in the autoclave will increase to approximately 3500 psi, which is well within the operating capabilities of the autoclave, but sufficiently high to contain the pressure within the quartz tubes at the fill levels described here. At the conclusion of the reaction, the autoclave is vented in a hood and opened in the normal fashion, and the quartz tubes examined for products and worked up as described above.

B. TRISILVER ANTIMONY TRISULFIDE (PYRARGYRITE; Ag_3SbS_3)

$$6Ag + Sb_2S_3 \rightarrow 2Ag_3SbS_3$$

Procedure

Silver powder (0.0545 g, 0.505 mmol), antimony(III) sulfide (0.0285 g, 0.0839 mmol), and ammonium sulfide (0.100 mL, 40 wt%) are added to a quartz ampoule, sealed, and worked up in the same fashion as the procedure described above (98.8% yield based on Ag).

Properties

The compound Ag_3SbS_3 is obtained as transparent red polyhedron-shaped crystals that are stable in air. In the presence of concentrated KOH solutions, crystals of pyrargyrite tarnish black.[6] Observed X-ray powder pattern spacings [Å Cu K_α ($I/I_{max}\%$ in parentheses)]: 3.957(12), 3.335(54), 3.215(46), 3.185(58), 2.779(100), 2.569(55), 2.535(69), 2.264(19), 2.126(18), 2.086(10), 1.981(4), 1.959(27), 1.866(18), 1.841(12), 1.769(7), 1.751(14), 1.684(15), 1.676(6), 1.669(12), 1.611(7), 1.598(10), 1.595(9), 1.548(4), 1.533(8), 1.529(7). (JCPDS card no. 21-1173).

C. HEPTASILVER ANTIMONY HEXASULFIDE (ARGRODITE; Ag_7SbS_6)

$$14Ag + Sb_2S_3 + 20S \rightarrow 2Ag_7SbS_6$$

Procedure

Silver powder (0.0545 g, 0.505 mmol), antimony(III) sulfide (0.0125 g, 0.0368 mmol), elemental sulfur (0.0425 g, 1.33 mmol), and ammonium sulfide (0.100 mL, 40 wt%) are added to a quartz ampoule, sealed, and worked up in the same fashion as the procedure described above (64.6% yield based on Ag).

Properties

Ag_7SbS_6 is obtained as shiny black polyhedron-shaped crystals that are stable in air. Observed X-ray powder pattern spacing [Å Cu K_α] ($I/I_{max}\%$ in parentheses)]: 11.746(6), 11.296(6), 10.920(6), 10.722(10), 10.274(9), 10.163(5), 9.764(7), 9.391(8), 9.158(7), 8.777(5), 8.648(6), 8.426(5), 8.355(5), 8.238(6), 7.622(6), 7.449(5), 7.161(6), 6.862(4), 6.430(5), 6.129(8), 5.286(6), 5.175(5), 5.035(4),

4.724(5), 4.308(4), 3.849(5), 3.533(15), 3.378(5), 3.351(17), 3.197(27), 3.059(100), 2.963(8), 2.940(25), 2.834(41), 2.653(9), 2.609(7), 2.603(8), 2.587(9), 2.498(31), 2.444(8), 2.434(15), 2.383(8), 2.313(19), 2.259(10), 2.209(4), 2.165(14), 2.121(7), 2.081(17), 2.064(4), 2.040(21), 1.967(9), 1.935(12), 1.873(32), 1.846(5), 1.819(7), 1.791(21), 1.766(6), 1.764(7), 1.719(7), 1.679(5), 1.656(4), 1.598(6), 1.582(5), 1.564(8), 1.558(6), 1.498(6), 1.459(6) (JCPDS card no. 21-1333).

D. DODECACOPPER TETRAANTIMONY DECATRISULFIDE (TETRAHEDRITE; $Cu_{12}Sb_4S_{13}$)

$$12Cu + 2Sb_2S_3 + 7S \xrightarrow{CaCl_2} Cu_{12}Sb_{14}S_{13}$$

Procedure

Copper powder (0.0520 g, 0.818 mmol), antimony(III) sulfide (0.0442 g, 0.130 mmol), elemental sulfur (0.0150 g, 0.467 mmol) and calcium chloride (0.0300 g, 0.270 mmol) are added to a quartz ampoule, sealed, and worked up in the same fashion as the procedure above. $Cu_{12}Sb_4S_{13}$ is obtained ($\sim$67% yield based on Cu for the clusters only) as shiny black polyhedron-shaped crystals that grow together as clusters of crystals. (These clusters can be easily physically separated from the rest of the single crystals on the filter paper with a microspatula.) Occasionally the product is contaminated with small amounts of famatinite (Cu_3SbS_4) that can be easily separated manually because tetrahedrite is shiny black, whereas famatinite has a characteristic bronze sheen.

Properties

The shiny black crystals are air-stable. Observed X-ray powder pattern spacings [Å Cu K_α (I/I_{max}% in parentheses)]: 5.148(4), 4.205(3), 3.105(63), 2.976(100), 2.755(10), 2.577(16), 2.430(11), 2.305(3), 2.197(3), 2.105(3), 2.022(8), 1.902(15), 1.883(8), 1.824(44), 1.768(4), 1.673(7), 1.630(2), 1.622(13), 1.555(20), 1.489(2), 1.459(3) (JCPDS card no. 24-1318).

E. COPPER INDIUM DISULFIDE (ROQUESITE; $CuInS_2$)

$$Cu + InS_3 \xrightarrow{CaCl_2} CaCl_2CuInS_2 + soluble\ materials$$

Procedure

Copper powder (0.0160 g, 0.252 mmol), indium(III) sulfide (0.0765 g, 0.235 mmol), and calcium chloride (0.100 g, 0.901 mmol) are added to a quartz ampoule, sealed, and worked up in the same fashion as the procedure described above (90.6% yield based on Cu).

Properties

$CuInS_2$ is obtained as a black homogeneous microcrystalline powder that is stable in air. Observed X-ray powder pattern spacings [Å Cu K_α (I/I_{max}% in parentheses): 11.472(1), 8.598(1), 8.307(1), 7.339(1), 5.119(1), 4.539(3), 3.176(100), 3.121(8), 2.958(5), 2.754(11), 2.610(2), 2.574(2), 2.469(2), 2.388(1), 2.185(2), 2.092(2), 1.950(54), 1.883(3), 1,720(1), 1.716(1), 1.664(29), 1.593(4), 1.534(2), 1.528(1), 1.524(1), 1.520(1), 1.514(2), 1.511(2) (JCPDS card no. 27-0159).

F. IRON DIINDIUM TETRASULFIDE (INDITE; $FeIn_2S_4$)

$$2Fe + In_2S_3 \xrightarrow{CaCl_2} FeIn_2S_4 + soluble\ materials$$

Procedure

Iron powder (0.0200 g, 0.358 mmol), indium(III) sulfide (0.0580 g, 0.178 mmol), and calcium chloride (0.100 g, 0.901 mmol) are added to a quartz ampoule, sealed, and worked up in the same fashion as the procedure above (87.7% yield based on Fe).

Properties

$FeIn_2S_4$ was obtained as a rust-colored powder and is air-stable. Observed X-ray powder pattern spacing [Å Cu K_α (I/I_{max}% in parentheses)]: 11.517(12), 11.339(11), 11.126(7), 10.683(11), 10.530(10), 10.360(13), 10.132(11), 9.636(11), 9.420(7), 9.224(10), 8.964(13), 8.829(7), 7.564(12), 7.375(10), 7.179(9), 7.093(10), 6.992(7), 6.707(7), 6.489(10), 6.202(8), 5.813(10), 5.757(7), 3.773(25), 3.218(91), 2.668(24), 2.539(15), 2.175(20), 2.050(60), 1.883(100), 1.685(8), 1.682(8), 1.678(8), 1.621(24), 1.538(7), 1.487(11) (JCPDS card no. 16-0170).

References

1. (a) R. A. Laudise, *Chem. Eng. News* 30 (Sept. 28, 1987); (b) R. A. Laudise, *Prog. Inorg. Chem.* **3**, 1 (1962).

2. G. Kullerud, in *Research Techniques for High Temperature and Pressure*, G. C. Ulmer (Ed.), Springer-Verlag, New York, 1971, p. 289.
3. (a)Y. Takeuchi and R. Sadanaga, *Z. Kristallogr.* **130**, 346 (1969); (b) L. G. Berry, *Am. Mineral.* **50**, 301 (1965); (c) E. Hellner, *J. Geol.* **66**, 503 (1958).
4. P. Ribbe (Ed.), *Sulfide Mineralogy*, Mineralogical Society of America, Blacksburg, VA, 1974.
5. (a) A. Rabenau and H. Rau, *Inorg. Synth.* **14**, 160 (1973); (b) A. Rabeneau, *Angew. Chem., Int. Ed. Engl.* **24**, 1026 (1983).
6. P. Ramdohr, *The Ore Minerals and Their Intergrowth*, Permagon Press, New York, 1969, p. 771.

21. OPEN-FRAMEWORK SOLIDS OF THE VANADIUM OXIDE–PHOSPHATE SYSTEM

Submitted by GRANT BONAVIA and JON ZUBIETA[*]
Checked by L. LARSON and M. KANATZIDIS[†]

The metal oxo unit (M=O) is a fundamental constituent of both soluble molecular clusters[1] and of complex solid materials.[2] The practical interest in the molecular species reflects applications to homogeneous catalysis, heterogeneous catalysis, photocatalysis, electrocatalysis, magnetic materials, and materials synthesis. Likewise, the solid metal oxides exhibit a remarkable range of properties, with applications to high-temperature ferroelectrics, frequency doubling nonlinear optics, electrode materials in solid-state batteries, high-temperature superconductors, catalysis, sorption, and ceramics.

The evolution of metal oxide chemistry is critically dependent on the synthesis of new solids possessing unique structures and properties. However, when considering synthetic materials containing metaloxo functionalities, solid-state phases are far outnumbered by molecular examples. This disparity reflects the tendency of synthetic chemists to study soluble molecular materials that can be prepared and crystallized using previously developed, rational synthetic methodologies and that can be characterized using conventional analytical techniques, such as mass spectroscopy and high-resolution NMR in solution. In contrast, for solid-state synthesis, kinetic control of the reaction parameters is often lost at the elevated temperatures required to achieve convenient reaction rates for the inter-diffusion of solid-state starting materials. Furthermore, since open-framework solids are metastable phases, the sytheses of new compositions requires intermediate temperature techniques, which exploit activation barriers to kinetically trap a new material with modified electronic, magnetic, or catalytic properties. While the rational synthesis of metal oxides remains a challenge, the techniques

[*] Department of Chemistry Syracuse University, Syracuse, NY 13244-4100.
[†] Department of Chemistry, Michigan State University, East Lansing, MI 48824.

of hydrothermal synthesis,[3] particularly in combination with the structure directing properties of organic components, including the coordination chemistry of metal/ligand subunits, and with multipoint hydrogen bonding of organic cations to encourage ordering the organic component, may be exploited in the preparation of solids that retain the structural elements of the precursors, so as to confer a primitive element of design to the solid-state synthesis.

The oxovanadium phosphates constitute a subclass of the general family of metal oxides, which display important catalytic properties and a remarkable structural diversity. The syntheses selected for this volume do no reflect a comprehensive attempt to provide the most important representatives of oxovanadium phosphates that have been described since the late 1990s. Rather, the focus is on the critical role of the organic component as a structure-directing agent and on the efficacy of the hydrothermal technique for the isolation of crystalline metastable phases.

The two oxovanadium phosphate phases and the oxovanadium phosphite material whose syntheses are described contain alkyldiammonium cations whose presence is essential to the isolation of the products. The structures of these materials, which are represented schematically in Figs. 1–3, reveal the importance of hydrophobic/hydrophilic interactions and multipoint hydrogen bonding.

A. 1,3-DIAMMONIUMPROPANE–BISAQUO-BISHYDROXO-BISPHOSPHATO-TRISVANADYL(IV) $\{[H_3N(CH_2)_3NH_3][(VO)_3(OH)_2(H_2O)_2(PO_4)_2]\}$[4]

$$0.75VO(H_2O)_5^{2+} + 4.1H_3PO_4 + 6.0(NH_2)_2C_3H_6 + xsH_2O \rightarrow$$
$$[H_3N(CH_2)_3NH_3][(VO)_3(OH)_2(H_2O)_2(PO_4)_2]$$

■ **Caution.** *Hydrothermal reactions are carried out at high temperatures and pressures. Appropriate shielding and specially designed autoclaves must be employed.*

Procedure

A 1.66 M vanadyl "VO^{2+}" solution is prepared by careful hydrolysis of 16.00 g (83 mmol) of vanadium tetrachloride (VCl_4) in 50 mL of water. A mixture of 0.45 mL (0.747 mmol) of this vanadyl solution, 0.28 mL of 85% phosphoric acid (0.472 g, 4.09 mmol), 0.50 mL (0.444 g, 5.99 mmol) of diaminopropane, and 10 mL (555 mmol) of water is placed in a 23-mL Parr acid digestion bomb.[5] The autoclave is placed in a furnace maintained at 200°C for 48 h. A monophasic product consisting of dark blue plates of $[H_3N(CH_2)_3NH_3][(VO)_3(OH)_2$ $(H_2O)_2$ $(PO_4)_2]$ is collected on a medium-porosity frit by suction and washed with three 20-mL portions of water. The product is then dried for 24 h in vacuo (0.1 torr). Yield: 0.107 g (0.20 mmol), 80% of theoretical based on vanadium.

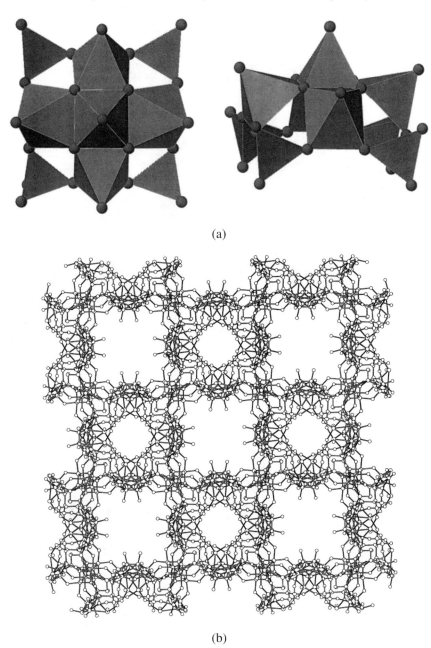

(a)

(b)

Figure 1. (a) Two polyhedral views of the pentanuclear $\{V_5O_9(PO_4)_{4/2}\}$ units of $[HN(CH_2CH_2)_3NH]K_{1.35}[V_5O_9(PO_4)_2] \cdot xH_2O$. (b) The V-P-O framework of the unit cell of $[HN(CH_2CH_2)_3NH]K_{1.35}[V_5O_9(PO_4)_2] \cdot xH_2O$ projected down the [100] direction.

Figure 2. A polyhedral view of the unit cell contents of $[H_3N(CH_2)_3NH_3]$ $[(VO)_3(OH)_2$ $(PO_4)_2(H_2O)_2]$ projected down [100] parallel to the propanediammonium filled tunnels.

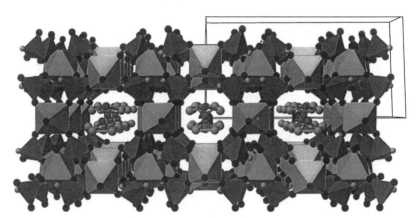

Figure 3. A polyhedral representation of the structure of $[H_2N(CH_2CH_2)_2NH_2]$ $[(VO)_3(HPO_3)_4(H_2O)_2]$ parallel to the *a* axis, showing the channel occupied by the organoammonium cations.

Anal. Calcd. for $C_3H_{18}N_2O_{15}P_2V_3$: C, 6.71; H, 3.35; N, 5.22; V, 28.46. Found: C, 6.65; H, 3.42; N, 5.32; V, 28.32.

Properties

Blue, crystalline $[H_3N(CH_2)_3NH_3][(VO)_3(OH)_2(H_2O)_2(PO_4)_2]$ is indefinitely stable when stored in vial under ambient conditions. The infrared spectrum exhibits characteristic features (KBr, 500–4000 cm^{-1}): 3545(m), 3149(m), 3047(m), 1029(vs), 940(sh), 763(m), 621(m).

B. $[HN(CH_2CH_2)_3NH]K_{1.35}[V_5O_9(PO_4)_2]x H_2O$[6]

$$9KVO_3 + 9N(CH_2CH_2)_3N + 3.3H_2(C_6H_5)PO_3 + 9H_3PO_4 + xsH_2O \rightarrow$$
$$[HN(CH_2CH_2)_3NH]K_{1.35}[V_5O_9(PO_4)_2] \cdot xH_2O$$

A mixture of 0.613 g (4.44 mmol) of potassium vanadate (KVO_3),1.00 g (8.88 mmol) of diaminobicyclooctane (DABCO), 0.92 ml (0.65 g, 8.89 mmol) of diethylamine, 0.526 g (3.33 mmol) of phenylphosphonic acid, 0.61 mL (1.028 g, 8.92 mmol) of phosphoric acid, and 8 mL (444 mmol) of water is placed in a 23 mL Parr acid digestion bomb and heated at 170°C for 4 days. The product is isolated as perfectly formed dark green-black rhombic dodecahedra. Yield: 0.498 g (0.62 mmol), 70% of theory based on vanadium (for $x = 2$).

Anal. Calcd. for $C_6H_{18}K_{1.35}N_2O_{19}P_2V_5$: C, 9.10 H, 2.27; N, 32.17. Found: C, 8.92; H, 2.32; N, 3.45; V, 31.86.

Properties

$[HN(CH_2CH_2CH_2)NH]K_{1.35}[V_5O_9(PO_4)2] \cdot xH_2O$ crystallizes as characteristic dark green-black rhombic dodecahedra or truncated tetrahedra that are indefinitely stable under ambient conditions. The infrared spectrum exhibits characteristic features (KBr, 500–4000 cm^{-1}): 3448(br), 1654(m), 1012(vs), 801(w), 648(m), 531(m).

C. $[H_2N(CH_2CH_2)_2NH_2][(VO)_3(HPO_3)_4(H_2O)_2]$[7]

$$VCl_4 + N_2C_4H_8 + H_3PO_3 + xsH_2O \rightarrow$$
$$[H_2N(CH_2CH_2)_2NH_2][(VO)_3(HPO_3)_4(H_2O)_2]$$

A mixture of 0.11 mL (0.18 mmol) of 1.66 M solution of VCl_4 in water, 0.040 g (0.464 mmol) of piperazine, 0.091 g (1.11 mmol) of phosphorus acid (H_3PO_3), and 6 mL (333 mmol) of water is placed in a 23-mL Parr acid digestion bomb and heated at 150°C for 6 days. The product was isolated as green plates. Yield: 0.052 g (0.08 mmol), 45% of theory based on vanadium.

Anal. Calcd. for $C_4H_{20}N_2O_{17}P_4V_3$: C, 7.44; H, 3.13; N, 4.34, V, 23.79. Found: C, 7.37; H, 4.55; N, 4.25; V, 23.66.

Properties

The compound is indefinitely stable under ambient conditions. It has a characteristic infrared spectrum (KBr, 500–4000 cm^{-1}): 3419(br), 2998(m), 1091(s, br) 961(m), 593(m).

References

1. A. Muller and M. T. Pope (Eds.), *Polyoxometalates: From Platonic Solids to Anit-Retroviral Activity*, Kluwer, Dordrecht, The Netherlands, 1993.
2. P. A. Cox, *Transition Metal Oxides*, Claredon Press, Oxford, UK, 1995.
3. J. Gopalakirshnan, *Chem Mater.* **7**, 1265 (1995).
4. V. Soghomonian, Q Chen, R. C. Haushalter, J. Zubieta, C. J. O'Connor, and Y. S. Lee, *Chem. Mater.* **5**, 1690 (1993).
5. Parr Instrument Corp., 211 53rd St., Moline, IL 61265.
6. M. I. Khan, L. M. Meyer, R. C. Haushalter, A. L. Schweizer, J. Zubieta, and J. L. Dye, *Chem. Mater.* **8**, 43 (1996).
7. G. Bonavia, J. DeBord, R. C. Haushalter, D. Rose, and J. Zubieta, *Chem. Mater.* **7**, 1995 (1995).

22. SULFUR-BRIDGED INCOMPLETE CUBANE-TYPE MOLYBDENUM AND TUNGSTEN AQUA IONS

Submitted by GENTA SAKANE* and TAKASHI SHIBAHARA*
Checked by JOHN H. ENEMARK,[†] JONATHAN McMASTER,[†]
and MICHAEL VALEK

Many reports on molybdenum and tungsten clusters with sulfur-bridged incomplete cubane-type cores, $M_3(\mu_3\text{-}S)(\mu\text{-}S)_3$ (M = Mo,W), have appeared in the literature.[1] The aqua ion $[Mo_3S_4(H_2O)_9]^{4+}$ (Mo3) is very useful for the

*Department of Chemistry, Okayama University of Science, Ridaicho, Okayama 700-0005, Japan.
[†]Department of Chemistry, University of Arizona, Tucson, AZ 85721.

preparation of cubane-type mixed-metal aqua ion clusters with Mo_3MS_4 cores (M = Fe,Co,Ni,Cu,In,Sn,Hg, etc.).[2]

We report here the preparation of the aqua clusters Mo3 and $[W_3S_4(H_2O)_9]^{4+}$ (W3), which are obtained by the reduction of $(NH_4)_2MS_4$ (M = Mo,W) with $NaBH_4$ in dilute hydrochloric acid.[3] Although several methods have been reported for the preparation of the aqua ions Mo3 and W3,[4] we believe that the method described here gives the most facile procedures for the preparation of these aqua ions. Crystalline samples of $[Mo_3S_4(H_2O)_9](CH_3C_6H_4SO_3)_4 \cdot 9H_2O$ (Mo3pts) and $[W_3S_4(H_2O)_9](CH_3C_6H_4SO_3)_4 \cdot 9H_2O$ (W3pts) have been isolated from Mo3 and W3, respectively, by the methods described here, and for these, crystal structures have been determined by X-ray diffraction.

General Procedure and Reagents

Anaerobic conditions or special equipment are not required. Sodium borohydride as well as most other reagents are commercially available reagents and are used as received. *p*-Toluenesulfonic acid monohydrate (Hpts $\cdot$ H_2O) is dissolved to make Hpts (4 or 2 M) solutions. $(NH_4)_2WS_4$ and $(NH_4)_2MoS_4$, are obtained by published procedures.[5] All the procedures described here can be scaled up.

A. SULFUR-BRIDGED INCOMPLETE CUBANE-TYPE MOLYBDENUM AQUA ION, $[Mo_3S_4(H_2O)_9]^{4+}$ (Mo3), IN SOLUTION AND $[Mo_3S_4(H_2O)_9)](CH_3C_6H_4SO_3)_4 \cdot 9H_2O$ (Mo3pts)

I. Nonaaquotrimolybdenum-tetrasulfide, $[Mo_3S_4(H_2O)_9]^{4+}$ (Mo3), in solution

$$6(NH_4)_2MoS_4 + 3NaBH_4 + 23HCl + 27H_2O \rightarrow$$
$$2\{[Mo_3S_4(H_2O)_9]^{4+}(Cl)_4\} + 3B(OH)_3 + 16H_2S$$
$$+ 6H_2 + 15Cl^- + 3Na^+ + 12NH_4^+$$

■ **Caution.** *Rapid addition of $NaBH_4$ will cause violent H_2 evolution. Hydrogen sulfide is toxic and a well-ventilated fume hood must be used until the end of addition of 6 M HCl.*

Procedure

Sodium borohydride (9.0 g, 0.24 mol) in H_2O (120 mL) and HCl (6 M, 120 mL) are pipetted alternately ($\sim$ 4 mL each) into a red solution of $(NH_4)_2MoS_4$ (9.0 g, 0.0346 mol) in H_2O (300 mL) with vigorous stirring at room temperature.

Another quantity of HCl (6 M, 480 mL) is added to the resultant dark brown suspension, through which air is passed above 90°C for 20 h, which need not be continuous. During aeration the suspension changes to solution and the color of this solution turns into dark green.

If the volume of the solution becomes less than ~ 300 mL during air oxidation, HCl (1 M) is added to the solution. After the solution is cooled in an ice-water bath, it is filtered by suction. The precipitate (mainly boric acid) is washed with a small amount of HCl (1 M) and the washing solution from the filter is combined with the filtrate. After the volume is reduced to ~ 100 mL using a rotary evaporator, a Sephadex G-10 column chromatography (diameter 4 cm, length 80 cm) is used to chromatograph the solution using HCl (1 M). A very small amount ($\sim 1\%$) of green $[Mo_3(\mu_3\text{-}S)(\mu\text{-}O)(\mu\text{-}S)_2 (H_2O)_9]^{4+}$ ($\lambda_{max} = 605$ nm in 1 M HCl)[4,6] is followed by the green aqua ion Mo3 ($\lambda_{max} = 620$ nm in 1 M HCl), which is collected. For concentration and further purification, the solution containing the aqua ion Mo3 is diluted 5 times its original volume with water, absorbed on a short Dowex 50W-X2 cation exchange column (diameter 2 cm, length 15 cm), and eluted with HCl (2 M) slowly. A trace amount of the yellow Mo(V) aqua dimer $[Mo_2O_2S_2(H_2O)_6]^{2+}$ is eluted first and then the aqua ion Mo3 is collected. Yield: 50%. Typically 57 mL of a 0.10 M (per Mo3 trimer) solution is obtained. Note that

1. Freshly prepared samples of $(NH_4)_2MoS_4$ (red) (or samples that have been stored under dinitrogen atmosphere) should be used in the synthesis of $[Mo_3S_4(H_2O)_9]^{4+}$. Old samples (black) do not dissolve well in water and therefore give lower yield.

2. The Sephadex G-10 column used can be refreshed by use of a small amount of diluted hydrogen peroxide ($\sim 3\%$) followed by thorough washing with water. Sephadex G-15 can also be used.

3. A dark green powder is obtained by evaporation of 2 M HCl solutions of the aqua ion (Mo3) to dryness. A rotary evaporator is used and the aqueous solution is heated above 90°C. Solidification of Mo3 is very convenient for storage or transportation.

Properties

The aqua ion Mo3 in HCl, Hpts, or dilute H_2SO_4 is very stable in the air; however, it becomes colorless in dilute HNO_3. Powdered Mo3 dissolves not only in dilute HCl to give the aqua ion Mo3 but also in some organic solvents: methanol (very well) ethanol (slightly). The powdered Mo3' is slightly deliquesent, but can be kept dry in a sealed bottle.

Electronic spectral data are shown in Table I.

TABLE I. Electronic Spectral Data for Aqua Trimers with M_3S_4 Cores $(M = Mo,W)^a$

Compounds	Solvent	Color	λ_{max}, nm (ε, M^{-1} cm^{-1})	Ref.
$[Mo_3S_4(H_2O)_9]^{4+}$	2 M Hpts	Green	367 (5190), 500(sh) (290), 602 (351)	3
	1 M HCl	Green	256 (10,350), 372 (5510), 620 (315)	3
$[W_3S_4(H_2O)_9]^{4+}$	2 M Hpts	Purple	314 (7040), 430(sh) (330), 560 (490)	3
	1 M HCl	Blue-violet	221 (15,000), 317 (6100), 430(sh) (310), 570(480)	3

^{a}Values per trimer.

II. Nonaaquotrimolybdenum-tetrasulfide-tetrakis-*p*-toluenesulfonate-nonahydrate, $[Mo_3S_4(H_2O)_9](CH_3C_6H_4SO_3)_4 \cdot 9H_2O$ (Mo3pts)

Procedure[7]

The aqua ion Mo3 (0.10 M per trimer, 100 mL) in HCl (2 M), is diluted to 10 times its original volume with water and absorbed on a Dowex 50W-X2 cation exchanger. The resin is washed with Hpts (0.1 M) to remove Cl^- ion, and following elution with Hpts (4 M), gives a green solution. Cooling of the eluted solution in a freezer (ca. $-10°C$) for a few days gives green plate crystals. Yield: 1.0 g (70% based on Mo3).

Anal. Calcd. for $C_{28}H_{64}O_{30}S_8Mo_3$: C, 23.59; H, 4.52; Mo, 20.20. Found: C, 23.94; H, 4.02; Mo, 20.81.

B. SULFUR-BRIDGED INCOMPLETE CUBANE-TYPE TUNGSTEN AQUA ION, $[W_3S_4(H_2O)_9]^{4+}$ (W3), IN SOLUTION AND $[W_3S_4(H_2O)_9](CH_3C_6H_4SO_3)_4 \cdot 9H_2O$ (W3pts)

■ **Caution.** Rapid addition of $NaBH_4$ will cause violent H_2 evolution. Hydrogen sulfide is toxic, and a well-ventilated fume hood must be used until the end of addition of HCl (6 M).

I. Nonaaquotritungsten-tetrasulfide, $[W_3S_4(H_2O)_9]^{4+}$ (W3), in solution

$$6(NH_4)_2WS_4 + 3NaBH_4 + 23HCl + 27H_2O \rightarrow$$
$$2\{[W_3S_4(H_2O)_9]^{4+}(Cl)_4\} + 3B(OH)_3 + 16H_2S + 6H_2$$
$$+ 15Cl^- + 3Na^+ + 12NH_4^+$$

Procedure

Solium borohydride (9.0 g, 0.24 mol) in H_2O (60mL) and HCl (6 M, 60 mL) are pipetted alternately ($\sim$4 mL each) into a yellow solution of $(NH_4)_2WS_4$ (9.0 g, 0.0258 mol) H_2O (150 mL) with vigorous stirring at room temperature. Another quantity of HCl (6 M, 240 mL) is added to the resultant dark brown suspension, through which air is passed above 90°C for 5 h, which need not be continuous. During aeration the suspension turns to solution and the color of the solution turns into black.

If the volume of the solution becomes less than 300 mL during air oxidation, HCl (1 M) is added to the solution. After the solution is cooled in an ice-water bath, it is filtered by suction. The precipitates (mainly boric acid) are washed with small amount of HCl (1 M) and the washing solution from the filter is combined with the filtrate and the volume is reduced to $\sim$100 mL by use of a rotary evaporator. The solution is filtered and charged on a Sephadex G-10 column chromatography column (diameter 4 cm, length 80 cm). the eluant used is HCl (1 M). The fourth eluate, which contains blue-violet W3 [λ_{max} = 570 nm in HCl (1 M)], is collected ($\sim$350 mL). Other eluates are as follows: (1) [W_3 (μ_3-S)(μ-O)$_2$(μ-S)(H$_2$O)$_9$]$^{4+}$ (red, 0.3%);[8] (2) [W_3(μ_3-S)(μ-O)(μ-S)$_2$(H$_2$O)$_9$]$^{4+}$ (red-purple, 5%);[9] (3) [$W_2O_2S_2(H_2O)_6$]$^{2+}$ (yellow, 10%);[10] and (4) the reduced form of W3 (brown). For concentration and purification, the solution obtained from the fourth band, containing the aqua ion W3, is diluted 5 times its original volume with water, absorbed on a short Dowex 50W-X2 cation exchange column (diameter 2 cm. length 10 cm), and eluted with HCl (2 M) slowly. A small amount of tungsten(V) dimer [$W_2O_2S_2(H_2O)_6$]$^{2+}$ is eluted first and then the aqua ion W3 is collected. To further purify the solution (ca. $\sim$30 mL) from the cation exchanger, a Sephadex G-10 column chromatography (diameter 2 cm, length 90 cm) is used again, with HCl (1 M) as eluant. A small amount of [$W_3(\mu_3$-S)(μ-O)(μ-S)$_2$(H$_2$O)$_9$]$^{4+}$ is eluted first and the main band containing W3 is collected; yield is 25% based on $(NH_4)_2WS_4$. Typically 165 mL of 0.013 M (per trimer) solution in HCl (1 M) is obtained.

II. Nonaaquotritungsten-tetrasulfide-tetrakis-*p*-toluenesulfonate-nonahydrate, [$W_3S_4(H_2O)_9$]$(CH_3C_6H_4SO_3)_4 \cdot 9H_2O$ (W3pts)

Procedure

The aqua ion W3 (0.013 M per trimer, 0.1 L) in HCl (1 M) is absorbed on a Dowex 50W-X2 cation exchanger. The resin is washed with Hpts (0.1 M) to remove Cl$^-$ ion, and a purple solution is obtained by slow elution with Hpts (4 M). Cooling of the eluted solution in a freezer (ca. -20°C) for a few days gives purple plate crystals. Yield: 0.47 g [21% based on in HCl (1 M)].

Anal. Calcd. for $C_{28}H_{64}O_{30}S_8W_3$: C, 19.91; H, 3.81%. Found: C, 20.21; H, 3.84%. Also

1. A dark violet powder of the aqua ion W3 in HCl (1 M) is obtained by evaporation to dryness by using a rotary evaporator at a temperature above 90°C. Solidification of W3 is very convenient for storage or transportation.
2. The brown solution from the fifth band obtained by the first Sephadex column chromatography turns blue-violet within several hours, and the resulting blue-violet solution containing W3 can be purified similarly to yield ca. 5% W3.

Properties

The aqua ion W3 in HCl, freshly prepared Hpts, and diluted H_2SO_4 solutions is stable in air. It decomposes to give a precipitate in aged Hpts solutions. The aqua in dilute HNO_3 solution becomes colorless. The powder W3′ dissolves not only in diluted HCl to give the aqua ion W3 but also in some organic solvents: methanol, ethanol, acetone, and acetonitrile. The powder W3′ is very deliquescent, but can be kept dry in a sealed bottle. The solid cluster W3pts dissolves in some organic solvents, acetone, acetonitrile, methanol, ethanol, and other compounds. Electronic spectral data are shown in Table I. The reactivity of the tungsten aqua ion W3 toward metals is lower than that of Mo3.

References

1. (a) F. A. Cotton, P. A. Kibala, and C. S. Miertschin, *Inorg. Chem.* **30**, 548 (1991); (b) V. P. Fedin, M. N. Sokolov, O. A. Geras'ko, B. A. Kolesov, V. Ye. Fedorov, A. V. Mironov, D. S. Yufie, Yu. L. Slovohotov, and Yu. T. Struchkov, *Inorg. Chim. Acta* **175**, 217 (1990); (c) T. Shibahara, K. Kohda, A. Ohtsuji, K. Yasuda, and H. Kuroya, *J. Am. Chem. Soc.* **108**, 2757 (1986).
2. G. Sakane and T. Shibahara, *Am. Chem Soc., Symp. Ser.* 653, *Transition Metal Sulfur Chemistry*, E. I. Stiefel and K. Matsumoto (Eds.), 1996, Chapter 13, p.225, and references cited therein.
3. T. Shibahara, M. Yamasaki, G. Sakane, K. Minami, T. Yabuki, and A. Ichimura, *Inorg. Chem.* **31**, 640 (1992).
4. T. Shibahara and H. Akashi, *Inorg. Synth.* **29**, 260 (1992), and references cited therein.
5. W. M. McDonald, G. D. Friesen, L. D. Rosenhein, and W. E. Newton, *Inorg. Chim. Acta* **72**, 205 (1983).
6. T. Shibahara, H. Miyake, K. Kobayashi, and H. Kuroya, *Chem. Lett.*, 139 (1986).
7. H. Akashi, T. Shibahara, and H. Kuroya, *Polyhedron* **9**, 1671 (1990).
8. T. Shibahara, A. Takeuchi, T. Kunimoto, and H. Kuroya, *Chem. Lett.*, 867 (1987).
9. T. Shibahara, A. Takeuchi, and H. Kuroya, *Inorg. Chim. Acta* **127**, L39 (1987).
10. T. Shibahara, Y. Izumori, R. Kubota, and H. Kuroya, *Chem. Lett.*, 2327 (1987).

23. SULFUR-BRIDGED CUBANE TYPE MIXED-METAL CLUSTERS WITH Mo₃MS₄ (M = Fe,Co,Ni,Cu,In,Sn) AND W₃NiS₄ CORES

Submitted by GENTA SAKANE[*] and TAKASHI SHIBAHARA[*]
Checked by JOHN H. ENEMARK,[†] JONATHAN MCMASTER,[†]
and MICHAEL VALEK[†]

Sulfide, disulfide, and thiolate ligands combine metal ions to give varied types of metal clusters, and much chemical and biological interest has been focused on sulfur-bridged cubane-type mixed-metal clusters.[1] The report[2] that the incomplete cubane-type sulfur-bridged molybdenum aqua cluster $[Mo_3S_4(H_2O)_9]^{4+}$ (Mo3) reacts with iron metal to give the molybdenum–iron mixed-metal cluster $[Mo_3FeS_4(H_2O)_{10}]^{4+}$ (Scheme 1) provided a ready route to mixed-metal clusters of the type Mo₃MS₄ (M = metal).[3] Three kinds of cubane-type cores are known as shown in Fig. 1. In addition to the method described here, NaBH₄ reduction of a solution of Mo3 and M^{2+} has also been used to prepare these compounds.[4] Other routes to clusters with Mo₃M'S₄ cores have also been reported.[5] Contrary to the high reactivity of Mo3 toward metal ions, the reactivity of the corresponding tungsten aqua cluster $[W_3S_4(H_2O)_9]^{4+}$ (W3) is relatively low, and only the reaction of this ion W3 with Ni[6] and Sn (or Sn^{2+})[7] has been reported thus far.

Other routes to the clusters with the W₃MS₄ cores (M = metal) are limited and only clusters with the W₃CuS₄ cores have been reported.[8] The uptake of ethylene by sulfur-bridged cubane-type molybdenum/tungsten-nickel clusters $[M_3NiS_4(H_2O)_{10}]^{4+}$ (M₃ = Mo₃, Mo₂W, MoW₂, W₃) and ¹H NMR spectra of the resultant ethylene clusters have been reported.[9] Herein we report metal incorporation reactions of Mo3 with Fe,[3a] Co,[10] Ni,[11] Cu,[12] In[13] and Sn[14] as well as that of W3 with Ni,[6] to give mixed-metal aqua clusters. All of the structures of the aqua clusters reported herein have been determined by X-ray crystallography. The electronic spectra of all of the clusters are summarized in Table I.

General Procedures

All the procedures described are carried out under a dinitrogen atmosphere unless otherwise stated, and the column used for the separation under dinitrogen

[*] Department of Chemistry, Okayama University of Science, Ridaicho, Okayama 700-0005, Japan.
[†] Department of Chemistry, University of Arizona, Tucson, AZ 85721.
[‡] The checkers found it difficult to control the flow of dinitrogen to various parts of the inert-atmosphere chromatography apparatus, described in Fig. 2. This can be rectified by using a dinitrogen supply from several outlets from a Schlenk line manifold.

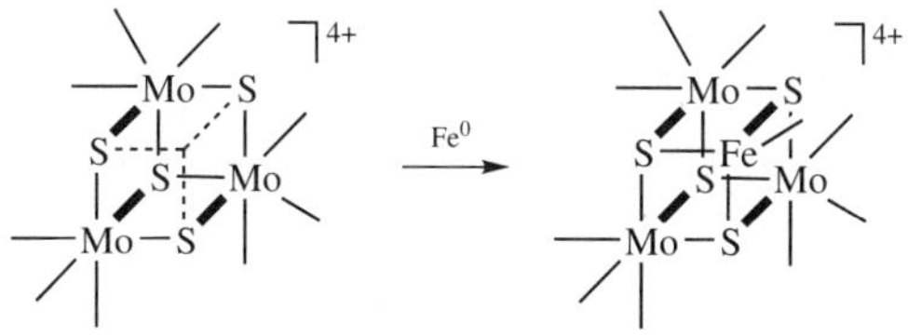

Scheme 1. Formation of molybdenum-iron $[Mo_3FeS_4(H_2O)_{10}]^{4+}$. Coordinated H_2Os are omitted for clarity.

atmosphere is shown in Fig. 2.[‡] Metals are analyzed by ICP emission spectrometry, and other elements are analyzed by standard microanalytical procedures; some solid samples of the aqua clusters easily lose their water of crystallization, and the determined analyses for hydrogen are low.

Source of Chemicals

The facile preparation of the starting materials Mo3 and W3 have been reported.[7a] The aqua cluster ions Mo3 and W3 are stable against air oxidation in dilute HCl and freshly prepared Hpts solution (Hpts = *p*-toluenesulfonic acids). However, they decompose to give precipitates in aged Hpts solutions. The Hpts solution is prepared by dissolving the monohydrate of Hpts $(C_7H_8O_3S \cdot H_2O)$. All chemicals except Mo3 and W3 are commercially available and can be used without further purification.

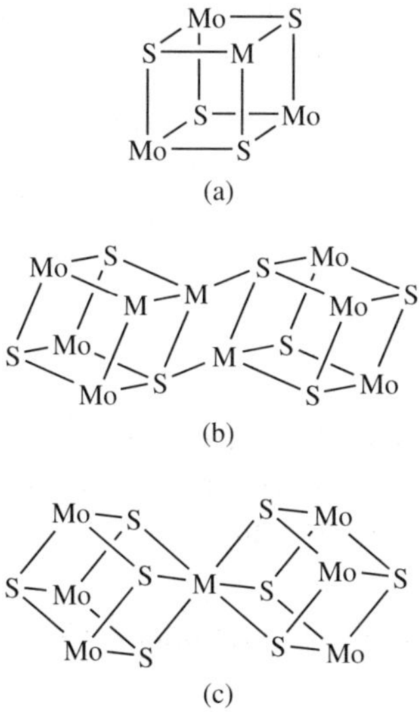

(a)

(b)

(c)

Figure 1. Three kinds of sulfur-bridged cubane-type mixed-metal cores: (a) single cubane type; (b) double cubane type; (c) sandwich cubane type.

TABLE I. Electronic Spectral Data for Aqua Clusters with Mo_3S_4, W_3S_4, Mo_3MS_4 (M = Fe,Co,Ni,Cu,In,Sn), and W_3NiS_4 Cores

Compounds	Solvent	Color	λ_{max}, nm (ε, M^{-1} cm^{-1})	Ref.
Mo3	1 M HCl	Green	256 (10,350), 372 (5510), 620 (315)	7a
Mo3	2 M Hpts	Green	367 (5190), 500(sh) (290), 602 (351)	7a
W3	1 M HCl	Purple	221 (15,000), 317 (6100), 430(sh) (310),570 (480)	7a
W3	2 M Hpts	Red-wine	314 (7040), 430(sh) (330), 560 (490)	7a
Mo3Fe	1 M HCl	Red-purple	464 (875), 512 (830), 600(sh) (535), 960 (177)	3a
Mo3Fe	2 M Hpts	Red-purple	470(sh) (765), 505 (846), 600 (507), 995 (141)	3a
Mo3Co-D[a]	2 M HCl	Red-purple	360 (10,740), 445 (8520), 796 (6720)	10
Mo3Co-D[a]	2 M Hpts	Red-purple	360 (13,860), 450 (8700), 790 (7020)	10
Mo3Ni	2 M Hpts	Blue-green	680 (610), 500 (200)	11
Mo3Ni	2 M HCl	Deep-green	800(sh) (360),686(488), 498(366)	11
Mo3Cu	1 M HCl	Red-brown	330(sh) (2760), 370(sh) (2175), 490 (1398), 710(sh) (201)	12
Mo3Cu-D[a]	2 M Hpts	Red-brown	325 (5254), 360(sh) (4376), 470 (2652), 620(sh) (740), 975 (588)	12
Mo3In	2 M Hpts	Red-brown	552 (192), 758 (510)	13
Mo3Sn-SW	2 M HCl	Red-purple	545 (14,550), 1010 (798)	14
Mo3Sn-Sw	1 M Hpts	Red-purple	545 (15,080), 960 (852)	14
Mo3Sn-OX	1 M HCl	Yellow-green	590 (348), 673 (447)	14
Mo3Sn-OX	3 M Hpts	Yellow-green	560 (345), 644(sh) (327)	14
W3Ni	1 M HCl	Green-blue	691 (565), 600(sh) (447), 520(sh) (384) 433 (730), 240(sh) (18,000), 221 (72,300)	6,9
W3Ni[b]	2 M Hpts	Blue-green	685 (605), 600(sh) (530), 520(sh) (390), 428 (657)	6,9

[a] Per double cubane.
[b] Per single cubane.

A. MOLYBDENUM-IRON CLUSTER AQUA ION, $[Mo_3FeS_4(H_2O)_{10}]^{4+}$ (Mo3Fe), IN SOLUTION AND $[Mo_3FeS_4(H_2O)_{10}]$ $(CH_3C_6H_4SO_3)_4 \cdot 7H_2O$ (Mo3Fepts)

I. Decaaquotrimolybdenum-iron-tetrasulfide, $[Mo_3FeS_4(H_2O)_{10}]^{4+}$ (Mo3Fe), in solution

$$[Mo_3S_4(H_2O)_9]^{4+} + Fe + H_2O \rightarrow [Mo_3FeS_4(H_2O)_{10}]^{4+}$$

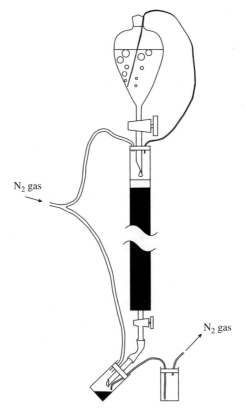

Figure 2. Apparatus for column chromatography under a dinitrogen atmosphere.

Procedure[2,3a]

Iron wire (3.0 g, 54 mmol) is introduced into a conical flask containing the aqua ion Mo3 (0.05 M per trimer in HCl (2 M), 25 mL).[7a] During stirring, the color of the solution changes from green to red-purple in a few hours. After the remaining iron is removed, the solution is filtered and diluted to 10 times its original volume with water, and the resultant solution is absorbed on a Dowex 50W-X2 cation exchanger (diameter 2.2 cm, length 80 cm). $[Fe(H_2O)_6]^{2+}$ is eluted with HCl (0.5 M) and a red-purple fraction containing Mo3Fe is obtained by elution with HCl (1 M): yield 78% (based on Mo3). A green band of the unreacted starting material Mo3 ($\sim 10\%$) follows the red-purple band.

II. Decaaquotrimolybdenum-iron-tetrasulfide-tetrakis-*p*-toluenesulfonate-heptahydrate, [Mo$_3$FeS$_4$(H$_2$O)$_{10}$](CH$_3$C$_6$H$_4$SO$_3$)$_4$·7H$_2$O (Mo3Fepts)

Procedure

Method 1. In order to obtain crystals of Mo3Fepts, the solution of Mo3Fe in HCl (1 M) is absorbed on a short cation exchanger, Dowex 50W-X2 (diameter 2.2 cm, length 5 cm). The resin is washed with Hpts (0.1 M) to remove chloride ion, and a red-purple solution is obtained by slow elution with Hpts (4 M). The eluate from the column is stored in a refrigerator (ca. −5°C). After one week black plate-shaped crystals deposit, which are collected by filtration, washed with ethyl acetate, and air-dried: yield 0.92 g (65% based on Mo3Fe in HCl (1 M).

Anal. Calcd. for Mo$_3$FeS$_8$O$_{29}$C$_{28}$H$_{62}$ (MW = 1462.98): Mo, 19.67; Fe, 3.82; C, 22.99; H, 4.27. Found: Mo, 19.56; Fe, 3.90; C, 22.92; H, 3.66.

Method 2. Iron powder (0.40 g, 7.2 mmol) is introduced into a conical flask containing the aqua ion Mo3 (0.025 M per trimer in Hpts (4 M), 25 mL). The mixture is heated at above 90°C for 1 h with stirring in a water bath, and then brought to room temperature and filtered to remove the remaining iron (∼0.2 g) and white precipitates (mainly Fe(pts)$_2$ *n*·H2O). Hpts·H$_2$O (3.0 g, 16 mmol) is added to the filtrate and kept in a refrigerator (ca. −10°C) for 4 days. The black crystals obtained are collected by filtreation, washed with ethyl acetate, and air-dried: yield 0.40 g (44% based on Mo3).

Properties[3a]

Solutions of Mo3Fe in acid are air-sensitive. Exposure to air gives back the Mo3 cluster together with [Fe(H$_2$O)$_6$]$^{2+}$ after several hours [Eq. (1)] Solid Mo3Fepts is more stable toward air oxidation and can be handled in air for brief periods of time (i.e., weighing a sample).

$$[Mo_3FeS_4(H_2O)_{10}]^{4+} + \frac{1}{2}O_2 + 2H^+ + 4H_2O \rightarrow$$
$$[Mo_3S_4(H_2O)_9]^{4+} + [Fe(H_2O)_6]^{2+} \quad (1)$$

The advantage of method 2 is that no column chromatography separation is required. However, using this procedure Mo3Fepts is contaminated by a few percent of the white iron salt of Fe(pts)$_2$·nH$_2$O.

B. MOLYBDENUM-COBALT CLUSTER AQUA ION, $[(H_2O)_9Mo_3S_4CoCoS_4Mo_3(H_2O)_9]^{8+}$ (Mo3Co-D), IN SOLUTION AND $[(H_2O)_9MO_3S_4CoCoS_4Mo_3(H_2O)_9(CH_3C_6H_4SO_3)_8 \cdot 18H_2O$ (Mo3Co-Dpts)

I. Bisnonaaquotrimolybdenum-cobalt-tetrasulfide, $[(H_2O)_9Mo_3S_4CoCoS_4Mo_3(H_2O)_9]^{8+}$ (Mo₃Co-D), in solution

$$2[Mo_3S_4(H_2O)_9]^{4+} + 2Co \rightarrow [(H_2O)_9Mo_3S_4CoCoS_4Mo_3(H_2O)_9]^{8+}$$

Procedure[10]

Cobalt powder (0.20 g, 3.4 mmol) is introduced into a conical flask containing the aqua ion Mo3 (0.05 M per trimer HCl (2 M), 25 mL).[7a] The color of the solution turns from green to brown within a couple of days. After a week the solution is filtered and chromatographed through a Dowex charged column (diameter 2.2 cm, length 80 cm). $[Co(H_2O)_6]^{2+}$ is eluted with HCl (0.5 M), the green aqua ion Mo3 (with HCl (1 M), and the dark brown Mo3Co-D [with HCl (2 M); yield 45%].

Anal. Mo/Co = 3.06.

II. Bisnonaaquotrimolybdenum-cobalt-tetrasulfide-octakis-*p*-toluenesulfonate-octadecahydrate, $[(H_2O)_9Mo_3S_4CoCoS_4Mo_3(H_2O)_9]$ $(CH_3C_6H_4SO_3)_8 \cdot 18H_2O$ (Mo3Co-Dpt)

Procedure

In order to obtain crystals of Mo3Co-Dpts the dark brown solution Mo3Co-D is absorbed on a short cation exchanger, Dowex 50W-X2 (diameter 2.2 cm, length 5 cm), and eluted with Hpts (4 M) The eluate from the column is kept in a refrigerator (ca. −5°C). After one week black-brown crystals deposit, which are collected by filtration, washed with ethyl acetate, and air-dried: yield 0.13 g (15% based on Mo3Co-D in HCl (1 M).

Anal. Calcd. for $Mo_6Co_2S_{16}O_{60}C_{56}H_{128}$ (MW = 2968.16): Mo, 19.39; Co, 3.97; C, 22.67; H, 4.35. Found: Mo, 19.71; Co, 4.20; C, 23.05; H, 3.48.

Properties

The aqua ion Mo3Co-D in HCl (2 M) or Hpts (2 M) is very air-sensitive and forms Mo3 and $[Co(H_2O)_6]^{2+}$ upon exposure to air.

C. MOLYBDENUM-NICKEL CLUSTER AQUA ION, $[Mo_3NiS_4(H_2O)_{10}]^{4+}$ (Mo3Ni) IN SOLUTION AND $[Mo_3NiS_4(H_2O)_{10}](CH_3C_6H_4SO_3)_4 \cdot 7H_2O$ (Mo3Nipts)

I. Decaaquotrimolybdenum-nickel-tetrasulfide, $[Mo_3NiS_4(H_2O)_{10}]^{4+}$ (Mo3Ni), in solution

$$[Mo_3S_4(H_2O)_9]^{4+} + Ni + H_2O \rightarrow [Mo_3NiS_4(H_2O)_{10}]^{4+}$$

Procedure[3a]

Concentrated HCl (10 mL) is added to a solution of Mo3 [0.080 M per trimer in HCl (2M), 10 mL]. Several pieces of nickel plates (1.25 g, 21.3 mmol), the surface of which has been washed with concentrated HCl, are introduced to a conical flask containing the aqua ion Mo3, and the mixture is allowed to stand for 5 days with stirring under a dinitrogen atmosphere at 50–55°C in a water bath. This is the only step that requires a dinitrogen atmosphere (see *Properties*). The color of the solution turns from green to dark brown. The nickel plates are removed by filtration. After the filtrate is diluted to 20 times, its original volume with water, is chromatographed through a Dowex 50W-X2 (2.5 × 45 cm). The first band containing $[Ni(H_2O)_6]^{2+}$ ion is eluted with HCl (0.5 M), and the second band (deep green) containing the mixed metal aqua cluster Mo3Ni is collected by elution with HCl (1 M): yield 75%. The third band contains small abounds of the starting material, Mo3. Uncharacterized brown bands remain at the upper part of the column.

II. Decaaquotrimolybdenum-nickel-tetrasulfide-tetrakis-*p*-toluenesulfonate-heptahydrate, $[Mo_3NiS_4(H_2O)_{10}]$ $(CH_3C_6H_4SO_3)_4 \cdot 7H_2O$ (Mo3Nipts)

Procedure

In order to obtain crystals of Mo3Nipts, a solution of Mo3Ni in HCl (1 M) is absorbed on a short cation exchanger, Dowex 50W-X2 (diameter 2.2 cm, length 5 cm). The resin is washed with Hpts (0.1 M) to remove the chloride ion, and slow elution with Hpts (4 M) gives bluish-green solution. The eluate from the column is stored in a refrigerator. After days, green crystals deposited are collected by filtration and washed with ethyl acetate and air-dried: yield 0.44 g [50% based on Mo3Ni in HCl (1 M)].

Anal. Calcd. for $Mo_3NiS_8O_{29}C_{28}H_{62}$ (MW = 1465.82): C, 22.94; H, 4.26. Found: C, 23.13; H, 4.21.

Properties

The cluster Mo3Ni in dilute HCl or Hpts is fairly stable toward air oxidtion and can be handled in the air; the solution of Hpts (2 M) is air-oxidized by only $\sim 15\%$ per month. However, if the solution is stored for a prolonged period, storage under a dinitrogen atmosphere is recommended. The crystals MoNipts dissolve in some organic solvents (e.g., acetonitrile, acetone, and *N,N*-dimethyl-folmamide).

Derivatives such as $Ca_{2.5}[Mo_3NiS_4(Hnta)(nta)_2Cl]\cdot 14H_2O$ (H_3nta = nitrilotri-acetic acid) can be obtained from Mo3Ni and H_3nta.[11]

D. MOLYBDENUM-COPPER CLUSTER AQUA ION DECAAQUO TRIMOLYBDENUM-COPPER-TETRASULFIDE, $[Mo_3CuS_4(H_2O)_{10}]^{4+}$ (Mo3Cu), AND BISNONAAQUOTRIMOLYBDENUM-COPPER-TETRASULFIDE, $[(H_2O)_9Mo_3S_4CuCuS_4Mo_3(H_2O)_9]^{8+}$ (Mo3Cu-D), IN SOLUTION AND BISNONAAQUOTRIMOLYBDENUM-COPPER-TETRASULFIDE-OCTAKIS-*p*-TOLUENESULFONATE-ICOSAHYDRATE, $[(H_2O)_9Mo_3S_4CuCuS_4Mo_3(H_2O)_9]$ $(CH_3C_6H_4SO_3)_8\cdot 20H_2O$ (Mo3Cu-Dpts)

$$[Mo_3S_4(H_2O)_9]^{4+} + Cu + H_2O \rightarrow [Mo_3CuS_4(H_2O)_{10}]^{4+}$$
$$2[Mo_3CuS_4(H_2O)_{10}]^{4+} \rightarrow [(H_2O)_9Mo_3S_4CuCuS_4Mo_3(H_2O)_9]^{8+} + 2H_2O$$

I. Bisnonaaquotrimolybdenum-copper-tetrasulfide-octakis-*p*-toluenesulfonate-icosahydrate, $[(H_2O)_9Mo_3S_4CuCuS_4Mo_3(H_2O)_9]$ $(CH_3C_6H_4SO_3)_8\cdot 20H_2O$ (Mo3Cu-Dpts)

Procedure[12]

Copper plates (0.2 g, 3.1 mmol) are introduced to a conical flask containing the aqua Mo3 cluster [0.05 M per trimer in HCl (2 M), 25 mL]. The color of the solution turns from green to green-brown in a few hours. After two days the resultant brown colution is filtered and chromatographed through a Dowex (diameter 2.2 cm, length 80 cm).

$[Cu(H_2O)_6]^{2+}$ * and a very light brown unknown species are eluted with HCl (0.5 M). The red-brown eluent, containing $[Mo_3CuS_4(H_2O)_{10}]^{4+}$, (Mo3Cu), is obtained by elution with HCl (1 M). A green band of the unreacted starting material, Mo3, follows the red-brown band. In order to obtain crystals of Mo3Cu-Dpts, the solution is absorbed on a short cation exchanger, Dowex 50W-X2 (diameter 2.2 cm, length 5 cm). The resin is washed with Hpts (0.1 M) to remove chloride ions, and slow elution with Hpts (4 M) gave a red-brown solution. The eluate from the column is a refrigerated (ca. $-5°C$). After several days black-brown crystals deposit, which are collected by filtration, washed with ethyl acetate, and air-dried: yield 0.24 g (13% based on Mo3).

Anal. Calcd. for $Mo_6Cu_2S_{16}O_{62}C_{56}H_{132}$ (MW = 3013.41); C, 22.32; H, 4.42. Found: C, 22.30; H, 3.87.

Properties

Column chromatography, X-ray structural analysis and electronic spectral changes[12] indicate that the molybdenum-copper species is present as either monomeric Mo3Cu (i.e, single cubane type) in HCl solution or dimeric Mo3Cu-D (i.e., double cubane type) in Hpts solution. The presence of monomeric Mo3Cu in $HClO_4$ (2 M) is indicated.[15] The compound Mo3Cu-Dpts dissolves sparingly in Hpts but easily in HCl solution. The aqua ion Mo3Cu, in dilute HCl or Mo3Cu-D in dilute Hpts are very air sensitive and turn to Mo3 and $[Cu(H_2O)_6]^{2+}$ upon exposure to air. The aqua ion Mo3 in dilute HCl or Hpts acts catalytically for the air oxidation of Cu metal to $[Cu(H_2O_6]^{2+}$ ion:

$$[Mo_3S_4(H_2O)_9]^{4+} + Cu + H_2O \rightarrow [Mo_3CuS_4(H_2O)_{10}]^{4+} \quad (2a)$$

$$[Mo_3CuS_4(H_2O)_{10}]^{4+} + \frac{1}{2}O_2 + 2H^+ + 4H_2O \rightarrow [Mo_3S_4(H_2O)_9]^{4+}$$
$$+ [Cu(H_2O)_6]^{2+} \quad (2b)$$

E. MOLYBDENUM-INDIUM CLUSTER AQUA ION DODECAAQUO TRIMOLYBDENUM-INDIUM-TETRASULFIDE, $[Mo_3InS_4(H_2O)_{12}]^{5+}$ (Mo3In), IN SOLUTION AND DI-*p*-TOLUENESULFONATO-DECAAQUO-TRIMOLYBDENUM-INDIUM-TETRASULFIDE-TRIS-*p*-TOLUENESULFONATE-TRIDECAHYDRATE, $[Mo_3InS_4(CH_3C_6H_4SO_3)_2(H_2O)_{10}]$ $(CH_3C_6H_4SO_3)_3·13H_2O$ (Mo3Inpts)

$$[Mo_3S_4((H_2O)_9]^{4+} + In + H^+ + 3H_2O \rightarrow [Mo_3InS_4(H_2O)_{12}]^{5+} + \frac{1}{2}H_2$$

I. Di-*p*-toluenesulfonato-decaaquo-trimolybdenum-tetrasulfide-tris-*p*-toluenesulfonate-tridecahydrate, [Mo$_3$InS$_4$(CH$_3$C$_6$SO$_3$)$_2$(H$_2$O)$_{10}$] (CH$_3$C$_6$H$_4$SO$_3$)$_3$ · 13H$_2$O (Mo3Inpts)

Procedure[13]

Indium plate (0.2 g, 5.4 mmol) is added to the green aqua ion Mo3 (0.027 M per trimer in Hpts (4 M), 20 mL) and is stirred for 2 days at room temperature. The resultant red-brown solution is filtered to remove unreacted indium metal and is stored in a refrigerator (ca. −5°C) for 2 days. Brown needle-like crystals are obtained: yield 0.55 g (57%). Anal. Calcd for Mo$_3$InS$_9$O$_{38}$C$_{35}$H$_{81}$ (MW = 1801.24): C, 23.34; H, 4.53. Found: C, 23.46; H, 4.57.

Properties

The compound Mo3Inpts in diluted HCl or Hpts is slowly oxidized to give Mo3 in a few days, while the solid sample Mo3Inpts is more stable and can be handled in the air: e.g. measurement of weight. Column chromatography indicates that no pts$^-$ is coordinated to indium atom of the Mo$_3$InS$_4$ core in Hpts solution.

F. MOLYBDENUM-TIN CLUSTER AQUA ION BIS(NONAAQUO TRIMOLYBDENUM-TETRASULFIDE)-TIN, [(H$_2$O)$_9$Mo$_3$S$_4$SnS$_4$Mo$_3$(H$_2$O)$_9$]$^{8+}$ (Mo3Sn-SW), AND DODECAAQUOTRIMOLYBDENUM-TIN-TETRASULFIDE, [Mo$_3$SnS$_4$(H$_2$O)$_{12}$]$^{6+}$ (Mo3Sn-OX), IN SOLUTION AND BIS(NONAAQUOTRIMOLYBDENUM-TETRASULFIDE)-TIN-OCTAKIS-*p*-TOLUENESULFONATE-HEXACOSAHYDRATE, [(H$_2$O)$_9$Mo$_3$S$_4$SnS$_4$Mo$_3$(H$_2$O)$_9$](CH$_3$C$_6$H$_4$SO$_3$)$_8$ · 26H$_2$O (Mo3Sn-SWpts)

I. Bis(nonaaquotrimolybdenum-tetrasulfide)-tin-octakis-*p*-toluenesulfonate-icosahexahydrate, [(H$_2$O)$_9$Mo$_3$S$_4$SnS$_4$Mo$_3$(H$_2$O)$_9$] (CH$_3$C$_6$H$_4$SO$_3$)$_8$ · 26H$_2$O (Mo3Sn-SWpts)

$$2[Mo_3S_4(H_2O)_9]^{4+} + Sn \rightarrow [(H_2O)_9Mo_3S_4SnS_4Mo_3(H_2O)_9]^{8+}$$

Procedure[14]

Addition of tin metal (3.0 g, 25 mmol) to Mo3 (0.05 M per trimer in HCl (2 M), 25 mL) rapidly changes the color of the solution from green to red-purple. After

2 days at room temperature, the unreacted metal is removed and the solution is chromatographed through a Dowex 50WX2 column (2.2 × 85m cm), The resin is washed with HCl (0.5 M), and a small amount of yellow-green single cubane $[Mo_3SnS_4(H_2O)_{12}]^{6+}$ (Mo3Sn-OX) is eluted with HCl (1 M). A small amount of unreacted Mo3 may follow this species. A red-purple solution of $[(H_2O)_9mMo_3S_4SnS_4Mo_3(H_2O)_9]^{8+}$ (Mo3Sn-SW), is obtained by use of HCl (2 M); yield 80%. The red-purple solution is absorbed on a short cation exchanger again and eluted with Hpts (4 M). The eluate from the column is refrigerated, and dark brown crystals of $[(H_2O)_9Mo_3S_4SnS_4Mo_3(H_2O)_9](CH_3C_6H_4$-$SO_3)_8 \cdot 26H_2O$ (Mo3Sn-SWpts) are obtained in a few days; yield 0.78 g (40% based on Mo3).

Anal. Calcd. for $Mo_6SnS_{16}O_{68}C_{56}H_{144}$(MW = 3113.12): Mo, 18.49; Sn, 3.81, C, 21.61; H, 4.66. Found: Mo, 18.49; Sn, 4.05; C, 21.64; H, 3.31.

II. Dodecaaquotrimolybdenum-tin-tetrasulfide, $[Mo_3SnS_4(H_2O)_{12}]^{6+}$ (Mo3Sn-OX), in solution

Procedure II[14]

The single-cubane-type aqua ion Mo3Sn-OX can also be prepared easily by the addition of $SnCl_2 \cdot 2H_2O$ or SnO to Mo3 [0.067 M per trimer HCl (2 M), 20 mL]. After filtration, Dowex 50W-X2 column chromatography is used (HCl (1 M) and the yellow-green solution is collected. Yield: 95% from Mo3.

Anal. Mo/Sn = 3.1. To prepare a Mo3Sn-OX solution in Hpts, the solution is absorbed on the cation exchanger again and eluted with Hpts (3 M).

Properties

The aqua ion Mo3Sn-SW is very air-sensitive, giving Mo3 and Mo3Sn-OX in the ratio of 1:1 on exposure to air. In contrast to Mo3Sn-SW, the aqua ion Mo3Sn-OX is fairly air-stable. Only a small change in absorbance per day is observed in the electronic spectrum of Mo3Sn-OX when solutions of the latter are exposed to air. The compound Mo3Sn-SWpts is fairly soluble in methanol, ethanol, and acetone; slightly soluble in acetonitrile and ethyl acetate; and insoluble in carbon tetrachloride. Solid Mo3Sn-SWpts easily loses some of its water of crystallization; thermal analysis shows that 22 water molecules are lost at 68°C under an argon stream. The derivative $K_3[Mo_3SnS_4(Hnta)_3Cl_3] \cdot 3H_2O$ has been isolated,[14] and the X-ray structure of $(Me_2NH_2)_6[Mo_3(SnCl_3)S_4(NCS)_9] \cdot 0.5H_2O$ has been determined.[16]

G. TUNGSTEN-NICKEL CLUSTER AQUA ION DECAAQUO TRITUNGSTEN-NICKEL-TETRASULFIDE, $[W_3NiS_4(H_2O)_{10}]^{4+}$ (W3Ni), IN SOLUTION AND BISNONAAQUOTRITUNGSTEN-NICKEL-TETRASULFIDE-OCTAKIS-*p*-TOLUENESULFONATE-ICOSAHYDRATE, $[(H_2O)_9W_3S_4NiNiS_4W_3(H_2O)_9]$ $(CH_3C_6H_4SO_3)_8 \cdot 20H_2O$ (W3Ni-Dpts)

$$[W_3S_4(H_2O)_9]^{4+} + Ni + H_2O \rightarrow [W_3NiS_4(H_2O)_{10}]^{4+}$$

$$2[W_3NiS_4(H_2O)_{10}]^{4+} \rightarrow [(H_2O)_9W_3S_4NiNiS_4W_3(H_2O)_9]^{8+} + 2H_2O$$

I. Bisnonaaquotritungsten-nickel-tetrasulfide-octakis-*p*-toluenesulfonate-icosahydrate $[(H_2O)_9W_3S_4NiNiS_4W_3(H_2O)_9](CH_3C_6H_4SO_3)_8 \cdot 20H_2O$ (W3Ni-Dpts)

Procedure[6]

Nickel plate (3.38 g, 57.6 mmol) is added to the violet aqua ion W3 (0.038 M per trimer in HCl (7 M), 50 mL), which is stirred for 17 h above 90°C in a water bath. The color of the solution turns from violet to green. The solution is filtered and the remaining nickel plates are removed. After the filtrate is diluted to 35 times its original volume, it is chromatographed through a Dowex 50W-X2 column (2.1 × 10 cm). The first band, containing the $[Ni(H_2O)_6]^{2+}$ ion, is eluted with HCl (0.5 M), and the second band (green), containing the mixed-metal cluster W3Ni, is collected using HCl (1 M) yield, $\sim$ 86% (based on W3). Unreacted W3 is eluted after W3Ni using HCl (1 M). In order to obtain crystals of W3Ni-Dpts, the solution of W3Ni in HCl (1 M) from the second band is absorbed on a short cation exchanger, Dowex 50W-X2 (2.1 × 1.8 cm). The resin was washed with Hpts (0.1 M) to remove chloride ion, and slow elution with Hpts (4 M) gave a bluish-green solution. Cooling of the elutate from the column in a freezer for several days gave green crystals. Yields: 1.85 g (64% based on the solution of W3Ni in HCl 1 M).

Anal. Calcd. for $W_6Ni_2S_{16}O_{62}C_{56}H_{132}$ (MW = 3531.11): C, 19.05; H, 3.77. Found: C, 19.18; H, 3.74.

Properties

The cluster W3Ni in solution is more air-sensitive than the corresponding molybdenum-nickel cluster Mo3Ni. Air oxidation of W3Ni ($\sim$ 1 mM) in Hpts (2 M) is completed in $\sim$ 15 h to give W3 and $[Ni(H_2O)_6]^{2+}$.

References

1. T. Shibahara, *Coord. Chem. Rev.* **123**, 73 (1993); (b) I. Dance and K. Fisher, *Prog. Inorg. Chem.* **41**, 637 (1994); (c) K. R. Tsai and H. L. Wan, *J. Cluster Sci.* **6**, 485 (1995); (d) K. D. Demadis, S. Malinak, and D. Coucouvanis, *Inorg. Chem.* **35**, 4038 (1996).

2. T. Shibahara, H. Akashi, and H. Kuroya, *J. Am. Chem. Soc.* **108**, 1342 (1986).
3. (a) T. Shibahara, G. Sakane, Y. Naruse, K. Taya, H. Akashi, A. Ichimura, and H. Adachi, *Bull. Chem. Soc. Jpn.* **68**, 2769 (1995); (b) G. Sakane and T. Shibahara, *Am. Chem. Soc. Symp. Ser. 653, Transition Metal Sulfur Chemistry*, E. I. Stiefel and K. Matsumoto (Eds.), 1996, Chapter 13, p. 225, and references cited therein.
4. D. M. Saysell and A. G. Sykes, *J. Cluster Sci.* **6**, 449 (1995).
5. (a) S.-F. Lu, H.-B Chen, J.-Q. Huang, Q.-J. Wu, Q.-L. Sun, J. Li, and J.-X. Lu, *Inorg. Chim. Acta* **232**, 43 (1995); (b) A. Deeg, H. Keck, A. Kruse, W. Kuchen, and H. Wunderlick, *Z. Naturforsch.* **43b**, 1541 (1988); (c) M. D. Curtis, *Appl. Organomet. Chem.* **6**, 429 (1992).
6. T. Shibahara, T. Yamamoto, and G. Sakane, *Chem. Lett.*, 1231 (1994).
7. (a) T. Shibahara, M. Yamasaki, G. Sakane, K. Minami, T. Yabuki, and A. Ichimura, *Inorg. Chem.* **31**, 640 (1992); (b) A. Müller, V. P. Fedin, E. Diemann, H. Bogge, E. D. Sölter, Krickemeyer, A. M. Giuliani, R. Barbieri, and P. Adler, *Inorg. Chem.* **33**, 2243 (1994).
8. (a) Y.-F. Zheng, H.-Q. Zhan, X.-T. Wu, and J.-X. Lu, *Trans. Met. Chem.* **14**, 161 (1989); (b) H.-Q. Zhan, Y.-F. Zheng, X.-T Wu, and J.-X Lu, *Inorg. Chim. Acta* **156**, 277 (1989).
9. T. Shibahara, G. Sakane, M. Maeyama, H. Kobashi, T. Yamamoto, and T. Watase, *Inorg. Chim. Acta* **251**, 207 (1996).
10. T. Shibahara, H. Akashi, M. Yamasaki, and K. Hashimoto, *Chem. Lett.* 689 (1991).
11. T. Shibahara, M. Yamasaki, H. Akashi, and T. Katayama, *Inorg. Chem.* **30**, 2693 (1991).
12. T. Shibahara, H. Akashi, and H. Kuroya, *J. Am. Chem. Soc.* **110**, 3313 (1988).
13. G. Sakane and T. Shibahara, *Inorg. Chem.* **32**, 777 (1993).
14. H. Akashi and T. Shibahara, *Inorg. Chem.* **28**, 2906 (1989).
15. M. Nasreldin, Y.-J. Li, F. E. Mabbs, and A. G. Sykes, *Inorg. Chem.* **33**, 4283 (1994).
16. J. E. Varey, G. J. Lamprecht, V. P. Fedin, A. Holder, W. Clegg, M. R. J. Elsegood, and A. G. Sykes, *Inorg. Chem.* **35**, 5525 (1996).

24. MOLYBDENUM AND TUNGSTEN CLUSTERS AS AQUA IONS $[M_3Q_4(H_2O)_9]^{4+}$ (M = Mo,W; Q = S,Se) AND RELATED CHALCOGEN-RICH TRINUCLEAR CLUSTERS

Submitted by VLADIMIR P. FEDIN[†] and A. GEOFFREY SYKES[*]
Checked by DIRK KUPPERT,[‡] KASPAR HEGETSCHWEILER,[‡]
STEPHAN APPEL-COLBUS,[§] and HORST PHILIPP BECK[§]

The trinuclear incomplete cuboidal cluster $[Mo_3S_4(H_2O)_9]^{4+}$, first prepared in the mid-1980's, combines a high degree of stability with quite unusual reactivity.[1–4] One route for the preparation of $[Mo_3S_4(H_2O)_9]^{4+}$ from $Na_2[Mo_2O_2S_2$ $(cys)_2] \cdot 4H_2O$ is now well established,[5,6] with yields typically $\sim 25\%$. In 1992

[†] Institute of Inorganic Chemistry, Russian Academy of Sciences, pr.Lavrentjeva 3, Novosibirsk 630090, Russia.
[*] Department of Chemistry, The University of Newcastle, Newcastle upon Tyne, NE1 7RU, UK.
[‡] Universitat des Saarlandes, Anorganische Chemie, Postfach 15 11 50 D-66041 Saarbrucken.
[§] Universitat des Saarlandes, Anorganische und Analytische Chemie und Radiochemie, Postfach 15 11 50 D-66041 Saarbrucken.

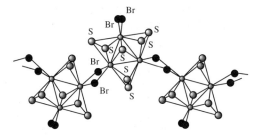

Figure 1

Shibahara prepared $[Mo_3S_4(H_2O)_9]^{4+}$ (yield 50%) and $[W_3S_4(H_2O)_9]^{4+}$ (yield 25%), from ammonium tetrasulfidomolybdate(VI) and tetrasulfidotungstate(VI), respectively, by reduction with sodium tetrahydroborate, $Na[BH_4]$, in 0.5 M HCl.[7] Although less extensively studied, the $[Mo_3Se_4(H_2O)_9]^{4+}$[8] and $[W_3Se_4(H_2O)_9]^{4+}$[9] clusters have also been prepared, but to date the yields are lower.

Solid-state sulfur and selenium containing clusters of molybdenum $Mo_3Q_7X_4$ (Q = S,Se; X = Cl,Br) were first obtained in the late 1960s by heating MoX_n (*n* = 2,3) with sulfur or selenium at 350–450°C[10]. They are polymeric solids $\{Mo_3Q_7X_4\}_x$, which contain chalcogen-rich $M_3(\mu_3-Q)(\mu_2-Q_2)_3$ clusters in chains formed via bridging halogen atoms, alternative formula $M_3Q_7X_2X_{4/2}$ (e.g. Fig. 1).[11,12] The amorphous solid material has a higher reactivity than does the crystalline material obtained at higher temperatures. Alternative routes for the preparation of $\{M_3Q_7Br_4\}_x$ (M = Mo,W; Q = S,Se) from the elements are now known.[9,13,14] From these polymeric compounds new procedures for the preparation of incomplete cuboidal molybdenum and tungsten aqua ions $[M_3Q_4(H_2O)_9]^{4+}$ (M = Mo,W; Q = S,Se) in different acids have been described. The yields are improved, and there is a net saving in time. In only one case with $[W_3Se_4(H_2O)_9]^{4+}$ is a single-step synthesis effective. In the other syntheses described sulfur-rich compounds $(NH_4)_2[Mo_3S_{13}] \cdot 2H_2O$, $(Et_4N)_2[Mo_3S_7Br_6]$, and $(NH_4)_2[W_3S_{16}(NH_3)_3] \cdot H_2O$ are first obtained as intermediates.

Materials

Molybdenum and tungsten powders with average particle size 10–20/µm (99.99%) have been used in these procedures. Such powders are generally free of oxide forms, but if old samples are used, they should be heated in a stream of H_2 (1000°C, 1 h) prior to use. Commercially available 99.5% liquid bromine is recommended to avoid the need to remove H_2O. Commercial 20% ammonium polysulfide $(NH_4)_2S_y$ can be used, or alternatively solutions can be prepared by passing H_2S through a suspension of sulfur (15 g) in 25% aqueous NH_3 (110 mL) until complete dissolution of the sulfur occurs. White crystalline 98.5%

TABLE I. UV–Visible Absorption Spectra, Peak Positions for $[M_3Q_4(H_2O)_9]^{4+}$ (M = Mo,W; Q = S,Se) and Related Clusters in Acidic Solutions

Compounds	Acid	λ, nm (ε, $M^{-1}cm^{-1}$)
$[Mo_3S_4(H_2O)_9]^{4+\ a}$	2M HCl	370 (4995), 616 (326)
	2M Hpts	366 (5550), 603 (362)
$[W_3S_4(H_2O)_9]^{4+\ a}$	2M HCl	317 (6100), 570 (480)
	2M Hpts	315 (6350), 560 (546)
$[Mo_3Se_4(H_2O)_9]^{4+\ a}$	2M HCl	433 (5250), 681 (580)
	2M Hpts	427 (5820), 646 (595)
$[W_3Se_4(H_2O)_9]^{4+\ a}$	2M HCl	360 (6950), 625 (500)
	2M Hpts	359 (6660), 618 (547)
$[Mo_4S_4(H_2O)_{12}]^{5+\ b}$	2M Hpts	635 (435), 1100 (122)
$[Mo_7S_8(H_2O)_{18}]^{8+\ c}$	2M Hpts	416 (7190), 480(sh) (9065), 518 (1.03×10^4), 635 (8860), 950 (4242)

[a] Absorption coefficients per M_3 clusters.
[b] Absorption coefficients per Mo_4 cube.
[c] Absorption coefficients per Mo_7 double cube.

p-toluenesulfonic acid $CH_3C_6H_4SO_3H \cdot H_2O$ (abbreviated Hpts) and Dowex 50W-X2 dry mesh size 200–400, a strongly acidic cation exchange resin, are used as supplied. Use of lower-grade pink-colored Hpts should be avoided. The resin is equilibrated in water overnight and then washed with water, 4–6 M HCl, and water until a neutral eluant is obtained from the column (typically 1.2 × 25 cm). Hypophosphorous acid (H_3PO_2), a strong reductant, is available as a 1 : 1 w/w aqueous solution. All other chemicals were of analytical-grade purity. Some of the polymer products contain small amounts of Mo or W metal detected by XRD (X-ray diffraction) powder diffractograms.

The aqua ions can be isolated as solids $[M_3Q_4(H_2O)_9]Cl_4 \cdot nH_2O$ after evaporation of solvent (preferably on a vacuum line). The yields were determined by UV–vis spectrophotometry (see Table I).

■ **Caution.** *Liquid bromine is an extremely caustic, oxidizing liquid that can inflict great bodily injury if not properly handled. Appropriate clothing and gloves must be worn when handling this element in a well-ventilated fume hood! Sealed tubes can explode even after a reaction has been completed. Exercise of extra caution is highly recommended.*

A. HEPTACHALCOGENOTETRABROMOTRIMETALLO COMPOUNDS, $\{M_3Q_7Br_4\}_x$, (M = Mo,W; Q = S,Se)

$$3M + 7Q + 2Br_2 \rightarrow M_3Q_7Br_4$$

The polymeric compounds $\{M_3Q_7Br_4\}_x$ (M = Mo,W; Q = S,Se) are prepared by the high-temperature reaction of the the metal, chalcogen and bromine (M : Q : Br = 3 : 7 : 4) in a sealed ampoule made from Pyrex (or quartz) glass (diameter 2.0–2.5 cm, volume $\sim$40 mL, walls $\sim$1 mm thick). Amounts used are in ratios according to the polymeric products, including tungsten (5.0 g; 27.2 mmol), sulfur (2.03 g; 63.3 mmol), and Br_2 (0.96 mL; 18.6 mmol). The tube and contents were cooled in liquid N_2, evacuated to a residual pressure of 0.01–1.0 Torr, and glass-sealed. The ampoule was contained in a steel tube for safety, placed in a variable-temperature oven, and gradually heated to 150°C ($\sim$3 h); 200°C ($\sim$3 h). The temperatures required were 400(10)°C for $\{Mo_3S_7Br_4\}_x$, 350(10)°C for $\{Mo_3Se_7Br_4\}_x$ and $\{W_3Se_7Br_4\}_x$, and 310(10)°C for $\{W_3S_7Br_4\}_x$. Heating was for 48 h, after which time the solid was thoroughly shaken and heated for another 72 h. The ampoule was then cooled and opened. Any solid lumps were ground prior to use. The solid was washed with $CHCl_3$ (3 × 30 mL), hot toluene (3 × 30 mL), and then dried under vacuum. Yields were in the range 95–98%, as in the case of $\{W_3S_7Br_4\}_x$ with amounts as indicated, 9.7 g (97%). The procedure can be scaled up or down and 2–20 g amounts of $\{M_3Q_7Br_4\}_x$ prepared. The solids can be stored indefinitely.

B. NONAAQUATETRASULFIDOTRIMOLYBDENUM(IV) $[Mo_3S_4(H_2O)_9]^{4+}$ (FIRST METHOD)[15]

$$Mo_3S_7Br_4 \xrightarrow[\text{conc. HCl}]{H_3PO_2} [Mo_3S_4(H_2O)_9]^{4+} + [Mo_4S_4(H_2O)_{12}]^{5+} + [Mo_7S_8(H_2O)_{18}]^{8+}$$

$$\sim 2\% \qquad\qquad \sim 13\% \qquad\qquad \sim 6\%$$

$$[Mo_4S_4(H_2O)_{12}]^{5+} \xrightarrow{\text{heat}} [Mo_3S_4(H_2O)_9]^{4+} + \{Mo\}$$

$$[Mo_7S_8(H_2O)_{18}]^{8+} \xrightarrow{\text{heat}} 2[Mo_3S_4(H_2O)_9]^{4+} + \{Mo\}$$

To polymeric $\{Mo_3S_7Br_4\}_x$ (1.0 g) and concentrated hydrochloric acid (20 mL, 11.3 M), excess H_3PO_2 (2 mL) is added and the mixture heated on a steam bath (90°C) under N_2 for 15 h. On cooling the solution is diluted 1 : 1 and filtered. Unreacted black polymer-related solid ($\sim$0.3 g) is recovered, and can be reused. The checker obtained more black solid (×2) at this stage. The filtrate is diluted to 0.3 M HCl and loaded onto a Dowex 50W-X2 cation exchange column. Four bands are eluted: $[Mo_3S_4(H_2O)_9]^{4+}$ and $[Mo_4S_4(H_2O)_{12}]^{5+}$ (both green) with 2 M HCl, the corner-shared double cube $[Mo_7S_8(H_2O)_{18}]^{8+}$ (violet), and a related brown form with 4 M HCl. Solutions of the hydrochloric acid eluates $[Mo_4S_4(H_2O)_{12}]^{5+}$, $[Mo_7S_8(H_2O)_{18}]^{8+}$, and related compounds are heated together or separately in air at 90°C (steam bath) for 1–2 days. Decay of the cube and double cube to $[Mo_3S_4(H_2O)_9]^{4+}$ is observed. The solutions are diluted to 0.3 M HCl and reloaded onto a Dowex 50W-X2 column and the single band of

$[Mo_3S_4(H_2O)_9]^{4+}$ eluted with 2 M HCl. Yield: 25–30%. The time required is more favorable than in procedure C below, but the yield is less.

C. [Mo$_3$S$_4$(H$_2$O)$_9$]$^{4+}$ (SECOND METHOD)

$$Mo_3S_7Br_4 + 3(NH_4)_2S_y + 2H_2O \rightarrow$$
$$(NH_4)_2[Mo_3S_{13}] \cdot 2H_2O + 4NH_4Br + (3y - 6)S$$
$$(NH_4)_2(Mo_3S_{13}) \cdot 2H_2O + 6HBr + 2Et_4NBr \rightarrow$$
$$(Et_4N)_2(Mo_3S_7Br_6) + 3H_2S + 3S + 2NH_4Br + 2H_2O$$
$$(Et_4N)_2[Mo_3S_7Br_6] + 3PPh_3 \rightarrow$$
$$[Mo_3S_4(H_2O)_9]^{4+} + 3SPPh_3 + 2Et_4NBr + 4Br^-$$

D. PREPARATION OF (NH$_4$)$_2$[Mo$_3$S$_{13}$]·2H$_2$O FROM {Mo$_3$S$_7$Br$_4$}$_x$[16]

The compound $\{Mo_3S_7Br_4\}_x$ (5.0 g) is added to ammonium polysulfide (100 mL), and the mixture boiled for 6 h. The precipitate is filtered and washed with water (4 × 50 mL), ethanol (2 × 30 mL). Hot toluene (4 × 30 mL) or ambient-temperature carbon disulfide are used to wash the precipitate free of sulfur. The red-brown solid that remains is dried under vacuum. Yield: 4.29 g (92%) of $(NH_4)_2[Mo_3S_{13}] \cdot 2H_2O$.

■   **Caution.**   *H_2S is toxic compound. All manipulations should be carried out in an efficient (well-ventilated) fume hood.*

The procedure of Müller and Krickemeyer[17] also can be used for the preparation of $(NH_4)_2[Mo_3S_{13}] \cdot 2H_2O$ by heating ammonium paramolybdate, $(NH_4)_6[Mo_7O_{24}] \cdot 4H_2O$, in an aqueous solution of ammonium polysulfide in air.

E. BISTETRAETHYLAMMONIUM-HEPTASULFIDOHEXABROMOTRIMOLYBDATE, (Et$_4$N)$_2$[Mo$_3$S$_7$Br$_6$][18]

The compound $(NH_4)_2[Mo_3S_{13}] \cdot 2H_2O$ (5.0 g, 6.43 mmol) is boiled in concentrated HBr (150 mL) for 30–40 min and the solution filtered off while hot. Addition of Et_4NBr (5.0 g, 23.8 mmol) to the solution gives a precipitate of orange-red $(Et_4N)_2[Mo_3S_7Br_6]$. The reaction mixture is kept at +5°C for 12 h, when the orange-red crystals are removed by filtration and dried in vacuum. Yield: ~6.5 g (80–85%) of $(Et_4N)_2[Mo_3S_7Br_6]$.

■   **Caution.**   *H_2S is toxic compound. All the manipulation should be carried in an efficient fume hood. Excess H_2S can be taken up in concentrated sodium hypochlorite solution. Concentrated HBr is highly corrosive. Breathing* *its vapors is extremely dangerous, and use of appropriate clothing, safety glasses, and a highly efficient fume hood is recommended.*

F. NONAAQUATRIMOLYBDENUM-TETRASULFIDE, [Mo₃S₄(H₂O)₉]⁴⁺[15]

The compound $(Et_4N)_2[Mo_3S_7Br_6]$ (2.70 g, 2.15 mmol) is dissolved in hot concentrated HCl (50 mL). The solution is stirred vigorously using a magnetic stirrer and solid PPh_3 (1.75 g; 6.65 mmol) added to the hot solution. (It is important to ensure that no solid is left before adding PPh_3.) The color begins to change immediately to the characteristic green of the product. A copious precipitate of $SPPh_3$ forms. Heating is continued for a further 10–15 min, after which time the reaction mixture is diluted (with stirring) with 4 times the volume of water, and the precipitate is filtered off. The filtrate is evaporated to dryness on a rotary evaporator, and the solid is taken up in 0.5 M HCl. The product is loaded onto a Dowex 50W-X2 cation exchange column (15 × 1.5 cm diameter) and washed with 0.5 M HCl or Hpts (100 mL), and a single green band is eluted with 2 M HCl or Hpts as required. Yield: 80–85%.

■ **Caution.** *Hot concentrated HCl is highly corrosive. Breathing its vapors is extremely dangerous; therefore use of appropriate clothing, safety glasses, and a highly efficient fume hood are highly recommended.*

G. [W₃S₄(H₂O)₉]⁴⁺ (FIRST PROCEDURE)[19]

$$W_3S_7Br_4 \xrightarrow[\text{conc.HCl}]{H_3PO_2} [W_3S_4(H_2O)_9]^{4+}$$

The polymer $\{W_3S_7Br_4\}_x$ (1.8 g) is heated with HCl (25 mL, 11.3 M) and H_3PO_2 (30 mL) at 90°C under N_2 for 15 h. The solution is diluted twofold and filtered, and after further dilution to <0.5 M HCl is loaded onto a Dowex 50W-X2 cation exchange column. After washing with 0.5M HCl or Hpts (150 mL) a single purple-violet band is eluted with 2M HCl or Hpts. Yield: 20%. The black polymer-related solid residue ($\sim$1.2 g) obtained on filtering can be recycled.

H. [W₃S₄(H₂O)₉]⁴⁺ (SECOND PROCEDURE)

$$W_3S_7Br_4 + 3(NH_4)_2S_y + H_2O + 3NH_3 \rightarrow$$
$$(NH_4)_2[W_3S_{16}(NH_3)_3]\cdot H_2O + 4NH_4Br + (3y-9)S$$
$$(NH_4)_2[W_3S_{16}(NH_3)_3]\cdot H_2O + 9HCl + 8H_2O \rightarrow$$
$$[W_3S_4(H_2O)_9]^{4+} + 4Cl^- + 3H_2S + 9S + 5NH_4Cl$$

I. $(NH_4)_2[W_3S_{16}(NH_3)_3]\cdot H_2O$ FROM $\{W_3S_7Br_4\}_x$ [13,20]

A solution of $(NH_4)_2S_y$ (100 mL) is added to $\{W_3S_7Br_4\}_x$ (2.20 g) and the mixture kept for a week at room temperature without stirring. A black crystalline solid forms and is filtered off and washed with ethanol (2×30 mL) and hot toluene (3×30 mL). Yield: 2.11 g (90%) of $(NH_4)_2[W_3S_{16}(NH_3)_3]\cdot H_2O$.

■ **Caution.** H_2S *is toxic compound. All manipulations should be carried in an efficient fume hood.*

J. $[W_3S_4(H_2O)_9]^{4+}$ [13,20]

The complex $(NH_4)_2[W_3S_{16}(NH_3)_3]\cdot H_2O$ (2.00 g) is refluxed in concentrated HCl (40 mL, 11.3 M) for 20–30 min. The solution is filtered while hot and the purple-violet solution in 0.5 M HCl loaded onto a Dowex 50W-X2 cation exchange column (15×1.5 cm diameter), washed with 0.5 M HCl (100 mL), and a single purple-violet band eluted with 2 M HCl or Hpts as required. Yield: 80–85%.

■ **Caution.** H_2S *is toxic compound. All manipulations should be carried in an efficient fume hood. Hot concentrated HCl is highly corrosive. Breathing its vapors is extremely dangerous; therefore use of appropriate clothing, safety glasses, and a highly efficient fume hood are highly recommended.*

K. $[Mo_3Se_4(H_2O)_9]^{4+}$

$$Mo_3Se_7Br_4 + 2PPh_4Br \rightarrow (PPh_4)_2[Mo_3Se_7Br_6]$$
$$(PPh_4)_2[Mo_3Se_7Br_6] + 4Hpts + 6H_2O \rightarrow$$
$$[Mo_3Se_7(H_2O)_6]^{4+} + 4pts^- + 2PPh_4Br + 4HBr$$
$$[Mo_3Se_7(H_2O)_6]^{4+} + 3PPh_3 + 3H_2O \rightarrow [Mo_3Se_4(H_2O)_9]^{4+} + 3SePPh_3$$

A mixture of $\{Mo_3Se_7Br_4\}_x$ (1.00 g, 0.86 mmol) and PPh_4Br (2.0 g, 4.76 mmol) is heated at 280–290°C for 4–5 h under N_2. A sand tray can be used in this heating process. The dark red melt is ground, washed free of PPh_4Br using ethanol (3×30 mL), and the product $(PPh_4)_2[Mo_3Se_7Br_6]$ stirred with 4 M Hpts (75 mL) for 24 h to substitute Br^- by H_2O. On filtration an orange solution is obtained which is diluted twofold with H_2O. A solution of PPh_3 (1.0 g, 3.81 mmol) in CH_2Cl_2 (30 mL) is added to the solution of $[Mo_3Se_7(H_2O)_6]^{4+}$ in 2 M Hpts, and the mixture stirred vigorously for 2–3 h. The organic layer is removed, the yellow-brown aqueous solution diluted fivefold with water, and the product loaded onto a Dowex 50W-X2 column (15 × 1.5 cm diameter). After washing with water (200 mL) and 0.5 M HCl (200 mL), a single yellow-brown band eluted with 2 M HCl or Hpts as required. Yield: 60–65%.

L. PREPARATION OF $[W_3Se_4(H_2O)_9]^{4+}$ FROM $\{W_3Se_7Br_4\}_x$ [19]

$$W_3Se_7Br_4 + 3H_3PO_2 + 12H_2O \rightarrow [W_3Se_4(H_2O)_9]^{4+} + 3H_2Se + 3H_3PO_3 + 4Br^-$$

■ **Caution.** *H_2Se has a very unpleasant smell and is very toxic in small amounts. All manipulations should be carried in an efficient fumehood. The H_2Se can be destroyed by bubbling the gas through a solution of HNO_3 when red Se forms.*

Polymeric $\{W_3Se_7Br_4\}_x$ (0.80 g) is heated (90°C) with H_3PO_2 (2 mL) and HCl (20 mL, 11.3 M) for 6–8 h in N_2. The green color of $[W_3Se_4(H_2O)_9]^{4+}$ is observed and H_2Se is released. The dark green solution is filtered, and the remaining solid reheated (90°C) with H_3PO_2 (2 mL) in HCl (20 mL, 11.3 M) for 6–8 h in N_2. The dark green solutions of $[W_3Se_4(H_2O)_9]^{4+}$ are combined, diluted with H_2O to conc. $H^+ = 4$ M, filtered, and diluted again to conc. $H^+ = 0.5$ M. The product is loaded onto a Dowex 50W-X2 cation exchange column (8 × 1.2 cm diameter), and washed successively with 0.5 M HCl (100 mL) and 1.0 M HCl or Hpts (100 mL). No other colored bands are observed. A green band of $[W_3Se_4(H_2O)_9]^{4+}$ is eluted with 2.0 M HCl or Hpts as required. Yield: 40–50%.

Properties

All aqua ions require acidic conditions to avoid H_2O ligand acid dissociation–oligomerization processes. The aqua ion $[Mo_3S_4(H_2O)_9]^{4+}$ is very stable (years!) toward air oxidation and does not require a nitrogen atmosphere for storage. Solutions of $[W_3S_4(H_2O)_9]^{4+}$, $[Mo_3Se_4(H_2O)_9]^{4+}$, and $[W_3Se_4(H_2O)_9]^{4+}$ are less stable (red Se is formed), but can be stored in air at +5°C for 2–3 months. Solutions of $[Mo_3Se_4(H_2O)_9]^{4+}$ and $[W_3Se_4(H_2O)_9]^{4+}$ are sensitive to light, and direct sunlight is best avoided. Spectrophotometric information is given in Table I. Procedures for the conversion to heterometallic (M′) containing cubes $[M_3M'Q_4(H_2O)_{10/12}]^{n+}$ and/or related double-cube forms have been described.[4,22–24] In particular, metallic state Fe, Co, Ni, Pd, Cu, Hg, Ga, In, Tl, Ge, Sn, Pb, Sb, and Bi are known to react directly (at different rates) with solutions of $[Mo_3S_4(H_2O)_9]^{4+}$, and in some cases (studies have been less extensive) with $[W_3S_4(H_2O)_9]^{4+}$, $[Mo_3Se_4(H_2O)_9]^{4+}$, and $[W_3Se_4(H_2O)_9]^{4+}$. The aqua ions $[Mo_3Q_7(H_2O)_6]^{4+}$ have been prepared from the compounds $(Et_4N)_2 [Mo_3Q_7Br_6]$.[21]

References

1. T. Shibahara, *Adv. Inorg. Chem.* **37**, 143 (1991).
2. T. Shibahara, *Coord. Chem. Rev.* **123**, 73 (1993).
3. T. Saito, in *Early Transition Metal Clusters with p-Donor Ligands*, M. H. Chisholm (Ed.), VCH, New York, 1995, p. 63.
4. D. M. Saysell, M. N. Sokolov, and A. G. Sykes, *ACS Symp. 653, Transition Metal Sulfur Chemistry*, E. I. Stiefel and K. Matsumoto (Eds.), ACS, 1996, pp. 216–224.
5. M. Martinez, B.-L. Ooi, and A. G. Sykes, *J. Am. Chem. Soc.* **109**, 4615 (1987).
6. T. Shibahara and H. Akashi, *Inorg. Synth.* **29**, 260–269 (1992).
7. T. Shibahara, M. Yamasaki, G. Sakane, K. Minami, T. Yabuki, and A. Ichimura, *Inorg. Chem.* **31**, 640 (1992) (see also Synthesis 23 in this volume).
8. M. Nasreldin, G. Henkel, G. Kampmann, B. Krebs, G. J. Lamprecht, C. A. Routledge, and A. G. Sykes, *J. Chem. Soc., Dalton Trans.*, 737 (1993).
9. V. P. Fedin, M. N. Sokolov, A. V. Virovets, N. V. Podberezskaya, and V. Ye. Fedorov, *Polyhedron* **11**, 2973 (1992).
10. A. A. Opalovskii, V. Ye. Fedorov, and K. A. Khaldoyanidi, *Dokl. Akad. Nauk SSSR* **182**, 1095 (1968).
11. J. Marcoll, A. Rabenau, D. Mootz, and H. Wunderlich, *Rev. Chim. Min.* **11**, 607 (1974).
12. F. A. Cotton, P. A. Kibala, M. Matsz, C. S. McCaleb, and R. B. W. Sandor, *Inorg. Chem.* **28**, 2693 (1989).
13. V. P. Fedin, M. N. Sokolov, O. A. Gerasko, B. A. Kolesov, V. Ye. Fedorov, A. V. Mironov, D. S. Yufit, Yu. L. Slovohotov, and Yu. T. Struchkov, *Inorg. Chim. Acta* **175**, 217 (1990).
14. V. P. Fedin, M. N. Sokolov, K. G. Myakishev, O. A. Gerasko, V. Ye. Fedorov, and J. Macicek, *Polyhedron* **10**, 1311 (1991).
15. M. N. Sokolov, N. Coichev, H. D. Moya, R. Hernandez-Molina, C. D. Borman, and A. G. Sykes, *J. Chem. Soc., Dalton Trans.*, 1863 (1997).
16. V. P. Fedin, B. A. Kolesov, Yu. V. Mironov, and V. Ye. Fedorov, *Polyhedron* **8**, 2419 (1989).
17. A. Müller and E. Krickemeyer, *Inorg. Synth.* **27**, 47 (1990).
18. V. P. Fedin, Yu. V. Mironov, M. N. Sokolov, B. A. Kolesov, S. V. Tkachev, and V. Ye. Fedorov, *Inorg. Chim. Acta* **167**, 39 (1990).
19. V. P. Fedin, M. N. Sokolov, A. V. Virovets, N. V. Podberezskaya and V. E. Fedorov, *Inorg. Chim. Acta* **269**, 292 (1998).
20. V. P. Fedin, M. N. Sokolov, O. S. Kibirev, A. V. Virovets, N. V. Podberezskaya, and V. Ye. Fedorov, *Russ. J. Inorg. Chem.* **38**, 1735 (1991).
21. D. M. Saysell, V. P. Fedin, G. J. Lamprecht, M. N. Sokolov, and A. G. Sykes, *Inorg. Chem.* **36**, 2982 (1997).
22. R. Hernandez-Molina, A. J. Edwards, W. Clegg, and A. G. Sykes, *Inorg. Chem.* **37**, 2989 (1998).
23. R. Hernandez-Molina, D. N. Dybtsev, V. P. Fedin, M. R. J. Elsegood, W. Clegg, and A. G. Sykes, *Inorg. Chem.* **37**, 2995 (1998).
24. V. P. Fedin, M. N. Sokolov, and A. G. Sykes, *J. Chem. Soc., Dalton Trans.*, 4089 (1996).

Chapter Four

COMPOUNDS OF GENERAL INTEREST

25. THE DIPHENYLMETHYLENETHIOPHOSPHINATE (MTP) LIGAND IN GOLD(I), PLATINUM(II), LEAD(II), THALLIUM(I), AND MERCURY(II) COMPLEXES, sym-Au$_2$(MTP)$_2$, (PPN)[Au(MTP)$_2$], Au$_2$Pt(MTP)$_4$, Au$_2$Pb(MTP)$_4$, AuTl(MTP)$_4$, Hg(MTP)$_2$, Hg(MTP)$_2$(AuCl)$_2$, AND Hg^{II}Au^{I}(MTP)$_2$Au^{III}Cl$_4$

Submitted by JOHN P. FACKLER, Jr.,[*,†] ESPERANZA GALARZA,[‡] GUILLERMO GARZÓN,[‡] ANTHONY M. MAZANY,[†] H. H. MURRAY,[§] MANAL A. RAWASHDEH OMARY,[†] RAPHAEL RAPTIS,[#] RICHARD J. STAPLES,[**] WERNER E. VAN ZYL[†], and SUNING WANG[††]
Checked by ELENA CERRADA and MARIANO LAGUNA[##]

The monoanionic compound diphenylmethylenethiophosphinate (MTP), CH$_2$P(S)Ph$_2^-$, was first described by Seyferth and co-workers [together with MOP, CH$_2$P(O)Ph$_2^-$] as lithium and tin salts to be used in organic synthesis[1]. The compound has been used as a ligand for the synthesis of a variety of mono-, di-, and trinuclear late-transition metal complexes. The ligand forms neutral (thio)ylide dinuclear complexes of gold[2] and silver[3] and cationic[4] dinuclear Hg^{II}-Au^{I} complexes. The ligand can bind in a monodentate fashion to form

[*] Author to whom correspondence should be addressed.
[†] Department of Chemistry, Texas A&M University, College Station, TX 77843-3255.
[‡] Department of Chemistry, Universidad del Valle, Cali, Colombia.
[§] Exxon Research and Engineering Company, Annandale, NJ 08801.
[#] Department of Chemistry, University of Puerto Rico.
[**] Department of Chemistry, Harvard University, Cambridge, MA 02138.
[††] Department of Chemistry, Queens University, Kingston, Ontario K7L 3N6 Canada.
[##] Departamento de Quimica Inorganica. E-50009 Zaragosa Spain.

the mononuclear complexes,[5] $[Au(MTP)_2]^-$ and $Hg(MTP)_2$,[4] and these complexes can also bind in a bidentate fashion to form organo(thio)heterobimetallic[4,6] and heterotrimetallic[5,7] complexes. Syntheses of the complexes *sym*-$Au_2(MTP)_2$, $[PPN][Au(MTP)_2]$, $Au_2Pt(MTP)_4$, $Au_2Pb(MTP)_4$, $AuTl$ $(MTP)_4$, $Hg(MTP)_2$, $Hg(MTP)_2(AuCl)_2$, and $Hg^{II}Au^I(MTP)_2Au^{III}Cl_4$ are described here. In addition to their interesting physical properties, these complexes have been used to investigate a variety of reactions such as insertion reactions,[8] oxidative addition reactions,[2,9] and Lewis acid activation,[10] as demonstrated in the related $[Au(ylide)]_2$ complexes.[2c]

Abbreviations Used

$MTP = CH_2P(S)Ph_2^-$;   [PPN] $Cl = [(Ph_3P)_2N]Cl = $ Bis(triphenylphosphoranylidene)ammonium chloride; THT = tetrahydrothiophene, SC_4H_8; THF = tetrahydrofuran, OC_4H_8.

General Procedures

Unless otherwise stated, all reactions and manipulations are carried out under a dinitrogen atmosphere using standard Schlenk techniques.[11] Solvents (THF and Et_2O) are distilled under dinitrogen over a Na/K benzophenone ketyl. Chlorinated solvents are distilled under dinitrogen from P_2O_5, and alcohols are distilled from Mg turnings. Methyl lithium is standardized[12] prior to use. $LiCH_3$, $P(S)Ph_3$, [PPN]Cl, $HgCl_2$, and THT are available from Aldrich Chemical Company and may be used without further purification. $Pb(NO_3)_2$ is available from Fisher Scientific Company, and Tl_2SO_4 is available from Alfa Chemicals. The compounds $Au(THT)Cl$, $PtCl_2(SEt_2)_2$, and $PhI\cdot Cl_2$ were prepared according to methods described in the literature.[13–15]

A. (DIPHENYLMETHYLENETHIOPHOSPHINATO) DI-GOLD(I), [AuCH₂P(S)Ph₂]₂

$$LiCH_3 + P(S)Ph_3 \rightarrow Li[CH_2P(S)Ph_2] + C_6H_6$$
$$2Li[CH_2P(S)Ph_2] + 2Au(THT)Cl^* \rightarrow [AuCH_2P(S)Ph_2]_2 + 2THT + 2LiCl$$

---

* Au(Ph₃As)Cl may be substituted for Au(THT)Cl.

Procedure

■ **Caution.**   *The synthesis must be performed in a well-ventilated fume hood. Avoid skin contact with LiCH₃, which is pyrophoric and corrosive. Tetrahydrothiophene has a foul smell, and can cause conjunctivitis. Benzene is a suspected carcinogen. All lithium organometallics are reducing and sensitive to air and moisture. Hence procedures are best performed under dry N_2.[9] The checkers used Au(Ph₃As)Cl in A instead of the more reactive Au(THT)Cl, to avoid some reduction of the gold.*

A 50-mL Schlenk tube provided with a Teflon-coated magnetic stirring bar, is charged with 1.17 g (4.0 mmol) of P(S)Ph₃ in 30 mL of THF under an inert atmosphere. It is important to measure the volume of THF since an approximate solution concentration is required for a later step in the synthesis. After dissolution of the solid, the solution is cooled to $-78°C$ (dry-ice/acetone bath), and 2.9 mL of 1.4 M methyllithium in ether (4.06 mmol) is added by syringe. The low-temperature bath is removed after 15 min and the solution is allowed to equilibrate to room temperature. The color of the solution gradually changes from light yellow to deep red after 1.5–2 h while stirring at room temperature under inert gas. The Li(MTP) complex should be used as quickly as possible.

In another 50-mL Schlenk tube, 0.48 g Au(THT)Cl[13] (1.5 mmol), is dissolved in 20 mL of THF, and cooled to approximately $-78°C$. To this solution 11.3 mL of the Li(MTP) solution is slowly added (1.5 mmol). Since Li(MTP) is a reducing agent, care must be taken not to greatly exceed the stoichiometric equivalent of the gold starting material present (a gold metal precipitate forms on reduction). The solution is stirred at about $-78°C$ for 15 min, and then allowed to equilibrate to room temperature and is stirred for an additional hour. The volume of the solution is reduced in vacuo to approximately 20 mL, and dry ether is added to induce precipitation of the yellow solid. The mixture is refrigerated for 12 h and filtered, and the solid is dried under vacuum for 2 h. This compound is not air-sensitive. Recrystallization from hot toluene yields light yellow crystals. Yield after recrystallization: 92%.

Anal. Calcd. for $C_{26}H_{24}P_2S_2Au_2$: C, 36.46; H, 2.82. Found: C, 36.67; H, 2.91.

Properties

The yellow solid *sym*-Au₂(MTP)₂ is air-stable with a melting point of 250°C. ¹H NMR (CDCl₃/TMS) $\delta CH_2 = 2.00$ ppm, d, CH₂, $J_{P-H} = 12$ Hz; $\delta = 7.3–7.9$ (*m*-C₆H₅), ³¹P{¹H} NMR δ 50.6 ppm (checkers found δ 51.6 ppm) relative to a 85% H₃PO₄ in D₂O standard. The structure has been determined by single-crystal X-ray diffraction.[2,9a]

B. BIS(TRIPHENYLPHOSPHORANYLIDENE)AMMONIUM (DIPHENYLMETHYLENETHIOPHOSPHINATO)GOLD(I), (PPN){Au[CH₂P(S)Ph₂]₂}

$$LiCH_3 + P(S)Ph_3 \rightarrow Li[CH_2P(S)Ph_2] + C_6H_6$$
$$2Li[CH_2P(S)Ph_2] + Au(THT)Cl \rightarrow Li\{Au[CH_2P(S)Ph_2]_2\} + LiCl + THT$$
$$Li\{Au[CH_2P(S)Ph_2]_2\} + [PPN]Cl \rightarrow (PPN)\{Au[CH_2P(S)Ph_2]_2\} + LiCl$$

Procedure

■   **Caution.**   *LiMe reacts explosively with water, and ignites spontaneously in moist air. [PPN]Cl is a harmful dust and inhalation must be avoided. Benzene is a suspected carcinogen.*

All manipulations are done under nitrogen.

A 25-mL Schlenk tube provided with a Teflon-coated magnetic stir bar is charged with 1.17 g of P(S)Ph₃ (4.0 mmol) in 15 mL THF. The solid is allowed to dissolve, and the solution is cooled to −78°C (dry-ice/acetone bath). A precipitate forms during cooling and redissolves upon warming. To this mixture 2.9 mL of 1.4 M LiMe in Et₂O (4.06 mmol) is slowly added via a syringe. After 15 min the cold bath is removed, and the light yellow mixture is allowed to equilibrate to room temperature. A clear pale yellow solution results. The color of the solution gradually intensifies to a deep red after 1.5–2 h of stirring at room temperature. The red-colored solution is cooled to −10°C, and 641 mg Au(THT)Cl (2.0 mmol) is added. The solution is stirred for one hour at this temperature, and then allowed to warm to ambient temperature over one hour, during which time the solution turns green. In a separate Schlenk tube, 1.146 g [PPN]Cl (2.0 mmol) is dissolved in 5 mL of THF and 5 mL of methanol, and slowly added to the solution containing the gold. Within 10 min a white precipitate is observed. The volume of the solvent is decreased by half under vacuum, and further product is obtained. The mixture is cooled to −10°C and the solid material separated by filtration. The product is washed with three 15-mL portions of dry ether and dried under vacuum for 2 h. Yield: 1.79 g (74%), the checkers obtained an 84–87% yield.

Anal. Calcd. for C₆₂H₅₄NP₄S₂Au: C, 62.16; H, 4.54; N, 1.17; S, 5.35. Found: C, 61.81; H, 4.64; N, 1.16; S, 5.04.

Properties

The product is a white solid that is stable at ambient temperature under nitrogen. Melting point 172–176°C (decomposition). ^{1}H NMR in $CDCl_3$ (relative to TMS) at 22°C shows $\delta(CH_2) = 1.52$ ppm, $J_{P-H} = 15.1$ Hz, $^{31}P\{^1H\}$ NMR (relative to a 85% H_3PO_4 standard), and δ 55.09 and 21.5 ppm (the checkers report δ 56.1 and 21.7 ppm). The structure has been determined by single-crystal X-ray crystallography.[5]

C. TETRAKIS(DIPHENYLMETHYLENETHIOPHOSPHINATO DIGOLD(I)PLATINUM(II), Au₂Pt(CH₂P(S)Ph₂)₄

$$2(PPN)\{Au[CH_2P(S)Ph_2]_2\} + cis\text{-}PtCl_2(SEt_2)_2 \rightarrow$$
$$Au_2Pt[CH_2P(S)Ph_2]_4 + 2PPNCl + 2SEt_2$$

Procedure

To a 50-mL Schlenk flask containing 1.60 g (1.34 mmol) of (PPN)[Au (CH₂P(S)Ph₂)₂] in 15–20 mL of THF, *cis*-PtCl₂(SEt₂)₂[14] (300 mg, 0.67 mmol) is added in one portion, resulting in a quick succession of color changes to an orange-red colored solution. After 2 h the solution becomes red-brown. The solvent volume is decreased to approximately half under reduced pressure. Dry ether (20 mL) then is added to induce precipitation of the PPNCl and the product. The solid material is removed by filtration, washed with five 8-mL portions of MeOH to remove the [PPN]Cl, and dried under reduced pressure for 2 h. The green-gray product should be recrystallized from a CH_2Cl_2–ether solvent mixture. Yield: 0.918 g (90%).

Anal. Calcd. for $C_{52}H_{48}P_4S_4Au_2Pt$: C, 41.25; H, 3.18. Found: C, 41.10 H, 3.20.

Properties

Red crystals of the product are obtained from recrystallization. M.p. 214–215°C (decomposition). It is air/moisture-stable at ambient temperature for several months. It is slightly soluble in chlorinated solvents and insoluble in diethyl ether and nonpolar solvents. Purification to obtain X-ray-quality crystals is achieved by recrystallization from CH_2Cl_2 by layering with hexane. The ^{1}H NMR in CD_2Cl_2 relative to CH_2Cl_2 at 22°C shows $\delta(CH_2) = 1.46$ ppm, $J_{P-H} = 11.2$ Hz, $\delta(Ph) = 7.76$–7.72 ppm, m; ^{31}P NMR (85% solution of H_3PO_4) δ 48.2 ppm. In

CD_2Cl_2 or $CDCl_3$ the solution changes color from an initial brown-red to orange-red, and a ^{31}P singlet simultaneously grows in at δ 62.76 ppm. This corresponds to the formation, in trace quantities, of the oxidative addition product $Au_2Pt[CH_2P(S)Ph_2]_4Cl_2$. The structure of $Au_2Pt[CH_2P(S)Ph_2]_4$ has been determined by single-crystal X-ray crystallography.[5]

D. TETRAKIS(DIPHENYLMETHYLENETHIOPHOSPHINATO) DI-GOLD(I)LEAD(II), $Au_2Pb[CH_2P(S)Ph_2]_4$

$$2(PPN)[Au(CH_2P(S)Ph_2)_2] + Pb(NO_3)_2 \rightarrow Au_2Pb[CH_2P(S)Ph_2]_4 + 2(PPN)NO_3$$

Procedure

■ **Caution.** *Lead nitrate is a harmful dust and may be fatal if swallowed.* A 25-mL Schlenk tube provided with a Teflon-coated magnetic stir bar is charged with 742 mg of $(PPN)[Au(MTP)_2]$ (0.70 mmol) in 10 mL of CH_2Cl_2. In a separate Schlenk tube 119 mg of $Pb(NO_3)_2$ (0.36 mmol) is dissolved in 10 mL of MeOH. This solution is slowly added to the gold-containing solution, which immediately becomes yellow. After approximately 15 min, bronze-colored microcrystals precipitate from the solution. Reducing the volume of the solvent to half induces further precipitation. After one hour the solid material is filtered off and washed 3 times with 10-mL portions of MeOH to extract the $[PPN]NO_3$ and the bronze-colored crystals are dried in vacuo for 2 h. Yield: 401.2 mg (73%).

Anal. Calcd. for $C_{52}H_{48}P_4S_4Au_2Pb$: C, 40.93; H, 3.14. Found: C, 40.62; H, 3.19.

Properties

The product is relatively air/moisture-stable and can be recrystallized from either a THF/ether or a CH_2Cl_2/ether solvent mixture at 0°C. The structure of the product crystallized from THF has been determined by single-crystal X-ray crystallography.[7] Bronze fibers form from CH_2Cl_2/CH_3OH, which strongly luminesces a bright red color (752 nm) at 298 K under UV excitation. The ^1H NMR spectrum shows no coupling of $-CH_2$ groups with Pb (^{207}Pb, $I = 2$ natural abundance 23%). In $CDCl_3$ (relative to TMS) the NMR spectrum is $\delta(CH_2) = 1.71$ ppm, d, $J_{P-H} = 12$ Hz; $\delta(C_6H_5) = 7.16$–7.38 ppm, m, $\delta C_6H_5 = 7.65$–7.82 ppm, m. ^{31}P NMR in $CDCl_3$, $\delta = 50.55$ ppm, $^2J_{Pb-P} = 16$ Hz.

E. BIS(DIPHENYLMETHYLENETHIOPHOSPHINATO) GOLD(I)THALLIUM(I), 2AuTl[CH$_2$P(S)Ph$_2$]$_2$

$$2PPN\{Au[CH_2P(S)Ph_2]_2\} + Tl_2SO_4 \rightarrow 2AuTl[CH_2P(S)Ph_2]_2 + [PPN]_2SO_4$$

Procedure

■ **Caution.** *Thallium salts are extremely toxic, and any contact should be avoided.*

All manipulations are carried out under N$_2$(g) using standard Schlenk procedures. Alternatively, nitrogen glove box or bag may be used.

In a dry 25-mL Schlenk nitrogen-filled flask containing a magnetic stirring bar are placed 119.8 mg (0.100 mmol) of (PPN)[Au(MTP)$_2$] and 5 mL of CH$_2$Cl$_2$. In a nitrogen-filled Schlenk tube, Tl$_2$SO$_4$ (25.24 mg, 0.05 mmol) is suspended in 3 mL of MeOH. This solution is added to the gold-containing solution. After approximately 10 min a yellow-orange precipitate is observed. The mixture is stirred for one hour, and the solid is collected by filtration under nitrogen and washed with small portions of MeOH to extract the (PPN)$_2$SO$_4$. The solid is vacuum dried for 2 h. Yield: 67.0 mg (53%).

Anal. Calcd. for C$_{26}$H$_{24}$P$_2$S$_2$AuTl: C, 36.14; H, 2.78. Found: C, 36.65; H, 2.53.

Properties

The air-stable product is a yellow solid and can be recrystallized from a warm CH$_2$Cl$_2$/ether solvent mixture. The ^{1}H NMR spectrum shows no coupling of δCH$_2$ groups with Tl (^{205}Tl, $I = 2$, natural abundance 70%). In CDCl$_3$ (relative to TMS) δ(CH$_2$) = 1.62 ppm, d, J_{P-H} = 12 Hz; δC$_6$H$_5$ = 7.20–7.40 ppm, m, δC$_6$H$_5$ = 7.65–7.80 ppm, m. ^{31}P NMR in CDCl$_3$ δ = 50.79 ppm. The structure has been determined by single-crystal X-ray crystallography.[7]

F. BIS(DIPHENYLMETHYLENETHIOPHOSPHINATO) MERCURY(II), Hg[CH$_2$P(S)Ph$_2$]$_2$

$$LiCH_3 + P(S)Ph_3 \rightarrow Li[CH_2P(S)Ph_2] + C_6H_6$$
$$2Li[CH_2P(S)Ph_2] + HgCl_2 \rightarrow Hg[CH_2P(S)Ph_2]_2 + 2LiCl$$

Procedure

■ **Caution.** *LiMe reacts explosively with water, and ignites spontaneously in moist air. Mercury(II) chloride is a highly toxic, light sensitive compound.*

All manipulations prior to product recrystallization are done under a dry $N_2(g)$ atmosphere.

A 50-mL Schlenk tube provided with a Teflon-coated magnetic stirring bar is charged with $P(S)Ph_3$ (1.17 g, 4.0 mmol) in 20 mL of THF. The solid is allowed to dissolve and is then cooled using a dry-ice/acetone bath to $-78°C$. To this solution 11.3 mL of 1.4 M LiMe (in ether) (4.06 mmol) is slowly added *via* a syringe. After 15 min the cold bath is removed and the light yellow solution allowed to equilibrate to room temperature. The color of the solution gradually intensifies from a light yellow to a deep red after 1.5–2 h of stirring at room temperature. The Li(MTP) complex generated should be used as soon as possible. The red solution is cooled to $-78°C$, and 500 mg of $HgCl_2$ (1.8 mmol) is added in one portion. The solution is stirred for one hour and gradually warmed to $0°C$, at which temperature it is stirred for an additional one hour, and then the product starts to precipitate. To increase the yield, further precipitation may be induced by reducing the volume of the solvent under reduced pressure. However, care must be taken to avoid coprecipitation of LiCl as a byproduct. The mixture is filtered and the precipitate washed with cold THF and cold EtOH. Yield: 780 mg (65%).

Anal. Calcd. for $C_{26}H_{24}P_2S_2Hg$: C, 47.09; H, 3.65. Found: C, 46.37; H, 3.34.

Properties

The product is a white solid that is air-stable for several months. Melting point $>230°C$. A gray discoloration of the solid indicates initial stages of decomposition. The ^1H NMR spectrum in $CDCl_3$ (relative to TMS) shows $\delta(CH_2) = 2.32$ ppm, $J_{P-H} = 12$ Hz, $J_{Hg-H} = 135$ Hz. ^{31}P NMR in $CDCl_3$ shows $\delta = 44.78$ ppm, $J_{Hg-P} = 148$ Hz. The structure has been determined by single-crystal X-ray crystallography.[8]

G. BIS(DIPHENYLMETHYLENETHIOPHOSPHINATO)MERCURY(II) DICHLORO-DI-GOLD(I), $Hg[CH_2P(S)Ph_2]_2(AuCl)_2$

$$Hg[CH_2P(S)Ph_2]_2 + 2Au(THT)Cl \rightarrow Hg[CH_2P(S)Ph_2]_2(AuCl)_2 + 2THT$$

Procedure

■ **Caution.** *Mercury compounds are highly toxic. Although the materials in this synthesis are not particularly air/moisture-sensitive, it is best to exercise caution in all organometallic chemistry and carry out manipulations under an inert atmosphere. Standard Schlenk procedures work well.*

A 25-mL Schlenk tube provided with a Teflon-coated magnetic stirring bar is charged with 20 mg $Hg(MTP)_2$ (0.030 mmol) in 5 mL of CH_2Cl_2. To the solution is added 20 mg of $Au(THT)Cl$ (0.062 mmol), which causes the suspended $Hg(MTP)_2$ to dissolve, yielding a clear, colorless solution. The solution is stirred under a N_2 atmosphere for 2 h at room temperature, after which time a white-colored precipitate is observed. Dry diethyl ether (5–8 mL) is added to induce further precipitation. The solution is carefully decanted, and the white solid is washed 3 times with 5mL portions of diethylether. The product is filtered and dried under reduced pressure for 2 h. Yield: 30 mg (88%)

Anal. Calcd. for $C_{26}H_{24}Cl_2P_2S_2Au_2Hg$: C, 27.68; H, 2.13. Found: C, 27.85 H, 1.91.

Properties

The product is a colorless solid, and is air-stable for several months. M.p. 178–180°C. It is slightly soluble in warm, chlorinated solvents. ^1H NMR in $CDCl_3$ (relative to TMS) shows $\delta(CH_2) = 2.46$ ppm, d, $J_{P-H} = 12$ Hz, $J_{Hg-H} = 166$ Hz; $\delta(C_6H_5) = 7.30$–7.60 ppm, m, $\delta C_6H_5 = 7.65$–7.90 ppm, m. The structure has been determined by single-crystal X-ray crystallography.[3]

H. BIS(DIPHENYLMETHYLENETHIOPHOSPHINATO)MERCURY (II)GOLD(I)-TETRACHLOROGOLD(III), $Hg^{II}Au^I[CH_2P(S)Ph_2]_2Au^{III}Cl_4$

$$Hg[CH_2P(S)Ph_2]_2(AuCl)_2 + PhI \cdot Cl_2 \rightarrow: Hg^{II}Au^I[CH_2P(S)Ph_2]_2Au^{III}Cl_4 + PhI$$

Procedure

A 25-mL Schlenk tube provided with a Teflon-coated magnetic stir bar is charged with 32 mg of $[Hg(MTP)_2](AuCl)_2$ (0.028 mmol) in 5 mL of CH_2Cl_2. To the white suspension 9 mg of $PhI \cdot Cl_2$ (0.033 mmol)[15] is added in one portion. The solution rapidly turns to a yellow color. After the solution is stirred for 1.5 h, excess diethyl ether is added to precipitate the product. The yellow solid is concentrated by centrifugation, the liquid phase decanted, and the product washed with diethyl ether. Recrystallization of the solid powder from a CH_2Cl_2/ether solvent mixture produces orange colored crystals. Yield: 23 mg (68%).

Anal. Calcd. for $C_{26}H_{24}Cl_4P_2S_2Au_2Hg$: C, 26.04; H, 2.00. Found: C, 26.19; H, 1.94

Properties

The product is an orange solid that is air- and moisture-stable at room temperature, m.p. 165°C. It is slightly soluble in chlorinated solvents at room temperature. ^1H NMR in $CDCl_3$ (relative to TMS) shows $\delta(CH_2) = 2.51$ ppm, $J_{P-H} = 12$ Hz, $J_{Hg-H} = 152$ Hz; $\delta(C_6H_5) = 7.40$–7.90 ppm, m. The solid-state structure has been determined by X-ray crystallography.

References

1. (a) D. Seyferth, D. E. Welch, and J. K. Heeren, *J. Am. Chem. Soc.* **85**, 642 (1963); (b) D. Seyferth, D. E. Welch, and J. K. Heeren, *J. Am. Chem. Soc.* **86**, 1100 (1964).
2. (a) A. M. Mazany and J. P. Fackler, Jr., *J. Am. Chem. Soc.* **106**, 801 (1984); (b) A. M. Mazany, Ph.D. dissertation, Case Western Reserve University, 1984; (c) A. Grohmann and H. Schmidbaur, in *Comprehensive Organometallic Chemistry II*, Vol. 3, J. Wardell, E. W. Abel, F. G. A. Stone, and G. Wilkinson (Ed. in Chief) (Eds.), Pergamon, 1995.
3. (a) S. Wang, J. P. Fackler, Jr., and T. F. Carlson, *Organometallics* **9**, 1973 (1990); (b) S. Wang and J. P. Fackler, Jr., *Organometallics* **9**, 111 (1990).
4. (a) S. Wang and J. P. Fackler, Jr., *Organometallics* **7**, 2415 (1988); (b) S. Wang and J. P. Fackler, Jr., *Inorg. Chem.* **28**, 2615 (1989).
5. H. H. Murray, D. A. Briggs, G. Garzón, R. G. Raptis, L. C. Porter, and J. P. Fackler, Jr., *Organometallics* **6**, 1992 (1987).
6. S. Wang, J. P. Fackler, Jr., C. King, and J. C. Wang, *J. Am. Chem. Soc.* **110**, 3308 (1988).
7. (a) S. Wang, G. Garzón, C. King, J. C. Wang, and J. P. Fackler, Jr., *Inorg. Chem.* **28**, 4623 (1989); (b) T. F. Carlson, J. P. Fackler, Jr., R. J. Staples, and R. E. P. Winpenny, *Inorg. Chem.* **34**, 426 (1995); (c) S. Wang and J. P. Fackler Jr., *Organometallics* **9**, 111 (1990).
8. S. Wang and J. P. Fackler, Jr., *Organometallics* **8**, 1578 (1989).
9. (a) J. P. Fackler, Jr., *Polyhedron* **16**, 1 (1997); (b) L. C. Porter and J. P. Fackler, Jr., *Acta Crystallogr.* **C43**, 587 (1987).
10. (a) C. King, D. D. Heinrich, G. Garzón, J. C. Wang, and J. P. Fackler, Jr., *J. Am. Chem. Soc.* **111**, 2300 (1989); (b) D. D. Heinrich, R. J. Staples, and J. P. Fackler, Jr., *Inorg. Chim. Acta* **229**, 61 (1995).
11. D. F. Shriver and M. A. Drezdzon, *The Manipulation of Air-Sensitive Compounds*, 2nd ed., Wiley, New York, 1986.
12. M. F. Lipton, C. M. Sorenson, A. C. Sadler, and R. H. Shapiro, *J. Organomet. Chem.* **186**, 155 (1980).
13. R. Usón and A. Laguna, in *Organometallic Syntheses*, R.B. King and J. J. Eisch (Eds.), Elsevier Scientific, Amsterdam, 1986, Vol. 3, p. 322.
14. G. B. Kauffman, D. O. Cowan, F. P. Dwyer, J. W. Hogarth, and A. M. Sargeson, *Inorg. Synth.* **6**, 211 (1960).
15. H. J. Lucas and E. R. Kennedy, *Organic Synthesis*, J Wiley, New York, 1955, Vol. 3, p. 482.

26. MESITYL-GOLD(I) COMPLEXES

Submitted by MARIANO LAGUNA,[*] JULIAN GARRIDO,[†] and MARIA CONTEL[†]
Checked by WERNER VANZYL,[‡] BELINDA PRIHODA[‡],
and JOHN P. PACKLER, Jr.[‡]

The original preparation of $\{[Au\,(mes)]_5\}$ is based on the method of Floriani and co-workers reacting [AuCl(CO)] with the corresponding Grignard reagent [Mg(mes)Br].[1,2] A later modification was the reaction between [AuCl(tht)] and $\{[Ag(mes)]_4\}$.[3] Both procedures require an inert atmosphere, total exclusion of light, and low temperatures. Furthermore, these methods demand the previous syntheses of [AuCl(CO)][4] or $\{[Ag(mes)]_4\}$,[2,5] which require special care and several hours of preparation.

The improved procedure described below is carried out in contact with air and at room temperature resulting in the rapid formation of $\{[Au(mes)]_5\}$. The method consists in displacing a chloro ligand from a chloromesityl-gold(I) complex[6] by using silver triflate:

$$BzPPh_3[Au(mes)Cl] + Ag(OSO_2CF_3) \xrightarrow[\substack{(1)\ -AgCl \\ (2)\ -BzPPh_3[SO_3CF_3]}]{RT} \frac{1}{5}\{[Au(mes)]_5\}$$

The gold(I) complex can be easily obtained from [Au(mes)(AsPh$_3$)][6] (whose preparation is also described below) by displacement of the triphenylarsine ligand with chloride using benzyltriphenylphosponium chloride.

A. MESITYLTRIPHENYLARSINE-GOLD(I) {[Au(mes)(AsPh₃)]}

$$[AuCl(AsPh_3)] + Mg(mes)Br \xrightarrow[-MgBrCl]{Et_2O,\ 0°C} [Au(mes)(AsPh_3)]$$

[*] Departamento de Quimica Inorganica, Instituto de Ciencia de Materiales de Aragon, Universidad de Zaragoza CSIC, E-50009 Zaragoza, Spain.
[†] Departamento de Quimica Aplicada, Universidad Publica de Navarra, E-31006 Pamplona, Spain.
[‡] Department of Chemistry, Texas A&M University, College Station, Texas 77843-3255.

Procedure

■ **Caution.** *Arsenic compounds are extremely toxic, solutions of Mg(mes)Br are pyrophoric, and silver trifluoromethanesulfonate is an irritant. They must be handled with the greatest care!*

A 100-mL, standard Schlenk flask, equipped with magnetic stirring bar is purged with nitrogen and then charged with 40 mL of freshly distilled diethyl ether and [AuCl(AsPh₃)] (0.538 g, 1 mmol). The resulting suspension is cooled to 0°C and a solution of [Mg(mes)Br] (1.5 mmol) in tetrahydrofuran (2 mL) is added [Mg(mes)Br] is easily prepared by standard Grignard procedures and is also commercially available (from Aldrich Chemical Co.)]. The mixture is stirred for 2 h at 0°C, and then one drop of distilled water is added to hydrolyze the excess Grignard reagent. Dark compounds could be observed in the reaction mixture. The solution is evaporated to dryness and the residue is redissolved in 30 mL of dichloromethane. The transparent solution is filtered through anhydrous MgSO₄ and concentrated to approximately 5 mL. Addition of 40 mL of *n*-hexane precipitates the compound as a white solid. The complex can be synthesized in larger quantities (up to 5 mmol) but with a slight decrease in the yield. Yield: 0.54 g, 87%. The crude product can be used without further purification.

Anal. Calcd. for $C_{27}H_{26}AuAs$: C, 52.1; H, 4.2. Found: C, 51.9; H, 4.1.

Properties

The compound [Au(mes)(AsPh₃)] is air-stable, in both solid state and solution. It can be stored without decomposition for several months at 0°C. At room temperature it decomposes after a few days. It is soluble in dichloromethane, acetone and chloroform; slightly soluble in diethyl ether; and insoluble in *n*-hexane. The infrared spectrum in Nujol mull shows asborptions corresponding to the mesityl ligand at 1582(w) and 840(m) cm⁻¹. The ¹H NMR spectrum in deuterated chloroform exhibits three singlets for the mesityl ligand at 6.91 (*m*-H), 2.59 (*o*-CH₃), and 2.25 (*p*-CH₃) ppm. This compound is an excellent starting material in the synthesis of new mesityl-gold(I) complexes, both mono- and polynuclear with the mesityl group acting as a terminal or bridging ligand.[6,7]

B. (BENZYL)TRIPHENYLPHOSPHONIUM-CHLOROMESITYLAURATE(I) {BzPPh₃[Au(mes)Cl]}

$$[Au(mes)(AsPh_3)] + [BzPPh_3]Cl \xrightarrow[-AsPh_3]{RT} BzPPh_3[Au(mes)Cl]$$

Procedure

To a dichloromethane solution (25 mL) of [Au(mes)(AsPh$_3$)] (0.622 g, 1 mmol) is added [BzPPh$_3$]Cl (commercially available from Aldrich) (0.389 g, 1 mmol) at room temperature, and the mixture is stirred for 20 min. Partial concentration under vacuum of the solution to $\sim$5 mL, and addition of diethyl ether (30 mL) gives the compound as a white solid. This crude product can be used without further purification and prepared in larger quantities (up to 5 mmol) without a significant change of yield. Yield: 0.620 g, 88%.

Anal. Calcd. for C$_{34}$H$_{32}$AuClP: C, 57.9; H, 4.7. Found: C, 58.3; H, 4.8.

Properties

The compound BzPPh$_3$[Au(mes)Cl] is an air-stable solid whose solution in acetone displays the conductivity typical of a 1 : 1 electrolyte.[6] It is more stable at room temperature than the starting material [Au(mes)AsPh$_3$], as it does not decompose over several months. BzPPh$_3$[Au(mes)Cl] is soluble in dichloromethane, acetone, and chloroform and insoluble in diethyl ether and *n*-hexane.

The infrared spectrum in Nujol exhibits absorptions due to the mesityl ligand at 1589(w) and 852(m) cm^{-1}, and the band assigned to the Au–Cl stretching vibration is approximately at 300 cm^{-1}. The ^{1}H NMR spectrum in deuterated chloroform shows the three singlets for the mesityl ligand at 6.69 (*m*-H), 2.31 (*o*-CH$_3$), and 2.16 (*p*-CH$_3$) ppm. A doublet assigned to the protons of the methylene group of the BzPPh$_3$ cation appears at 5.07 ppm ($^2J_{P–H} = 14.5$ Hz).

C. PENTAMESITYL-GOLD(I) {[Au(mes)]$_5$}

$$\text{BzPPh}_3[\text{Au(mes)Cl}] + \text{Ag(OSO}_2\text{CF}_3) \xrightarrow[\substack{(1)\ -\text{AgCl} \\ (2)\ -\text{BzPPh}_3[\text{SO}_3\text{CF}_3]}]{\text{RT}} \frac{1}{5}\{[\text{Au(mes)}]_5\}$$

Procedure

■ **Caution.** *The first step must be carried out while avoiding light exposure until the silver chloride is removed. This is achieved simply by wrapping the glassware in aluminum foil.*

To a dichloromethane solution (40 mL) of BzPPh$_3$[Au(mes)Cl] (0.353 g, 0.5 mmol), a solution of Ag(OSO$_2$CF$_3$) (commercially available from Aldrich) (0.128 g, 0.5 mmol) in diethylether (10 mL) is added. The reaction is instantaneous and the silver chloride starts to precipitate. The mixture is allowed to stir another 5–10 min and then is filtered through Celite. The yellow solution is concentrated to $\sim$2 mL, and 30 mL of diethylether is added to precipitate (BzPPh$_3$)SO$_3$CF$_3$ as a white solid (0.216 g, yield 86%) and is removed by filtration through a glass frit. The clear, bright yellow filtrate is evaporated to $\sim$5 mL, and 30 mL of *n*-hexane is added. The yellow complex {[Au(mes)]$_5$} precipitates and can be separated (as a pure product) by filtration through a glass frit (0.042 g). Later concentration of the mother liquor and storing at $-20°$C (during 24 h) yields another 0.033 g of the compound. Total yield: 0.075 g, 47%.

Anal. Calcd. for C$_{45}$H$_{55}$Au$_5$: C, 34.2; H, 3.5. Found: C, 34.7; H, 3.4.

Properties

The compound {[Au(mes)]$_5$} is air/moisture-stable, both as a solid and in solution. It can be stored at 0°C for several months. It is soluble in dichloromethane, acetone, chloroform, and diethylether and partially soluble in *n*-hexane and *p*-dioxane. Vibrations due to the mesityl ligand are observed at 1594(s) and 847(s) cm^{-1}, and the absorption assigned to $\nu_{Au–C}$ appears at 581(w) cm^{-1}. The ^{1}H NMR spectrum in deuterated chloroform shows the three singlets for the mesityl ligand at 6.70 (*m*-H), 2.58 (*o*-CH$_3$), and 2.08 (*p*-CH$_3$) ppm. Further characterization is given by Floriani and co-workers,[1,2] and its reactivity has been described.[1-3] {[Au(mes)]$_5$} reacts with phosphines, phosphites, and salts of general formula QX to give both mononuclear and dinuclear derivatives (neutral or anionic) with the mesityl group acting as a terminal ligand.

References

1. S. Gambarotta, C. Floriani, A. Chiesi-Villa, C. Guastini, *J. Chem. Soc., Chem. Commun.*, 1304 (1983).
2. E. M. Meyer, C. Floriani, A. Chiesi-Villa, and C. Guastini, *Organometallics* **8**, 1067 (1989).
3. R. Uson, A. Laguna, E. J. Fernandez, M. E. Ruiz Romero, P. G. Jones, and J. Lautner, *J. Chem. Soc., Dalton Trans.*, 2127 (1989).
4. D. Belli-Dell'Amico and F. Calderazzo, *Gazz. Chim. Ital.* **103**, 1099 (1973).
5. S. Gambarotta, C. Floriani, A. Chiesi-Villa, and C. Guastini, *J. Chem. Soc., Chem Commun.*, 870 (1984).
6. M. Contel, J. Jimenez, P. G. Jones, A. Laguna, and M. Laguna, *J. Chem. Soc., Dalton Trans.*, 2515 (1994).
7. M. Contel, J. Garrido, M. C. Gimeno, P. G. Jones, A. Laguna, and M. Laguna, *Organometallics* **15**, 4939 (1996).

27. AN IMPROVED SYNTHESIS OF *cis*-DITHIOCYANATO-BIS (4,4′-DICARBOXY-2,2′-BPY)Ru(II) SENSITIZER

Submitted by M. K. NAZEERUDDIN*,† and M. GRÄTZEL*
Checked by D. PAUL RILLEMA‡

Dye-sensitized nanocrystalline TiO_2 solar cells are presently under intensive investigation.[2-4] So far, the best performing charge transfer sensitizer employed for such an application is the *cis*-dithiocyanato-bis(4,4′-dicarboxy-2,2′-bipyridine)ruthenium(II) complex, yielding solar-to-electric power conversion efficiency of 10% under standard solar conditions.[1] These cells exhibit a remarkable stability, making practical applications feasible.

Herein the synthesis and characterization of this sensitizer is reported. The procedure allows the synthesis of ≤ 15-g quantities, by proportionately selecting the reactants. For a small batch of ~ 0.1 g, the minimum required DMF solvent is 30 mL. The total time necessary for the synthesis and workup is 3 days. All the reactions are carried out under argon and reduced light.

Materials

The ligand 4,4′-dicarboxy-2,2′-bipyridine is synthesized by using the literature procedure.[5] $RuCl_3 \cdot 3H_2O$ is obtained from Johnson Matthey, Wayne, PA.

A. *cis*-DICHLORO-BIS(4,4′-DICARBOXY-2,2′-BPY)Ru(II) [RuL₂(Cl)₂], 1

$$RuCl_3(H_2O)_3 + 2L \xrightarrow[N_2 \ or \ Ar]{DMF} RuL_2(Cl)_2 \qquad (L = 4,4'\text{-dicarboxy-2}, 2'\text{-bipyridine})$$

■ **Caution.** *The synthesis and all the manipulations should be carried out in reduced light to avoid possible trans isomerization.*

Reagent-grade N,N'-dimethylformamide (500 mL) is added to a 2000-mL, three-necked, round-bottomed flask fitted with a reflux condenser, a magnetic stirring

*Laboratory for Photonics and Interfaces, Swiss Federal Institute of Technology, CH-1015 Lausanne, Switzerland.
†Email: MdKhaja.Nazeeruddin@epfl.ch.
‡Department of Chemistry, Wichita State University, Wichita, KS 67260-0051.

bar, and a gas inlet. Then 5.23 g (0.02 mol) of $RuCl_3(H_2O)_3$ is dissolved in DMF under argon. After stirring for 15 min another 500 mL of DMF is added. To this solution 9.52 g (0.039 mol) of 4,4′-dicarboxy-2,2′-bipyridine ligand as a solid is added. The flask is wrapped completely with aluminum foil and then immersed in an oil bath maintained at 170–180°C. The reaction mixture is refluxed with vigorous stirring for 3 h in an oil bath at 180°C. (*The temperature is important; at lower temperatures the reaction time significantly increases.*) The progress of the reaction is monitored by UV–vis spectroscopy. A small aliquot of the reaction mixture is sampled and diluted with an absolute ethanol. The UV–vis absorption spectrum of this solution shows maxima at 565, 414, and 316. At the end of the reaction the relative intensities of these maxima are 1 : 1.05 : 3.33. After reaching these ratios (≈ 3 h) refluxing is suspended, while maintaining stirring for an additional 1 h. The reaction mixture is passed through a sintered-glass crucible. The solvent DMF is evaporated completely on a rotary evaporator under vacuum. The resulting air-stable solid product is stirred in 100 mL of 2 M HCl for 4 h in darkness and filtered through a sintered* glass crucible. After drying, the yield weighs 11.4 g (86%).

Anal. for $[RuL_2(Cl)_2]$: C, 24; H, 16; N, 4; O, 8; C, 12. Calcd. in (%) C, 43.67; H, 2.44; N, 8.48; Cl, 10.74. Found: C, 44.02, H, 2.58, N, 8.90; Cl, 10.74.

B. *cis*-DITHIOCYANATO-BIS(4,4′-DICARBOXY-2,2′-BPY) Ru(II) $[RuL_2(NCS)_2]$, 2

$$RuL_2(Cl)_2 + KNCS \xrightarrow[\text{N}_2 \text{ or Ar}]{\text{DMF, H}_2\text{O}} RuL_2(NCS)_2$$

$$(L = 4, 4'\text{-dicarboxy-2}, 2'\text{-bipyridine})$$

Procedure

An amount of KNCS [58.2 g (0.6 mol)] is dissolved in 100 mL of distilled water and transferred into a 2000-mL three-necked flask. To this solution, 500 mL of DMF is added and purged with Ar for 15 min. Then 11.0 g (0.017 mol) of $[RuL_2(Cl)_2]$ as a solid, is introduced into the flask in darkness, followed by another 500 mL of DMF solvent. The flask is covered with aluminum foil and refluxed (oil bath temperature 150°C) for 5 h. The reaction mixture is allowed to

---

Checkers' comment: The product was obtained after cooling the reaction mixture to room temperature and filtering it. Additional product was obtained by adding acetone to the filtrate. The combined products yielded 77% formulated as Ru(4,4′-dicarboxy-bpy)$_2$Cl$_2$-2-DMF.

cool and then filtered through a sintered-glass crucible, grade 4. An amount of a DMF-insoluble but water-soluble byproduct (0.3 g) is obtained. The absorption and emission spectral properties of this compound suggest that it could be a dimer.[*] The DMF and water solvents are removed using a rotary evaporator under high vacuum. To the resulting viscous liquid is added 50 mL of water. To this aqueous solution is added $\sim$100 mL of 0.5 M NaOH to give a dark purple-red homogeneous solution. During the addition of NaOH the solution pH is monitored by a pH meter and kept below 10. The solution is filtered, and the pH of this solution is lowered to 3.3 with a 0.5 M HNO_3 or CF_3SO_3H solution, giving a dense precipitate. The flask is placed in a refrigerator for 12 h at $-4°C$. After allowing it to reach room temperature, the solid is collected on a sintered-glass crucible by suction filtration. It is then washed (3×20 mL) with pH 3.5 water, which is prepared with the same acid as used previously and air-dried. Yield: 9.0 g, 85%.

Anal. for $[RuL_2(NCS)_2]·4(H_2O)$ (C, 26; H, 24; N, 6; O, 12; S, 2). Calcd. in (%): C, 40.15; H, 3.11; N, 10.80; S, 8.24. Found: C, 39.65; H, 3.13; N, 10.53; S, 8.67.

Physical Properties

In solid state both complexes are stable in air at 35°C and can be stored in a drawer for months. However, in solution complex **1** $[RuL_2(Cl_2)]$ undergoes trans isomerization under light. Complex **2** $[RuL_2(NCS)_2]$ is soluble in DMF, DMSO, methanol, and alkaline water. The chloride ligands in complex **1** are labile and form diaqua complex in water and solvent complexes in coordinating solvents such as DMSO.

Electronic Spectra

The absorption spectral data of complexes **1** and **2**, in the UV–vis region are presented in Table I. The absorption spectrum of **2** in ethanol shows two intense visible MLCT (metal–ligand charge transfer) bands at 538 and 398 nm. In the UV, the complex shows bands at 314 nm with a shoulder at 304 nm shifted when compared to the analogous unsubstituted bpy complex.

Complex **2** has an emission peak at 850 nm, with a short lifetime of 50 ns. The emission quantum yield (ϕ_{em}) of **2** is 0.0002. Cyclic voltammogram of complex **2** is measured in DMSO solvent with 0.1 mol tetrabutylammonium-trifluoro-methanesulfonate as supporting electrolyte. In the potential range of $+1.2$ to

---

[*] *Checkers' comment*: The reaction was carried out in water after adjusting the pH with 0.5 M NaOH until the dicarboxylate was soluble (pH $\sim$ 10). The reactants were then refluxed for 5 h, and no precipitate was observed.

TABLE I. Electronic Spectral Data of Complexes 1 and 2 in Solution at Room Temperature[a]

Complex	Solvent	MLCT	MLCT	$\pi - \pi^*$ CT
[RuL$_2$(Cl)$_2$] (**1**)	H$_2$O[b]	535 (1.0)	400 (1.04)	314 (3.60)
	C$_2$H$_5$OH	565 (1.0)	414 (0.98)	316 (3.30)
	DMSO	580 (1.0)	430 (0.94)	320 (3.65)
[RuL$_2$(NCS)$_2$] (**2**)	H$_2$O[b]	500 (1.0)	372 (1.03)	308 (3.91)
	C$_2$H$_5$OH	538 (1.0)	398 (0.98)	314 (3.70)
	DMSO	542 (1.0)	400 (0.91)	318 (2.92)

[a] Wavelengths are in nanometers; the values in parentheses are relative intensities normalized to the lowest MLCT band.
[b] Dissolved in pH 10 (NaOH) water

TABLE II. ^{1}H NMR Chemical Shifts for Free Ligand and Complex 2 and Coordination-Induced Shifts for the H6 proton (CIS = δ, complexed, $-\delta$, Free)[a]

Complex or Ligand[b]	6	6'	5	5'	3	3'	CIS
4,4'-COOH-2,2'bpy	8.78	—	7.87	—	8.40	—	—
2	9.55	7.82	8.22	7.52	8.94	8.77	0.77

[a] In ppm with respect to TMS. Positive CIS values refer to downfield shifts.
[b] Complex **2** and the ligand 4,4'-COOH-2,2'-bpy were recorded in D$_2$O containing 0.01 M NaOD.

-1.5 V, complex **B** shows an oxidation wave at $+0.85$ V and one reduction wave at -1.17 V versus SCE. The IR spectrum of complex **B** shows an intense absorbance at 2120 cm^{-1}. The resonance Raman spectrum of complex **B** obtained by excitation at 415.44 nm (krypton laser as excitation source) shows vibrational modes due to 4,4'-dicarboxy-2,2'-bpy ligand at 1612, 1538, 1477, 1295, 1269, 1045, and 1030 cm^{-1}. The broad and weak vibrational mode at 2140 cm^{-1} is due to coordinated NCS ligand. The ^{1}H NMR spectral data of **B** are given in Table II. The primed notation indicates the protons of the pyridine ring trans to the NCS ligand. The proton decoupled ^{13}C NMR spectrum of complex **B** shows 12 resonance peaks (6 pairs) corresponding to two different bpy ligands and one single peak at 132.84 ppm due to the carbon of *N*-coordinated NCS.

Acknowledgments

This work was supported by grants from the Swiss Federal Institute for Energy (OFEN) and the Institut für Angewandte Photochemie (INAP, Germany).

References

1. (a) M. K. Nazeeruddin, A. Kay, I. Rodicio, R. Humphry-Baker, E. Muller, P. Liska, N. Vlachopoulos, and M. Grätzel, *J. Am. Chem. Soc.* **115**, 6382 (1993); (b) P. Liska, N. Vlachopoulos, Md. K. Nazeeruddin, P. Compte, and M. Grätzel, *J. Am. Chem. Soc.* **110**, 3686 (1988).
2. M. Grätzel, *Platinum Metals Rev.* **4**, 151 (1994).
3. K. Murakoshi, G. Kano, Y. Wada, S. Yanagida, H. Miyazaki, M. Matsumoto, and S. Murasawa, *J. Electroanal. Chem.* **396**, 27 (1995).
4. R. Argazzi, C. A. Bignozzi, T. A. Heimer, and G. J. Meyer, *Inorg. Chem.* **36**, 2 (1997).
5. M. K. Nazeeruddin, K. Kalyanasundaram, and M. Grätzel, *Inorg. Synth.* **32**, 181 (1998).

28. DIMETHYLSULFOXIDE COMPLEXES OF PLATINUM(II): K[PtCl$_3$(Me$_2$SO)], *cis*-[PtCl$_2$L(Me$_2$SO)] (L = Me$_2$SO, MeCN), [PtCl(μ-Cl)(Me$_2$SO)]$_2$, AND [Pt(Me$_2$SO)$_4$](CF$_3$SO$_3$)$_2$

Submitted by VADIM YU. KUKUSHKIN, [*] **ARMANDO J. L. POMBEIRO,** [†]
CRISTINA M. P. FERREIRA, [†] **and LARS I. ELDING** [§]
Checked by RICHARD J. PUDDEPHATT [‡]

Since the early 1970s there has been intense interest in the synthesis,[1-3] reactivity,[1-4] structural features,[5] kinetic behavior,[3] cis–trans[6,7] and linkage[8] isomerization, and antitumor activity[9] of dimethylsulfoxide complexes of platinum. These complexes are also useful as synthons for preparation of antitumor malonato platinum compounds[10] as well as versatile Pt(II) and Pt(IV) dimethylsulfoxide and dimethylsulfide complexes via substitution,[1-3] oxidation of the central ion,[7] or deoxygenation reactions of coordinated sulfoxide ligands.[4] We describe syntheses and properties of the parent compounds, K[PtCl$_3$\{(CH$_3$)$_2$SO\}], *cis*-[PtCl$_2$\{(CH$_3$)$_2$SO\}$_2$], [PtCl(μ-Cl)\{(CH$_3$)$_2$SO\}]$_2$, and *cis*-[Pt\{(CH$_3$)$_2$SO\}$_2$ \{(CH$_3$)$_2$SO\}$_2$](SO$_3$CF$_3$)$_2$, which can be used for preparation of Pt complexes of different types. Synthesis of the starting material for obtaining the bridged dimer [PtCl(μ-Cl)\{(CH$_3$)$_2$SO\}]$_2$, namely, the acetonitrile complex

[*] Department of Chemistry, St. Petersburg State University, Universitetsky Pr., 2, 198904 Stary Petergof, Russian Federation; fax +7-(812)-428-6939; email kukushkin@VK2100.spb.edu.
[†] Centro de Química Estrutural, Complexo I, Instituto Superior Técnico, Av. Rovisco Pais, 1049-001 Lisbon, Portugal.
[§] Inorganic Chemistry, Chemical Center, University of Lund, PO Box 124, SE-221 00 Lund, Sweden.
[‡] Department of Chemistry, University of Western Ontaio, London, Ontario, Canada N6A 5B7.

cis-[PtCl$_2$(CH$_3$CN){(CH$_3$)$_2$SO}], is also included. The latter compound has been successfully employed for simple preparation of *trans*-[PtCl$_2$(PPh$_3$)$_2$].[11]

S-Coordinated dimethylsulfoxide, exhibiting a strong trans effect,[3,12] facilitates substitution trans to the sulfur. Therefore, the water-soluble sulfoxide complex K[PtCl$_3${(CH$_3$)$_2$SO}] is useful for preparation of *trans*-[PtCl$_2$(R$_2$SO)L′] compounds through reaction between K[PtCl$_3${(CH$_3$)$_2$SO}] and water-soluble L′ in aqueous media. The simple synthesis of K[PtCl$_3${(CH$_3$)$_2$SO}] reported below is based on treatment of K$_2$(PtCl$_4$) with 1 equiv of dimethylsulfoxide; an alternative method includes displacement of a dimethylsulfoxide in *cis*-[PtCl$_2${(CH$_3$)$_2$SO}$_2$] by Cl$^-$.[13] Metathesis of K[PtCl$_3${(CH$_3$)$_2$SO}] with phosphonium or alkylammonium chlorides, Q$^+$Cl$^-$, in aqueous solution leads to precipitation of Q[PtCl$_3${(CH$_3$)$_2$SO}] (e.g., Q = Ph$_3$PCH$_2$Ph[14]), which is soluble in most common organic solvents.

Reaction of K[PtCl$_3${(CH$_3$)$_2$SO}] with 1 equiv of dimethylsulfoxide or interaction between K$_2$(PtCl$_4$) and 2 equiv of Me$_2$SO in water results in precipitation of *cis*-[PtCl$_2${(CH$_3$)$_2$SO}$_2$].[13,15] Other routes for obtaining this complex, such as reaction of PtCl$_2$ with Me$_2$SO[16] or trans–cis isomerization of *trans*-[PtCl$_2${(CH$_3$)$_2$SO}$_2$],[17] are not recommended as preparative methods because of rather complicated procedures and availability of much simpler methods.

Reaction in nonaqueous media of [PtX(μ-X)(R$_2$SO)]$_2$ (X = Cl, Br; R = alkyl) with various ligands L′ soluble in organic solvents splits the bridge, giving the mononuclear complex *trans*-[PtX$_2$(R$_2$SO)L′], which is not always easily obtained by other methods. The parent compound of the series, the bridged dinuclear complex [PtCl(μ-Cl){(CH$_3$)$_2$SO}]$_2$, has been prepared by thermal solid-state elimination of ethylene from *cis*-[PtCl$_2$(ν^2-C$_2$H$_4$){(CH$_3$)$_2$SO}][18] via halide abstraction from K[PtCl$_3${(CH$_3$)$_2$SO}] by AgNO$_3$ in water,[19] and by treatment of *cis*-[PtCl$_2${(CH$_3$)$_2$SO}$_2$] with PtCl$_2$ in naphthalene at 165°C.[17] We describe a facile high-yield synthesis of [PtCl(μ-Cl){(CH$_3$)$_2$SO}]$_2$ that proceeds via thermal solid-state elimination of CH$_3$CN from *cis*-[PtCl$_2$(CH$_3$CN){(CH$_3$)$_2$SO}].[20]

The homoleptic dimethylsulfoxide complex *cis*-[Pt{(CH$_3$)$_2$SO}$_2${(CH$_3$)$_2$SO}$_2$][SO$_3$CF$_3$]$_2$ has been synthesized by reaction of Pt(OH)$_2$·xH$_2$O with trifluoromethanesulfonic acid in dimethylsulfoxide.[21] We report a much simpler method for preparation of this complex via halide abstraction from *cis*-[PtCl$_2${(CH$_3$)$_2$SO}$_2$] by 2 equiv of AgSO$_3$CF$_3$ in dimethylsulfoxide. This method has been previously used for synthesis of the perchlorate [Pt{(CH$_3$)$_2$SO}$_2${(CH$_3$)$_2$SO}$_2$](ClO$_4$)$_2$.[22]

■ **Caution.** *Potassium tetrachloroplatinate(II) is known as a sensitizing agent. Dimethylsulfoxide and silver trifluoromethanesulfonate are irritants. Acetonitrile is a lachrymatory and flammable liquid. All other organic solvents used are toxic. Contact with liquids and their vapors should be avoided. Appropriate precautions must be taken, and an efficient hood must be used.*

A. POTASSIUM TRICHLORO(DIMETHYLSULFOXIDE)
PLATINATE(II) (K[PtCl₃{(CH₃)₂SO}])

$$K_2[PtCl_4] + (CH_3)_2SO \rightarrow K[PtCl_3\{(CH_3)_2SO\}] + KCl$$

Procedure

An amount of $K_2[PtCl_4]^*$ (3.00 g, 7.23 mmol; purchased from Degussa, New Jersey, USA) is placed in a 100-mL beaker and dissolved in water (50 mL) at room temperature. Dimethylsulfoxide (0.52 mL, 7.33 mmol; purchased from Carlo Erba, Milan, Italy) is added, and the reaction mixture is stirred with a Teflon-coated magnetic bar at 20–25°C for 12 h. The yellow solution that is formed is filtered off from a small amount of pale yellow *cis*-[PtCl₂{(CH₃)₂SO}₂] through a paper filter (Schleicher & Schuell 589³ Blauband filter was used) to another 100-mL beaker. The filtrate is evaporated at 90–95°C on stirring with a Teflon-coated magnetic bar to a volume of 10–12 mL, when yellow-orange crystals start to precipitate. The mixture is cooled to room temperature and left without stirring for 1 h, whereafter the precipitate is collected on a Hirsch funnel, washed with two 0.5-mL portions of water, two 5-mL portions of ethanol, and two 5-mL portions of diethyl ether, and dried in air at 20–25°C. Yield of K[PtCl₃{(CH₃)₂SO}]: 2.49 g, 82%.

Anal. Calcd. for K[PtCl₃{(CH₃)₂SO}]: Cl, 25.4; Pt, 46.6. Found: Cl, 25.3; Pt, 46.6. FAB⁻-MS (NBA matrix): 379 (calcd. 379 for [PtCl₃{(CH₃)₂SO}]⁻).

Properties

K[PtCl₃{(CH₃)₂SO}] is crystallized from water as needle-like yellow-orange crystals in the monoclinic space group C2/m with $a = 14.969(7)$, $b = 8.116(4)$, $c = 7.746(4)$ Å; $\beta = 105.10(4)°$; $V = 908.6(8)$ Å³; $\rho_{exp} = 3.07(1)$ g/cm³ (flotation); and $Z = 4$.[23] Crystal structures of other complexes of (cation)[PtX₃(R₂SO)] type have been reviewed.[5] The complex is soluble in water and insoluble in dichloromethane, diethylether, ethanol, and acetone. ¹H NMR in D₂O, δ, ppm: 3.50 ($J_{PtH} = 23.4$ Hz); an additional peak at 3.57 ppm is due to the hydration product [PtCl₂(H₂O){(CH₃)₂SO}]. ¹⁹⁵Pt NMR, δ, ppm: -2990 (in water[24]) and -2959 (in Me₂SO-d_6[28]). IR spectrum in KBr, selected bands, cm⁻¹: 1102 (vs) $\nu_{S=O}$ all peaks in the ν_{Pt-Cl} region are 350(s), 336(w), 330(w), and 313(m) [*lit.*[25] 1102 (vs) $\nu_{S=O}$, 346(s), 331(m), and 309(s) ν_{Pt-Cl}].

---

*In many instances, commercially available K₂(PtCl₄) contains admixtures of K₂(PtCl₆). These impurities do not affect the reaction but slightly decrease its yield.

B. *cis*-DICHLOROBIS(DIMETHYLSULFOXIDE)PLATINUM(II) (*cis*-[PtCl$_2${(CH$_3$)$_2$SO}$_2$])

$$K_2(PtCl_4) + 2(CH_3)_2SO \rightarrow \textit{cis-}[PtCl_2\{(CH_3)_2SO\}_2] + 2KCl$$

An amount of K$_2$(PtCl$_4$) (1.00 g, 2.41 mmol; purchased from Degussa) is placed in a 30-mL beaker and dissolved in water (20 mL) at room temperature.* Dimethylsulfoxide (0.69 mL, 9.64 mmol; purchased from Carlo Erba) is added and the reaction mixture is stirred with a Teflon-coated magnetic bar at 20–25°C for 4 h. The pale yellow precipitate that forms is collected on a Hirsch funnel, washed with three 1-mL portions of water, and dried in air at 20–25°C. Yield of *cis*-[PtCl$_2${(CH$_3$)$_2$SO}$_2$]: 0.66 g, 65%.

Anal. Calcd. for *cis*-[PtCl$_2${(CH$_3$)$_2$SO}$_2$]: C, 11.4; H, 2.9; Cl, 16.8; Pt, 46.2. Found: C, 11.6; H, 3.1; Cl, 17.1; Pt, 46.3. FAB$^+$-MS (NBA matrix): 423 (M+H)$^+$.

Properties

The complex crystallizes from the reaction mixture as pale yellow rod-shaped crystals in the monoclinic space group $P2_1$/c with $a = 8.653(3)$, $b = 13.580(5)$, $c = 10.914(4)$ Å; $\beta = 123.43(3)°$; $V = 1070.3(7)$ Å^3; $\rho_{exp.} = 2.59(3)$ g/cm^3 (flotation); and $Z = 4$.[26] Crystal structures of other complexes of the *cis*-[PtX$_2$(R$_2$SO)$_2$] type have been reviewed.[5] ^{1}H NMR in CD$_2$Cl$_2$, δ, ppm: 3.47 (J_{PtH} 23.5 Hz) [*lit.*[27] 3.53 (J_{PtH} 23.3 Hz) in CD$_3$NO$_2$]. ^{13}C{^{1}H} NMR in CD$_3$NO$_2$, δ, ppm: 45.43 (J_{PtC} 52.1).[27] ^{195}Pt NMR in Me$_2$SO-d_6, δ, ppm: −3450.[28] IR spectrum in KBr, selected bands, cm^{-1}: 1155(vs) and 1131(vs) $\nu_{S=O}$, 334(s) and 309(s) ν_{Pt-Cl} [*lit.*[22] IR spectrum in Nujol: 1157 and 1134 $\nu_{S=O}$, 334 and 309 ν_{Pt-Cl}].

C. *cis*-DICHLORO(ACETONITRILE)(DIMETHYLSULFOXIDE) PLATINUM(II) (*cis*-[PtCl$_2$(CH$_3$CN){(CH$_3$)$_2$SO}])

$$K[PtCl_3\{(CH_3)_2SO\}] + CH_3CN \rightarrow \textit{cis-}[PtCl_2(CH_3CN)\{(CH_3)_2SO\}] + KCl$$

An amount of K[PtCl$_3${(CH$_3$)$_2$SO}] (0.50 g, 1.94 mmol) is placed in a 30-mL beaker is dissolved in water (15 mL) at room temperature. Acetonitrile (1.00 mL, 19.15 mmol; purchased from Aldrich) is added to the solution, the beaker is covered with Parafilm M and the reaction mixture is left at 20–25°C for 12 h. The needle-like yellow crystals formed are collected on a Hirsch funnel,

---

*To avoid contamination of the product with impurities of K$_2$(PtCl$_6$) from commercial K$_2$(PtCl$_4$), the solution of K$_2$(PtCl$_4$) should be filtrated before performing this reaction.

washed with two 3-mL portions of water, and dried in air at room temperature. Yield of *cis*-[PtCl$_2$(CH$_3$CN){(CH$_3$)$_2$SO}]: 0.37 g, 81%.

Anal. Calcd. for *cis*-[PtCl$_2$(CH$_3$CN){(CH$_3$)$_2$SO}]: C, 12.5; H, 2.4; N 3.6; Cl, 18.4; Pt, 50.7. Found: C, 12.1; H, 2.6; N 3.6; Cl, 18.3; Pt, 50.8. FAB$^+$-MS (NBA matrix): 386 (M+H)$^+$.

The complex *cis*-[PtCl$_2$(CH$_3$CN){(CH$_3$)$_2$SO}] can also be obtained from K[PtCl$_3${(CH$_3$)$_2$SO}] prepared in situ from K$_2$(PtCl$_4$) and 1 equiv of dimethyl-sulfoxide. However, the yield in this case is lower (61%[29]). The bromide com-plex *cis*-[PtBr$_2$(CH$_3$CN){(CH$_3$)$_2$SO}] can be synthesized analogously.[30]

Properties

cis-[PtCl$_2$(CH$_3$CN){(CH$_3$)$_2$SO}] precipitates as needle-like yellow crystals from the reaction mixture. They crystallize in the monoclinic space group $P2_1/n$ with $a = 9.566(4)$, $b = 9.966(4)$, $c = 9.995(3)$ Å; $\beta = 93.18(3)°$; $V = 954.3(6)$ Å^3; $\rho_{calcd.} = 2.681$ g/cm^3; $Z = 4$.[30] When the compound was prepared from K[PtCl$_3${(CH$_3$)$_2$SO}] obtained in situ from K$_2$(PtCl$_4$) and 1 equiv of DMSO, another polymorphic modification was obtained [the triclinic space group $P\bar{1}$ with $a = 7.346(1)$, $b = 8.865(1)$, $c = 14.886(2)$ Å; $\alpha = 90.58(1)°$, $\beta = 96.10(1)°$, $\gamma = 87.44(2)°$; $V = 962.9(3)$ Å^3; $\rho_{calcd.} = 2.65$ g/cm^3, $Z = 4$[30,32]]. In both polymorphs the DMSO is coordinated to Pt(II) through the sulfur.

On slow heating of the solid complex ($\sim$2–3°C/min), the nitrile ligand dis-sociates at 145–150°C, forming [PtCl(μ-Cl){(CH$_3$)$_2$SO}]$_2$, which decomposes at 235°C [*lit.*[31] m.p. 170°C decomposition]. ^{1}H NMR in acetone-d_6, δ, ppm: 2.68 (J_{PtH} 14.2 Hz, 3H) and 3.49 (J_{PtH} 22.5 Hz, 6H). IR spectrum in KBr, selected bands, cm^{-1}: 2339(m) and 2312(mw) $\nu_{C\equiv N}$, 1147(vs) $\nu_{S=O}$; all peaks in the ν_{Pt-Cl} region are 359(m), 318(s), and 313(sh).

D. DI-μ-CHLORO-DICHLORO-BIS(DIMETHYLSULFOXIDE) DIPLATINUM(II) ([PtCl(μ-Cl){(CH$_3$)$_2$SO}]$_2$)

$$2 \ cis\text{-}[PtCl_2(CH_3CN)\{(CH_3)_2SO\}] \rightarrow [PtCl(\mu\text{-Cl})\{(CH_3)_2SO\}]_2 + 2CH_3CN$$

The finely ground complex *cis*-[PtCl$_2$(CH$_3$CN){(CH$_3$)$_2$SO}] (1.00 g) is spread in a thin layer on a Petri dish. It is kept in an oven in open air at 145°C for $\sim$6 h. During the thermosynthesis, the color of the sample turns from pale yellow to orange; at the end of the synthesis the surface of the compound is slightly gray-ish. The progress of the reaction is controlled by monitoring the weight of the sample; the process is completed when mass loss is 10.7%. Yield of [PtCl(μ-Cl){(CH$_3$)$_2$SO}]$_2$ is quantitative (0.89 g).

Anal. Calcd. for [PtCl(μ-Cl){(CH$_3$)$_2$SO}]$_2$: C, 7.0; H, 1.8; Cl, 20.6; Pt, 56.7. Found: C, 6.7; H, 2.1; Cl, 20.5; Pt, 56.8. FAB$^+$-MS (NBA matrix): 688 (M$^+$).

Properties

The compound forms an orange powder slightly soluble in hot chloroform[19] and in dichloromethane. m.p. = 235°C decomposition; (Kofler tables) [*lit.*[19] m.p. 215°C]. ^{1}H NMR in CD$_2$Cl$_2$, δ, ppm: 3.46 (J_{PtH} 26.5 Hz) [*lit.*[17] 3.47 (J_{PtH} 27 Hz)]. IR spectrum in KBr, selected bands, cm^{-1}: 1152(vs), 1115(s), and 1099(s) $\nu_{S=O}$; all peaks in $\nu_{Pt-Cl,}$ and $\nu_{Pt-\mu-Cl}$ regions are 364(s), 309(m), 289(m), and 259(m) [*lit.*[17] 1155 $\nu_{S=O}$, 370 $\nu_{Pt-Cl,}$, and 264 $\nu_{Pt-\mu-Cl}$]. Observation of J_{PtH} coupling constants in ^{1}H NMR spectrum, and position of the $\nu_{S=O}$ stretching vibrations in the IR spectrum indicate *S*-coordination of the sulfoxide ligands in both solution and the solid state.[1-3] Although an X-ray structure of [PtCl(μ-Cl){(CH$_3$)$_2$SO}]$_2$ is yet unreported, its diethylsulfoxide analog, *trans*-[PtCl(μ-Cl){(C$_2$H$_5$)$_2$SO}]$_2$, has been characterized by X-ray diffraction.[20] In this complex, the (C$_2$H$_5$)$_2$SO ligands are coordinated to Pt through sulfur.

E. TETRAKIS(DIMETHYLSULFOXIDE)PLATINUM(II) BIS(TRIFLUOROMETHANESULFONATE) [Pt{(CH$_3$)$_2$SO}$_2$(CH$_3$)$_2$SO$_2$][SO$_3$CF$_3$]$_2$

cis-[PtCl$_2${(CH$_3$)$_2$SO}$_2$] + 2AgSO$_3$CF$_3$ + 2(CH$_3$)$_2$SO →
[Pt{(CH$_3$)$_2$SO}$_2${(CH$_3$)$_2$SO$_2$}][SO$_3$CF$_3$]$_2$ + 2AgCl

All operations should be done using standard inert gas (N$_2$) and high-vacuum Schlenk techniques. Silver trifluoromethanesulfonate (0.154 g, 0.60 mmol; purchased from Aldrich) is added to a solution of *cis*-[PtCl$_2${(CH$_2$)$_2$SO}$_2$] (0.125 g, 0.30 mmol) in anhydrous dimethyl sulfoxide (5 mL; purchased from Carlo Erba) placed in a 20-mL Schlenk tube. The mixture is stirred with a Teflon-coated magnetic bar for 20 h in the dark. The precipitate of AgCl formed is filtered off on a grade 4 glass filter. The yellow filtrate, contained in a 30-mL Schlenk tube, is washed thoroughly on stirring with three 10-mL portions of Et$_2$O and filtered again through a grade 4 glass filter. After each washing and separation of the two liquid phases by syringing out the lighter diethyl ether phase, the remaining heavier phase is placed under vacuum for ~2 min in 20-mL Schlenk tube. Dry CHCl$_3$ (5 mL) is added to the final residual solution, leading to precipitation of a white solid that is filtered off (grade 4 glass filter), washed with dry *n*-pentane (3 mL), and dried in vacuo at 20–25°C. Yield of *cis*-[Pt{(CH$_3$)$_2$ SO}$_2${(CH$_3$)SO}$_2$][SO$_3$CF$_3$]$_2$: 0.096 g, 40%.

Anal. Calcd. for [Pt{(CH$_3$)$_2$SO}$_4$][SO$_3$CF$_3$]$_2$: C, 14.9; H, 3.0. Found: C, 15.1; H, 2.9. FAB$^+$-MS (NBA matrix), m/z: 672 ([Pt{(CH$_3$)$_2$SO}$_4$][SO$_3$CF$_3$]$^+$ + O), 657 ([Pt{(CH$_3$)$_2$SO}$_4$][SO$_3$CF$_3$]$^+$), 578 ([Pt{(CH$_3$)$_2$SO}$_3$][SO$_3$CF$_3$]$^+$), and 500 ([Pt{(CH$_3$)$_2$SO}$_2$][SO$_3$CF$_3$]$^+$).

Properties

The compound forms colorless crystals that are stable in contact with supernatant DMSO solution but rapidly decompose when filtered and exposed to air.[21] The complex crystallizes from DMSO solution in the triclinic space group $P\bar{1}$ with $a = 8.630(2)$, $b = 9.557(3)$, $c = 16.659(3)$ Å; $\alpha = 73.33(2)°$, $\beta = 77.38(2)°$, $\gamma = 79.19(3)°$; $V = 1272.8$ Å^3; $\rho_{calcd.} = 2.10$ g/cm^3; $Z = 4$. Pt coordinates two O and two S atoms from different monodentate Me$_2$SO ligands in a cis arrangement.[21] ^{1}H NMR (300 MHz) in CD$_3$NO$_2$, δ, ppm: 3.63 (broad ^{195}Pt satellites overlapping with the main signal; J_{PtH} 21–27 Hz) and 3.02(s, br) [*lit.* 3.60 (J_{PtH} 28 Hz), 3.00, and 2.52 (free DMSO) in CD$_3$NO$_2$;[33] 3.59 (J_{PtH} 29.2 Hz) and 3.12(br) in CD$_3$NO$_2$ for [Pt{(CH$_3$)$_2$SO}$_2${(CH$_3$)$_2$SO}$_2$][BF$_4$]$_2$[27]]. ^{13}C{^{1}H} NMR in CD$_3$NO$_2$, δ, ppm: 39.77 (J_{PtC} 40.4) and 32.48(s, br) [*lit.* 43.15 (J_{PtC} 48.8) and 35.85(s, br) for [Pt{(CH$_3$)$_2$SO}$_2${(CH$_3$)$_2$SO}$_2$][BF$_4$]$_2$[27]]. ^{195}Pt NMR in Me$_2$SO-d_6, δ, ppm: −3231 for [Pt{(CH$_3$)$_2$SO}$_2${(CH$_3$)$_2$SO}$_2$][ClO$_4$]$_2$].[28] An ^{1}H NMR kinetic study of DMSO exchange at variable temperature and pressure has been published, indicating widely different exchange rates of the O- and S-bonded DMSO ligands.[33] IR spectrum in KBr, selected bands, cm^{-1}: 1160(vs) and 1140(vs) $\nu_{S=O}$, 1030(vs) and 1010(vs) $\nu_{S=O}$.

References

1. J. A. Davies, *Adv. Inorg. Chem. Radiochem.* **24**, 115 (1981).
2. H. B. Kagan and B. Ronan, *Rev. Heteroatom Chem.* **7**, 92 (1992).
3. Yu. N. Kukushkin, *Koord. Khim.* **23**, 163 (1997), and references cited therein.
4. V. Yu. Kukushkin, *Coord. Chem. Rev.* **139**, 375 (1995).
5. M. Calligaris and O. Carugo, *Coord. Chem. Rev.* **153**, 83 (1996).
6. Yu. N. Kukushkin, *Platinum Met. Rev.* **35**, 28 (1991).
7. V. Yu. Kukushkin, *Zh. Neorg. Khim.* **33**, 1905 (1988); *Russ. J. Inorg. Chem.* (Engl. transl.) **33**, 1085 (1988).
8. J. A. Davies, C. M. Hockensmith, V. Yu. Kukushkin, and Yu. N. Kukushkin, *Synthetic Coordination Chemistry: Principles and Practice*, World Scientific, Singapore, 1996, p. 370.
9. N. Farrell, D. M. Kiley, W. Schmidt, and M. P. Hacker, *Inorg. Chem.* **29**, 397 (1990).
10. P. Bitha, G. O. Morton, T. S. Dunne, E. F. Delos Santos, Y. Lin, S. R. Boone, R. C. Haltiwanger, and C. G. Pierpont, *Inorg. Chem.* **29**, 645 (1990).
11. V. Yu. Kukushkin, *Koord. Khim.* **15**, 1439 (1989).
12. L. I. Elding and Ö. Gröning, *Inorg. Chem.* **17**, 1872 (1978).
13. Yu. N. Kukushkin, Yu. E. Vyazmenski, and L. I. Zorina, *Zh. Neorg. Khim.* **13**, 3052 (1968).
14. S. S. Sotman, V. S. Fundamensky, V. Yu. Kukushkin, and E. Yu. Pankova, *Zh. Obsch. Khim.* **58**, 2297 (1988); *Russ. J. Gen. Chem.* (Engl. transl.) **58**, 2044 (1988).
15. Yu. N. Kukushkin, Yu. E. Vyazmenski, L. I. Zorina, and Yu. L. Pazukhina, *Zh. Neorg. Khim.* **13**, 1595 (1968).

16. P. Khodadad and R. Ceolin, *J. Term. Anal.* **30**, 1141 (1985).
17. G. Annibale, M. Bonivento, L. Canovese, L. Cattalini, G. Michelon, and M. L. Tobe, *Inorg. Chem.* **24**, 797 (1985).
18. I. V. Pakhomova, Yu. N. Kukushkin, L. V. Konovalov, and V. V. Strukov, *Zh. Neorg. Khim.* **29**, 1000 (1984), and references cited therein.
19. P.-C. Kong and F. D. Rochon, *Inorg. Chim. Acta* **37**, L457 (1979).
20. V. Yu. Kukushkin, V. K. Belsky, V. E. Konovalov, R. R. Shifrina, A. I. Moiseev, and R. A. Vlasova, *Inorg. Chim. Acta* **183**, 57 (1991).
21. L. I. Elding and Å. Oskarsson, *Inorg. Chim. Acta* **130**, 209 (1987).
22. J. H. Price, A. N. Williamson, R. F. Schramm, and B. B. Wayland, *Inorg. Chem.* **11**, 1280 (1972).
23. R. Melanson, J. Hubert, and F. D. Rochon, *Acta Crystallogr., Sect. B* **32**, 1914 (1976).
24. Ö. Gröning, T. Drakenberg, and L. I. Elding, *Inorg. Chem.* **21**, 1820 (1982).
25. R. Romeo and M. Tobe, *Inorg. Chem.* **13**, 1991 (1974).
26. R. Melanson and F. D. Rochon, *Can. J. Chem.* **53**, 2371 (1975).
27. J. D. Fotheringham, G. A. Heath, A. J. Lindsay, and T. A. Stephenson, *J. Chem. Res., Synop.* 1986, 82.
28. L. G. Marzilli, Y. Hayden, and M. D. Reily, *Inorg. Chem.* **25**, 974 (1986).
29. V. Yu. Kukushkin, E. Yu. Pankova, T. N. Fomina, and N. P. Kiseleva, *Koord. Khim.* **14**, 1110 (1988); *Sov. J. Coord. Chem.* **14**, 625 (1988).
30. V. K. Belsky, V. E. Konovalov, V. Yu. Kukushkin, and A. I. Moiseev, *Inorg. Chim. Acta* **169**, 101 (1990).
31. F. D. Rochon, P. C. Kong, and R. Melanson, *Inorg. Chem.* **29**, 1352 (1990).
32. V. Yu. Kukushkin and V. E. Konovalov, *Phosph. Sulf. Silicon* **69**, 305 (1992).
33. Y. Ducommun, L. Helm, A. E. Merbach, B. Hellquist, and L. I. Elding, *Inorg. Chem.* **28**, 377 (1989).

29. TETRASULFUR-TETRANITRIDE (S₄N₄)

Submitted by A. MAANINEN,* J. SIIVARI,* R. S. LAITINEN,* and T. CHIVERS[†]
Checked by J. D. LAWRENCE and T. B. RAUCHFUSS[‡]

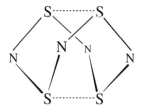

Figure 1. Tetrasulfur-tetranitride.

Tetrasulfur-tetranitride (S_4N_4) (Fig. 1) is widely used as a starting material for the preparation of other cyclic and acyclic SN derivatives.[1] The standard

* Department of Chemistry, P.O. Box 3000, 90014 University of Oulu, Finland.
[†] Department of Chemistry, The University of Calgary, Calgary, Alberta, Canada T2N 1N4.
[‡] Department of Chemistry, University of Illinois, Urbana-Champaign. Illinois, 61801-3792.

synthesis involves the treatment of S$_2$Cl$_2$ with Cl$_2$ gas in CCl$_4$ followed by the reaction of the product with NH$_3$ gas.[2] An aqueous workup procedure is necessary to remove NH$_4$Cl. A procedure has also been described by using CH$_2$Cl$_2$ as the solvent instead of CCl$_4$.[3] S$_4$N$_4$ may also be prepared by the reduction of various NSCl compounds.[4]

For example, the treatment of (NSCl)$_3$ with iron turnings in monoglyme produces S$_4$N$_4$ in 67% yield.[4a] The reaction of [(Me$_3$Si)$_2$N]$_2$S with an equimolar mixture of SCl$_2$ and SO$_2$Cl$_2$ in CH$_2$Cl$_2$ provides an alternative route to excellent yields of S$_4$N$_4$, which avoids the use of gaseous reactants and employs a simple purification procedure:[5]

$$[(Me_3Si)_2N]_2S + SCl_2 + SO_2Cl_2 \rightarrow S_4N_4 + 4Me_3SiCl + SO_2 \qquad (1)$$

Also reported herein is the synthesis of bis[bis(trimethylsilyl)amino]sulfide, [(Me$_3$Si)$_2$N]$_2$S. In an adaptation of the method of Wolmershäuser et al.,[6] [(Me$_3$Si)$_2$N]$_2$S is prepared by treatment of (Me$_3$Si)$_2$NH with n-BuLi followed by the reaction with SCl$_2$. Other sulfur–nitrogen species that can be prepared using [(Me$_3$Si)$_2$N]$_2$S include S$_4$N$_2$ and SeS$_3$N$_2$,[7] S$_3$N$_2$Cl$_2$,[5] and (SSe$_2$N$_2$Cl)$_2$.[6]

A. BIS[BIS(TRIMETHYLSILYL)AMINO]SULFIDE {[(Me$_3$Si)$_2$N]$_2$S}

Procedure

Diethyl ether is dried over Na/benzophenone under a nitrogen atmosphere prior to use. All reactions and manipulations are carried out under an argon atmosphere dried with P$_4$O$_{10}$. (Me$_3$Si)$_2$NH (Aldrich), and n-butyllithium (2.5 M in hexane, Aldrich) are used without further purification.

Bis(trimethylsilyl)imide (8.1 g, 0.05 mol) is dissolved in 100 mL of diethyl ether in a 500-mL round-bottomed flask. The solution is cooled in an ice bath and 20 mL (0.05 mol) of n-BuLi is added. The resulting mixture is stirred for one hour, during which time a white precipitate of (Me$_3$Si)$_2$NLi forms. A solution of sulfur dichloride (2.57 g, 0.025 mol) in 50 mL of diethyl ether is added to the resulting suspension at -78°C. After the addition is complete, the cooling bath is removed, and the stirred reaction mixture is allowed to warm to room temperature over the course of 12 h. The orange solution is filtered by use of a filter syringe to remove the lithium chloride precipitate, and the filtrate is evaporated. [(Me$_3$Si)$_2$N]$_2$S is purified by distillation (92°C at 0.007 Torr). The yield of [(Me$_3$Si)$_2$N]$_2$S is 5.7 g (65%).

Properties

$[(Me_3Si)_2N]_2S$ (m.p. 64°C) (cf. 66°C[8]) decomposes slowly in air and must therefore be stored in dry oxygen-free atmosphere. The IR spectrum of $[(Me_3Si)_2N]_2S$ shows bands at 2960(m), 2900(w), 1266(s), 1258(w), 1250(s), 905(vs), 846(vs), 792(s), 760(m), 728(m), 674(m), 645(w), 617(w), 488(w), and 460(vw) cm^{-1},[8] and the ^1H chemical shift is 0.252 ppm (CH_2Cl_2).[6] The compound is soluble in nonpolar organic solvents, but reacts with protic solvents.

B. TETRASULFUR TETRANITRIDE (S_4N_4)

Procedure

■   **Caution.**   *Samples of S_4N_4 are explosive under the influence of heat or mechanical stress. The dry recrystallized product should not be removed from the flask with a metal spatula. Use of Teflon spatula or glass rod is recommended. It is highly recommended that S_4N_4 not be prepared in amounts greater than 0.5 g at any given time.*

Dichloromethane is dried by distillation over P_4O_{10} under a nitrogen atmosphere prior to use. All reactions and manipulations are carried out under an argon atmosphere passed through P_4O_{10}. SCl_2 (Fluka Chemie AG) is distilled over PCl_3 and SO_2Cl_2 (Aldrich) over P_4O_{10} in an argon atmosphere.

Bis[bis(trimethylsilyl)amino]sulfide (17.6 g, 0.05 mol) is dissolved in 500 mL of dichloromethane in a 1000-mL round-bottomed flask, and a mixture of sulfur dichloride (5.15 g, 0.05 mol) and sulfuryl chloride (6.75 g, 0.05 mol) in 250 mL of dichloromethane is added dropwise at −78°C. The stirred reaction mixture is allowed to warm to room temperature over the course of 12 h. The brown solution is filtered by use of filter syringe in argon, and the filtrate is evaporated to give an orange-brown powder. The crude product is dissolved in 50 mL of toluene at 60°C and recrystallized at −20°C to form pale orange crystals. The total yield of S_4N_4 is 75% (3.45 g, 0.0188 mol).

Characterization of S_4N_4 is carried out using ^{14}N NMR, IR, and Raman spectroscopy. The ^{14}N NMR spectrum shows only one resonance at −256 ppm.[9] The characteristic IR absorptions of S_4N_4 occur at 928(vs), 768(w), 727(m), 700(vs), 630(w), 553(vs), 548(vs), and 529(w,m) cm^{-1} and Raman vibrations at 763(< 1), 719(4), 559(5), 519(1), 347(2), 215(9), and 197(10). Both IR and Raman spectra are in good agreement with earlier reported results.[10,11]

An analogous synthesis of Se_4N_4 from $[(Me_3Si)_2N]_2Se$ and $SeCl_4$ has been reported.[12] Se_4N_4 is extremely explossive when heated or subjected to mechanical stress. If it must be made, the preparation should be limited to amounts

< 0.5 g and the product should be handled with greater care than suggested for S_4N_4. A polycarbonate blast shield should be employed, and the experimenter should wear heavy-duty gloves and ear protection. Se_4N_4 must be stored under a hydrocarbon solvent.

References

1. (a) T. Chivers, in *The Chemistry of Inorganic Homo- and Heterocycles*, Academic Press, London, 1987, Vol. 2, pp. 793–869; (b) R. T. Oakley, *Prog. Inorg. Chem.* **36**, 299 (1988).
2. M. Villena-Blanco and W. L. Jolly, *Inorg. Synth.* **9**, 98 (1967).
3. P. A. G. O'Hare, E. G. Awere, S. Parsons, and J. Passmore, *J. Chem. Thermodyn.* **21**, 153 (1989).
4. (a) A. J. Banister, A. J. Fielder, R. G. Hey, and N. R. M Smith, *J. Chem. Soc., Dalton Trans.*, 1457 (1980); (b) M. P. Berthet, H. Vincent, and Y. Monteil, *Z. Naturforsch.* **35b**, 329 (1980).
5. T. Chivers, R. S. Laitinen, A. Maaninen, S. Pyykkönen, and J. Siivari, *Univ. Oulu Rep. Ser. Chem.* **42**, P18 (1993).
6. G. Wolmershäuser, C. R. Brulet, and G. B. Street, *Inorg. Chem.* **17**, 3586 (1978).
7. A. Maaninen, J. Siivari, R. J. Svontamo, J. Konu, R. S. Laitinen, and T. Chivers, *Inorg. Chem.* **36**, 2170 (1997).
8. G. Schubert and G. Gattow, *Z. Anorg. Allg. Chem.* **574**, 165 (1989).
9. T. Chivers, R. T. Oakley, O. J. Scherer, and G. Wolmershäuser, *Inorg. Chem.* **20**, 914 (1981).
10. I. S. Butler and T. Sawai, *Can. J. Chem.* **55**, 3838 (1977).
11. P. K. Gowik and T. M. Klapötke, *Spectrochim. Acta* **46A**, 1371 (1990).
12. J. Siivari, T. Chivers, and R. S. Laitinen, *Inorg. Chem.* **32**, 1519 (1993).

30. HYDROTRIS(METHIMAZOLYL)BORATE

Submitted by LUIS FERNANDO SOARES* and ROSALICE MENDONÇA SILVA*
Checked by JASON SMEE† and MARCETTA DARENSBOURG†

The cyclopentadienyl anion, $C_5H_5^-$ (Cp), is a very useful supporting ligand and has been used as such with almost all transition metals. However, in some cases, synthesis of the CpM starting material is very difficult, especially when M = early transition metal, mainly because of the instability of the intermediates that originate the metallocenes. The hydrotris(pyrazolyl)borate ligand, $HB(pz)_3^-$, Tp, belongs to a novel class of tridentate ligands, isolobal with Cp, that is a similar supporting ligand. For many applications it is easier to handle, more stable and cheaper, than is Cp.[1] The pyrazolyl nitrogen atoms are hard donors, and

* Departamento de Química, Universidade Federal de Minas Gerais, 31270-901 Belo Horizonte, MG, Brasil.
† Department of Chemistry, Texas A&M University, College Station, TX 77843-3255.

analogs of Tp, which have softer donor atoms, like sulfur, are of interest. One derives from 2-sulfanyl-1-methylimidazol, which as *methimazole*, which exists as a tautomer:[2]

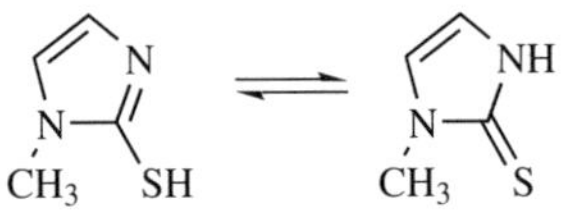

Reglinski et al.[2] reacted methiamazole with $NaBH_4$, following the same procedure that was reported by Trofimenko to prepare the Tp ligand,[1] obtaining the Tm ligand:

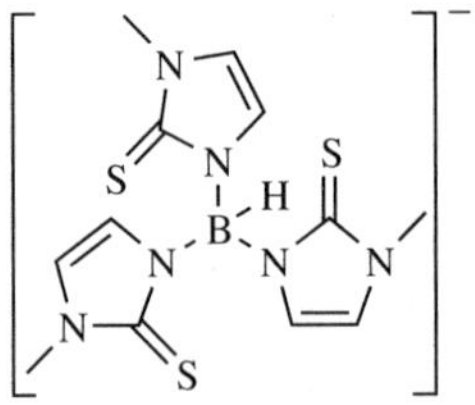

Tm

The synthesis described below for the Tm ligand is based on Trofimenko's literature reports of trispyrazolyl borate[1] and the trismethimazole borate synthesis of Lobbia et al.,[3] with modifications in temperature and solvent washes. Hydrogen evolution is monitored at a lower temperature, starting around 100°C. In this way we obtain very pure Tm in better yields.

A. HYDROTRIS(METHIMAZOLYL)BORATE

$$KBH_{4(s)} + 4C_4H_6N_2S_{(s)} \rightarrow KTm_{(s)} + 3H_{2(g)}$$

■ **Caution.** *Since large amounts of hydrogen are evolved, this reaction should be run in an efficient hood, and open flames or sparks are to be avoided.*

A mixture of 2.70 g (23.64 mmol) of methimazole* and 0.32 g of KBH_4 (5.53 mmol) is placed together with a magnetic stirring bar, into a 100-mL

---

*Aldrich Chemical Company, Milwaukee, WI 53233.

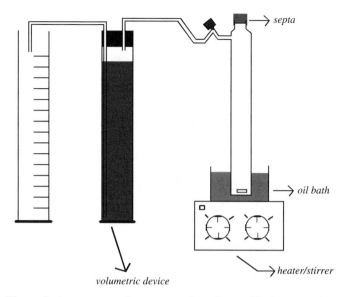

Figure 1. Apparatus used to measure the volume of hydrogen evolved.

Schlenk flask that is placed into an oil bath, with a thermometer, resting on a heating and stirring plate, under air. The Schlenk flask is connected to a volumetric device that measures the volume of hydrogen evolved by the displacement of water (Fig. 1). While stirring, the mixture is warmed gently, and the evolution of hydrogen starts at $\sim 100^\circ$C. The temperature of the oil bath is allowed to rise up to 140°C, and is kept there, with continuous stirring, until the desired amount of hydrogen has been evolved, 0.4 L (17.85 mmol; 400 mL of water displaced). The melt is allowed to cool down to room temperature, and 40 mL of THF (not dried) is added to it.* After stirring for 3 min a white solid settles on the bottom of the flask, leaving a light blue solution. This mixture is filtered through a glass-fritted funnel, in the open, and the white solid thus obtained is washed with three more 40 mL portions of THF. After drying under vacuum 1.73 g of KTm (4.43 mmol) is obtained. Yield: 75%.

Checkers' comment: Solidification of the product toward the end of the reaction, causing the stir bar to get stuck, might require use of overhead stirrer. The solid product should be thoroughly pulverized before washing with THF so as to completely wash out the starting material.

Anal. Calc. for $C_{12}H_{16}N_6S_3B$: K, 390.4; C, 36.96; H, 4.14; N, 21.53. Found: C, 37.01; H, 4.01; N, 21.45.

Properties

KTm is a white solid that decomposes at 220°C. It is very soluble in H_2O and insoluble in acetone and THF. IR (cm^{-1}, CsI, Nujol): ν_{B-H} = 2440; ^{1}H NMR (δ, D_2O): 6.89 [d, NC*H*=CHN(CH₃)], 6.14 [d, NCH=C*H*N(CH₃)], 3.46 (s, CH₃). ^{13}C NMR {^{1}H} (δ, D_2O): 158.23 (C_{quat}); 120.65 (CH), 120.28 (CH), 34.74 (CH₃).

Reglinski et al. showed that the Tm ligand produces a Zn complex utilizing all three thione sulfur atoms, TmZnBr, whose molecular structure is analogous to that of the pyrazolyl borate analog.[2] Using a mixed pyrazolyl/thioimidazolyl ligand, Parkin et al. developed a zinc biomimetic complex where two thioimidazolyl are *S*-complexed to the metal.[4] Lobbia et al. also reported a bidentade bonding mode for the $Ag(PPh_3)^+$ derivative as well as the possibility of Tm serving as a binucleating ligand.[3]

References

1. (a) S. Trofimenko, *J Am. Chem. Soc.* **89**, 3170 (1967); S. Trofimenko, *Inorg. Synth.* **99** (1970); (b) N. Armaroli, G. Accorsi, F. Barigelletti, S. M. Couchman, J. S. Fleming, N. C. Jeffery, J. C. Mann, L. V. Karen, J. A. McCleverty, L. H Rees, S. R. Starling, and M. D. Ward, *Inorg. Chem.* **38**, 5769, (1999); (c) A. A. Eagle, G. N. George, E. R. Tiekink, and C. G. Young, *J. Inorg. Biochem.* **76**, 39 (1999); (d) S. Tsuji, D. C. Swenson, and R. F. Jordan, *Organometallis* **18**, 4758 (1999); (e) G. Jia and C.-P. Lau, *Coord. Chem., Rev.* **19**, 83–102 (1999); (f) C. Santini, C. Pettinari, M. Pellei, G. Lobbia, A. Pifferi, M. Camalli, and A. Mele, *Polyhedron* **18**, 2255 (1999); (g) A. J. Canty, H. Jin, B. W. Skelton, and A. H. White, *Aust. J. Chem.* **52**, 417 (1999); (h) C. Janiak, L. Braun, and F. Girgsdies, *J. Chem. Soc., Dalton Trans.* **17**, 3133 (1999); (i) K. Reddy, A. Domingos, A. Paulo, and I. Santos, *Inorg. Chem.* **38**, 4278 (1999); (j) I. Lopes, G. Y. Lin, A. Domingos, R. McDonald, N. Marques, and J. Takats, *J. Am. Chem. Soc.* **121**, 8110 (1999); (k) C.-L. Lee, Y.-Y. Wu, C.-P. Wu, J.-D. Chen, T.-C. Keng, and J.-C. Wang, *Inorg. Chim. Acta* **292**, 182 (1999); (l) G. M. Ferrence, R. McDonald, and J. Takats, *Angew. Chem., Int. Ed. Engl.* **38**, 2233 (1999); (m) J. Reglinski, M. Garner, I. D. Cassidy, P. A. Slavin, M. D. Spicer, and D. R. Armstrong, *J. Chem. Soc., Dalton Trans.* **13**, 2119 (1999); (n) M. P. Campello, A. Domingos, A. Galva, A. de Matos, and A. I. Santos. *J. Organomet. Chem.* **579**, 5 (1999); (o) K. D. Demadis, E.-S. El-Samanody, T. J. Meyer, and P. S. White, *Polyhedron* **18**, 1587 (1999); (p) C. Slugovc, K. Mereiter, R. Schmid, and K. Kirchner, *Eur. J. Inorg. Chem.* **7**, 1141 (1999); (q) T. B. Gunnoe, M. Sabat, Harman, and W. Dean, *J. Am. Chem. Soc.* **121**, 6499 (1999); (r) D. D. DuMez, T. O. Northcutt, Y. Matano, and J. M. Mayer, *Inorg. Chem.* 38, 3309 (1999); (s) M. Leonor, M. P. Campello, A. Domingos, I. Santos, and R. Andersen, *J. Chem. Soc., Dalton Trans.* **12**, 2015 (1999).
2. M. Garner, J. Reglinski, I. Cassidy, M. D. Spicer, and A. R. Kennedy, *Chem. Commun.* 1975 (1996).
3. C. Santini, G. G. Lobbia, C. Pettinari, and M. Pellei, *Inorg. Chem.* **37**, 890 (1998).
4. C. Kimblim, T. Hascall, and G. Parkin, *Inorg. Chem.* **36**, 5680 (1998).

31. ARSENIC(III)BROMIDE

Submitted by FRANCISCO J. ARNAIZ[*] and MARIANO J. MIRANDA[*]
Checked by ARNOLD L. RHEINGOLD[†]

Arsenic(III) halides are important starting materials for the preparation of a variety of arsenic species in nonaqueous media.[1] Most synthetic procedures based on these halides make use of $AsCl_3$, which can be conveniently prepared by reacting arsenic(III) oxide with thionyl chloride.[2] However, it should be noted that many arsenic compounds derived substitutionally from $AsCl_3$ can similarly be achieved from $AsBr_3$, and that this is superior to the chloride in many instances, since at ambient temperature it is a solid that can be handled in air for a short time without noticeable decomposition. Traditionally, $AsBr_3$ is prepared by reacting arsenic with bromine.[3–5] More recently it has been prepared by halogen exchange between boron tribromide and arsenic trichloride.[6] The procedure described here makes use of the reaction of arsenic oxide with concentrated hydrobromic acid and is based on the displacement to the right of the equilibrium below in excess HBr, the limited solubility of $AsBr_3$ in aqueous HBr, and its high solubility in petroleum.

This reaction allows for the preparation of 50–60 g of pure $AsBr_3$ in less than 2 h and can be scaled without difficulty.

A. ARSENIC-TRIBROMIDE ($AsBr_3$)

$$As_4O_6 + 12HBr \rightleftarrows 4AsBr_3 + 6H_2O$$

Procedure

■ **Caution.** *Arsenic compounds are poisonous and hydrobromic acid is corrosive. All manipulations should be conducted in an efficient fume hood, and gloves should be worn.*

In a 250-mL Erlenmeyer flask are placed 20 g (0.051 mol) of As_4O_6 powder,[‡] 160 mL of 48% HBr, and a magnetic stirring bar. The mixture is heated with stirring at 120°C for 10–20 min until all the arsenic oxide is dissolved. The mixture is cooled in an ice bath and a significant amount of white solid separates. The aqueous solution is decanted, and the solid is treated with 100 mL of hexane at room temperature (in case two phases are produced, the minor aqueous phase

[*] Laboratorio de Chimica Inorganica, Universidad de Burgos, 09001 Burgos, Spain.
[†] Department of Chemistry, University of Delaware, Newark, DE 19716.
[‡] As_2O_3 obtained from Aldrich Chemicals.

is decanted and added to the previously obtained aqueous solution). The resulting aqueous solution is extracted with three 100-mL portions of hexane that are combined with the first, and the resulting solution is vigorously stirred with 30 g of anhydrous $MgSO_4$ for 15 min. After filtration the dry hexane solution is immersed in an ice-water bath and evaporated to dryness in vacuo. Then 51 g (80%) of pure arsenic(III)bromide, in the form of colorless neddles are obtained. The yield will be reduced if the evacuation is prolonged after removal of the hexane, since the vapor pressure of $AsBr_3$ is significant.

Anal. Calcd. for $AsBr_3$: As, 23.8; Br, 76.2. Found: As, 23.6; Br, 76.1.

Properties

$AsBr_3$ is a colorless crystalline solid that melts at 32°C and hydrolyzes slowly in air developing a white color. It is very soluble in common aprotic solvents, including hydrocarbons and ethers. The IR spectrum, KBr dispersion, has characteristic ν_{AsBr} bands at 272 and 287 cm^{-1}.

[*Note*: It is worth conserving the hexane distillate for further preparations because it contains a substantial amount of arsenic bromide (~ 10–15% of the $AsBr_3$ initially present in the parent solution). In this way, the cost of the disposal of toxic wastes is reduced and the yield in subsequent preparations can be noticeably increased.]

References

1. G. O. Doak and L. D. Freedman, in *Organometallic Compounds of Arsenic, Antimony and Bismuth*, Wiley-Interscience, New York, 1970.
2. S. K. Pandey, A. Steiner, and H. B. Roesky, *Inorg. Synth.* **31**, 148 (1997).
3. W. Wallace, *J. Prakt. Chem.* **78**, 119 (1859).
4. J. Nickles, *J. Pharm. Chim.* **41**, 142 (1862).
5. P. Baxter, *J. Am. Chem. Soc.* **55**, 1054 (1933).
6. P. M. Druce and M. F. Lappert, *J. Chem. Soc.*, 3595 (1971).

32. DICHLORODIOXO-BIS(TRIPHENYLPHOSPHINE OXIDE) URANIUM(VI) [UO₂Cl₂(OPPh₃)₂]

Submitted by FRANCISCO J. ARNAIZ*
Checked by CAROL J. BURNS†

$UO_2Cl_2(OPPh_3)_2$ is a valuable starting material for preparing other uranyl species in nonaqueous media.[1] The compound was first obtained by reacting

* Laboratorio de Química Inorgánica, Universidad de Burgos, 09001 Burgos, Spain.
† Los Alamos National Laboratory, Mail Stop D453, Los Alamos, NM 87545.

$UCl_4(OPPh_3)_2$ with H_2O_2,[2] and then by treating UO_2Cl_2 with $OPPh_3$ in anhydrous ethanol.[3] Both procedures require significant effort since the starting materials are prepared from the hygroscopic, not readily available, UCl_4.

More recently the preparation, starting from $UO_2(CH_3CO_2)_2.2H_2O$, has been described,[4] but care is required to avoid contamination with acetate complexes.

The procedure described here allows the preparation in excellent yield of pure $UO_2Cl_2(OPPh_3)_2$ by reacting $UO_2Cl_2 \cdot xH_2O$ with $OPPh_3$ in acetone. The procedure can be scaled up and down without significant difficulty. The synthesis of $UO_2Cl_2(OPPh_3)_2$ is based on the tendency of $OPPh_3$ to displace numerous ligands, including water, from many metallic centers. The high solubility of the reactants and the low solubility of the product in acetone further facilitate the process.

The equation for this procedure is as follows:

$$UO_2Cl_2 \cdot xH_2O + 2OPPh_3 \xrightarrow{CH_3COCH_3} [UO_2Cl_2(OPPh_3)_2] + xH_2O$$

Procedure

■ **Caution.** *Uranium compounds are toxic. Gloves should be worn during the synthesis and while handling the reactants and products.*

$UO_2Cl_2 \cdot H_2O^*$ [25.0 g (0.069 mol)] is dissolved in a mixture of 60 g of acetone and 3 g of concentrated hydrochloric acid, giving a clear yellow solution. This solution is then added dropwise to a warm (50–60°C) solution of 40 g (0.144 mol) of triphenylphosphine oxide dissolved in 150 g of acetone. During the addition a yellow precipitate forms. The resulting mixture is allowed to cool to room temperature while stirring. The solid product is collected by vacuum filtration, washed with three 50-mL portions of acetone, and dried under vacuum. Yield: 60 g (95%).

Anal. Calcd. for $UO_2Cl_2(OPPh_3)_2$: U, 26.52; Cl, 7.90; C, 48.18; H, 3.37. Found: U, 26.3; Cl, 7.8; C, 47.9; H, 3.4.

Properties

Yellow $UO_2Cl_2(OPPh_3)_2$ melts at 298–299°C. The IR spectrum, taken as a Nujol mull, has the characteristic band $\nu_{as}(UO_2)$ at 919 cm^{-1}. The ^{31}P NMR spectrum

---

*$UO_2Cl_2 \cdot H_2O$ is obtained from Rhône-Poulenc. If unavailable, a product suitable for this synthesis is prepared as follows. In a bath maintained at 110–120°C, heat to dryness a solution of $UO_2(NO_3)_2 \cdot 6H_2O$ (from Aldrich) in concentrated hydrochloric acid [HOOD]. Repeat this operation two additional times. This is necessary to insure complete destruction of the nitrate.

in $DCCl_3$ exhibits a peak at δ 50.2 ppm. It is insoluble in water, diethylether, and hydrocarbons; slightly soluble in methanol, dichlorometane, and chloroform; moderately soluble in acetonitrile; and very soluble in dimethylformamide and dimethylsulfoxide (DMSO). The product cannot be recovered readily from DMSO. It is very stable at room temperature and can be manipulated in air without special precautions [after storage for 2 years in the dark over KOH (potassium hydroxide), it remains unchanged, and after boiling with 6 M HCl for 5 min the compound is recovered without noticeable alteration].

References

1. C. J. Burns, D. C. Smith, A. P. Sattelberger, and H. B. Gray, *Inorg. Chem.* **31**, 3724 (1992).
2. P. Gans and B. C. Smith, *J. Chem. Soc.*, 4172 (1964).
3. J. P. Day and L. M. Venanzi, *J. Chem. Soc. A*, 1363 (1966).
4. F. J. Arnáiz and M. J. Miranda, *J. Chem. Educ.* **75**, 1457 (1998).

33. CHLOROHYDRO-TRIS(PYRAZOL-1-YL)BORATO-BIS (TRIPHENYLPHOSPHINE)RUTHENIUM(II) {RuCl[κ^3-HB(pz)$_3$](PPh$_3$)$_2$} (pz = pyrazol-1-yl)*

Submitted by ANTHONY F. HILL* and JAMES D. E. T. WILTON-ELY
Checked by THOMAS B. RAUCHFUSS and DANIEL E. SCHWARTZ[†]

Since its preparation,[1] reactions of {RuCl[k^3-HB(pz)$_3$](PPh$_3$)$_2$} have been investigated in a number of papers.[2] The lability of both the phosphines and the chloride ligand allow functional group transformations to be carried out in order to provide a wide range of ruthenium hydro-tris(pyrazol-1-yl)borate complexes. A straightforward preparation of the complex on a large scale (10 g) is described here.

Materials and General Procedures

All manipulations are carried out under nitrogen using solvents as obtained commercially but that are degassed before use. Dichlorotris(triphenylphosphine) ruthenium(II) is prepared by the literature method,[3] as is the hydrotris(pyrazol-1-yl) borate potassium salt.[4] The title compound is mildly air-sensitive in

* Department of Chemistry, Imperial College of Science, Technology and Medicine, South Kensington, London SW7 2AY England; email: a.hill@ic.ac.uk.
† Department of Chemistry, University of Illinois, Urbana—Champaign, IL 61801-3792.

solution but can be worked up under aerobic conditions without greatly compromising the yield. The exclusion of light and an atmosphere of nitrogen is recommended for long-term storage, although not essential.

The equation for this procedure is as follows:

$$[\text{RuCl}_2(\text{PPh}_3)_3] + \text{K}[\text{HB}(\text{pz})_3] \rightarrow [\text{RuCl}\{\kappa^3\text{-HB}(\text{pz})_3\}(\text{PPh}_3)_2] + \text{KCl} + \text{PPh}_3$$

Procedure

■ **Caution.** *Dichloromethane is harmful if inhaled or absorbed through the skin. It should be used in a well-ventilated fume hood; protective gloves and goggles should be worn. Hydrotris(pyrazol-1-yl) borate potassium salt is an irritant and should not be inhaled. Gloves are recommended during weighing and subsequent use of this ligand.*

Amounts of $[\text{RuCl}_2(\text{PPh}_3)_3]$ (10.0 g, 10.4 mmol) and $\text{K}[\text{HB}(\text{pz})_3]$ (2.89 g, 11.5 mmol) are dried under vacuum in a Schlenk tube and placed under an atmosphere of nitrogen. Degassed diclorometane (150 mL) is added and the solution stirred for 1 h. The resulting cloudy yellow solution is filtered through diatomaceous earth to remove KCl and placed in a single-necked round-bottomed flash. Reduction of solvent volume under reduced pressure (rotary evaporator) to ~ 80 mL is followed by addition of ethanol (100 mL). Further reduction of solvent volume results in precipitation of the yellow title complex. When the solvent volume has been reduced to ~ 60 mL, the product is filtered and washed with ethanol (2×50 mL), and hexane (50 mL) and dried under vacuum. Yield: 7.56 g (82%). A further crop of product can be isolated from the filtrate after being cooled overnight at $-20°\text{C}$. The product can be recrystallized aerobically from dichloromethane and ethanol to give yellow crystals. The yields of the synthesis are essentially invariant for scales between 0.5 and 10 mmol.

Properties

The complex $[\text{RuCl}\{\kappa^3\text{-HB}(\text{pz})_3\}(\text{PPh}_3)_2]$ is stable under air as a solid over a period of days but shows signs of surface decomposition over longer periods. The compound is indefinitely stable if stored under nitrogen. In solution this process is much more rapid, resulting in the development of a green colouration after only a few hours. The complex shows characteristic infrared activity (Nujol) due to the pyrazolylborate at 1307, 1215, 1120, and 1043 cm^{-1} as well as a ν_{BH}-associated absorption at 2467 cm^{-1} (2476 cm^{-1} in CH_2Cl_2). The ^1H NMR spectrum CDCl$_3$ (25°C) displays resonances due to the pyrazole protons at 5.15(t), 5.65(t), 6.86(d), and ~ 7.0(d) ppm; the remaining pyrazolyl protons are obscured by the phosphine resonances. The ^{31}P-$\{^1$H$\}$ NMR spectrum is a singlet at 42.9 ppm, as is the resonance at 3.67 in the ^{11}B NMR spectrum. A mass ion is observed in the FAB mass spectrum (NBA matrix) at m/z 874 as well as fragmentations due to loss of the chloride (839), a phosphine (612), and both these ligands (576). Microanalytical data match those obtained previously.[1]

References

1. N. W. Alcock, I. D. Burns, K. S. Claire, and A. F. Hill, *Inorg. Chem.* **31**, 2906 (1992).
2. (a) I. D. Burns, A. F. Hill, and D. J. Williams, *Inorg. Chem.* **33**, 2685 (1996); (b) W.-C. Chan, Y.-Z. Chen, Y.-Q. Fang, G. Jia, C.-P. Lau, and S.-M. Ng, *Organometallics* **16**, 34, (1997); (b) C. Slugoovc, K. Mereiter, E. Zobertz, R. Schmid, and K. Kirchner, *Organometallics* **15**, 5275 (1996).
3. P. S. Hallman, T. A. Stephenson, and G. Wilkinson, *Inorg. Synth.* **12**, 237 (1970).
4. S. Trofimenko, *Inorg. Synth.* **3**, 102 (1969).

34. OCTACARBONYL-DI-μ-HYDRIDODIRHENIUM [Re$_2$(μ-H)$_2$(CO)$_8$]

Submitted by MATTHEW C. COMSTOCK* and JOHN R. SHAPLEY*
Checked by RICHARD D. ADAMS† and JOSEPH L. PERRIN†

The dimetallic compound Re$_2$(μ-H)$_2$(CO)$_8$ is isolobal to an alkene,[1] undergoes anion-induced oligomerization,[2] and is a useful synthon in heterometallic cluster synthesis.[3] While the original synthesis from Re$_2$(CO)$_8$(H$_2$SiPh$_2$) provides Re$_2$(μ-H)$_2$(CO)$_8$ in 78% yield,[4] the most convenient method of preparation involves photolysis of Re$_2$(CO)$_{10}$ with dihydrogen, leading in hexane to a mixture of Re$_2$(μ-H)$_2$(CO)$_8$, ReH(CO)$_5$, and Re$_3$H(CO)$_{14}$, in yields of 35%, 15–25%, and 20%, respectively.[5] In the saturated hydrocarbon solvent the relatively insoluble Re$_3$H(CO)$_{14}$ tends to precipitate during the reaction. We have found that conducting the photolysis in benzene, while still producing a mixture of the same products, provides Re$_2$(μ-H)$_2$(CO)$_8$ in 60–70% yield.

Procedure

The equation for this procedure is as follows:

$$Re_2(CO)_{10} \xrightarrow[C_6H_6]{h\nu/H_2} Re_2(\mu-H)_2(CO)_8 + ReH(CO)_5 + Re_3H(CO)_{14}$$

* Department of Chemistry, University of Illinois, Urbana, IL 61801.
† Department of Chemistry, University of South Carolina, Columbia, SC 29208.

■ **Caution.** *Volatile metal carbonyls are highly toxic and benzene is a human carcinogen. These operations should be conducted with the appropriate precautions in a well-ventilated fume hood.*

A 100-mL quartz Schlenk tube (a clear Pyrex tube can also be used, but the reaction time is longer) is charged with $Re_2(CO)_{10}$ (Strem, Newburyport, MA 01950; 100 mg, 0.153 mmol), benzene (50 mL, distilled from sodium/benzophenone under nitrogen), and a Teflon-coated stirring bar. The colorless solution is stirred and purged for 5 min with H_2, then irradiated with a Hanovia 450-W mercury lamp for 1 h, with continued passage of a slow stream of H_2. During this time the solution turns bright yellow. The reaction can be monitored by observing the decrease of an IR peak at 2071 cm^{-1} for $Re_2(CO)_{10}$. On completion of the reaction, the solvent and $ReH(CO)_5$ are distilled under vacuum to a liquid-nitrogen-cooled trap. The yield of $ReH(CO)_5$ (~2 mg, 0.006 mmol, 2%) is determined from the IR spectrum.[6] The solid residue is dissolved in 30 mL of hexane and filtered through a column (20 × 1 cm) of Florosil (Fisher) in air. Further addition of hexane elutes a broad yellow band that is collected. Solvent removal under vacuum gives yellow, solid $Re_2(\mu\text{-H})_2(CO)_8$ (57 mg, 0.095 mmol, 62%). Continued elution of the column with dichloromethane provides a second yellow band, from which yellow, solid $Re_3H(CO)_{14}$ (13 mg. 0.014 mmol, 14%) is recovered and identified by IR spectroscopy.[6,7]

Anal. Calcd. for $Re_2(\mu\text{-H})_2(CO)_8$: C, 16.05; H, 0.34. Found: C, 16.21; H, 0.41.

The reaction can be conducted on a larger scale; however, the photolysis time must be increased. For example, 200 mg of $Re_2(CO)_{10}$ in 50 mL of benzene requires 2.5 h for completion and provides 5 mg of $ReH(CO)_5$ (3%), 116 mg of $Re_2(\mu\text{-H})_2(CO)_8$ (63%), and 30 mg of $Re_3H(CO)_{14}$ (15%).

Properties

Yellow, solid $Re_2(\mu\text{-H})_2(CO)_8$ can be handled in air, but for extended periods it should be stored under nitrogen. The infrared spectrum in hexane shows peaks at 2094, 2022, 2002, and 1980 cm^{-1} (*lit.*[4] in cyclohexane: 2093, 2020, 2000, and 1979 cm^{-1}). The ^{1}H NMR spectrum in CDCl$_3$ exhibits a singlet at δ −9.04.[4] Bands at 1275 and 1382 cm^{-1} for the stretching modes of the bridging hydride ligands have been identified,[5] and the exchange with D$_2$O to produce $Re_2(\mu\text{-D})_2(CO)_8$ has been studied.[5]

References

1. (a) R. Hoffmann, *Angew. Chem., Int. Ed. Engl.* **21**, 711 (1982); (b) F. G. A. Stone, *Angew. Chem., Int. Ed. Engl.* **23**, 89 (1984).
2. M. Bergamo, T. Beringhelli, G. D'alfonso, P. Mercandelli, M. Moret, and A. Sironi, *J. Am. Chem. Soc.* **120**, 2971 (1998), and references cited therein.

3. (a) T. Beringhelli, A. Ceriotti, G. D'Alfonso, R. D. Pergola, G. Ciani, M. Moret, and A. Sironi, *Organometallics* **9**, 1053 (1990); (b) P. Antognazza, T. Beringhelli, G. D'Alfonso, A. Monoja, G. Ciani, M. Moret, and A. Sironi, *Organometallics* **11**, 1777 (1992); (c) T. Beringhelli, G. Ciani, G. D'Alfonso, L. Garlaschelli, M. Moret, and A. Sironi, *J. Chem. Soc., Dalton Trans.* 1865 (1992);   (d) M. C. Comstock, T. Prussak-Wieckowska, S. R. Wilson, and J. R. Shapley, *Inorg. Chem.* **36**, 4397 (1997).
4. M. J. Bennet, W. A. G. Graham, J. K. Hoyano, and W. L. Hutcheon, *J. Am. Chem. Soc.* **94**, 6232 (1972).
5. M. A. Andrews, S. W. Kirtley, and H. D. Kaesz, *Inorg. Chem.* **16**, 1556 (1997).
6. M. A. Urbancic and J. R. Shapley, *Inorg. Synth.* **26**, 77 (1989).
7. W. Fellmann and H. D. Kaesz, *Inorg. Nucl. Chem. Lett.* **2**, 63 (1966).

35. ZERO-VALENT BINUCLEAR NICKEL COMPLEXES

Submitted by J. K. GONG,* C. A. WRIGHT, D. L. DELAET,† and C. P. KUBIAK‡
Checked by MATTHEW MILLER and MARCETTA Y. DARENSBOURG§

Zero-valent binuclear nickel complexes bridged by diphosphine ligands exhibit interesting ligand- and metal-centered chemistry.[1-3] They have been used as starting complexes to form nickel clusters.[4-6] The synthesis of $Ni_2(CO)_3(dppm)_2$ was first reported in 1988.[7]

We describe here a simple procedure to prepare the complex in high yield.

Materials and General Procedures

$Ni(COD)_2$ (COD cyclooctadiene, C_8H_{12}] is commercially available (Strem) or can be prepared following the literature synthesis.[8] Dppm is availbale commercially (Aldrich, Strem) and was used as received. Carbon monoxide, (available from various sources) was used as received from Aldrich. All solvents are freshly distilled from the appropriate drying agents and deoxygenated prior to use.

All reactions and sample preparations are carried under an inert atmosphere using standard Schlenk techniques

A. DI(CARBONYL)-(μ-CARBONYL)BIS[μ-BIS (DIPHENYLPHOSPHINO)METHANE]DINICKEL(0)
[Ni₂(μ-CO)(CO)₂(μ-dpmm)₂]

$$2Ni(C_8H_{12})_2 + 2dppm^\# + 3CO \rightarrow Ni_2(CO)_3(dppm)_2 + 4C_8H_{12}$$

* Department of Chemistry, Towson University, Towson, MD 21252.
† Ethyl Corporation, Baton Rouge, LA 70802.
‡ Department of Chemistry and Biochemistry, University of California, San Diego, CA 92093.
§ Department of Chemistry Texas A & M University, College Station, TX 77843.
# Dppm = bis(diphenylphosphino)methane, $(C_6H_5)_2PCH_2P(C_6H_5)_2$].

Procedure

■ **Caution.** *During the preparation, there is a possibility of forming a trace amount of Ni(CO)₄, a colorless, highly volatile (b.p. 42.4°C), highly toxic liquid. Manipulations should be carried out in an efficient fume hood. The reaction solution is stored in the freezer to prevent any Ni(CO)₄ from escaping. The volatile byproducts are collected in a liquid-nitrogen-cooled trap between the reaction flask and vacuum line. Any Ni(CO)₄ that may be present is disposed of by treatment of the trap contents with bromine water until the solution turns orange.*

Freshly prepared Ni(COD)$_2$ (2 g, 7.27 mmol) is dissolved in 50 mL of THF, to which is added 1.1 equiv of dppm (3.07 g) in a minimal amount of THF. The mixture is stirred for about 20 min, and the color of the solution changes from yellow to orange-yellow. With the flask under a slight vacuum, 1.5–1.6 equiv of CO gas were added via syringe (volume depending on room temperature and pressure). The mixture is stirred for 30 min and stored at $-20°C$ overnight. A small amount of yellow precipitate forms overnight. The reaction mixture is warmed to 45°C and stirred for 30 min.

The solution is allowed to cool to room temperature Volatiles are removed under vacuum until a large amount of yellow precipitate with a small amount of solvent is left. The yellow precipitate is filtered, washed with ether, and dried under vacuum, to give a yield of 2.82 g(80%).

Anal. Calcd. for C$_{53}$H$_{44}$Ni$_2$O$_3$P$_4$: C, 64.80; H, 4.52; Ni, 11.66; P, 12.54. Found: C, 64.80; H, 4.52; Ni, 11.66; P, 12.53.

Properties

The formula weight is 970.22 g mol^{-1}. It displays the following spectroscopic properties: IR (KBr): ν_{CO} 1970, 1948, 1781 cm^{-1}, ^{31}P{^{1}H} NMR CD$_2$Cl$_2$) δ 22.92(s).

B. DI(ISOCYANOMETHANE)-(μ-ISOCYANOMETHANE)BIS [μ-BIS(DIPHENYLPHOSPHINO)METHANE]DI-NICKEL(0) [Ni$_2$(μ-CNCH$_3$)(CNCH$_3$)$_2$(μ-dppm)$_2$]

$$Ni(\eta^5 - C_5H_5)_2 + 4CNCH_3 \rightarrow Ni(CNCH_3)_4$$
$$2Ni(CNCH_3)_4 + 2dppm \rightarrow Ni_2(CNCH_3)_3(dppm)_2 + 5CNCH_3$$

Materials and General Procedures

Bis(η^5-cyclopentadienyl)nickel is available commercially (Aldrich, Strem) or can be prepared from literature methods.[9] Methylisocyanide was freshly prepared according to the literature.[10] Dppm is available commercially (Aldrich, Strem) and is used as received. All solvents ware freshly distilled from the appropriate drying agents and deoxygenated prior to use. All reactions and sample preparations were carried out under an inert atmosphere using standard Schlenk techniques.

■ **Caution.** *Isocyanides are very toxic materials having an extremely obnoxious odor. Manipulations should be carried out in an efficient fume hood. Methylisoyanide has a very high vapor pressure and will also polymerize at room temperature over a period of time. It should be stored in glass (it reacts with rubber septa) in a freezer.*

To a solution of bis((η^5-cyclopentadienyl)nickel (1 g, 5.29 mmol) in ether is added methylisocyanide (1.6 mL, 26.89 mmol) at a temperature of 0°C. This reaction leads to the formation of $Ni(CNCH_3)_4$ as a light yellow solid. After 1 h, the solid is collected by vacuum filtration, rinsed with hexane, and dried under vacuum, giving a yield of 0.94g (80%).

To a suspension of $Ni(CNCH_3)_4$ (0.94 g, 4.22 mmol in 30 mL of benzene is added 1 equiv of dppm predissolved in a minimal amount of benzene. The reaction solution is stirred for several hours. The solvent is reduced to two-thirds of the original volume under vacuum in order to remove the liberated $CNCH_3$. More benzene is added to replace the volume removed, and stirring is continued. This procedure is repeated at least 3 times before the solution is concentrated to two-thirds the original volume and 40 mL, of hexanes is added to initiate precipitation. The solution is cooled to -20°C overnight. $Ni_2(CNCH_3)_3(dppm)_2$ is isolated the next day by vacuum filtration and washed with hexane and then acetonitrile to remove any impurities. Yield: 12.7–14.8 g, 60–70%.

Anal. Calcd. for $C_{56}H_{53}N_3Ni_2P_4$: C, 66.64; H, 5.29; N, 4.16. Found: C, 66.60; H, 5.23; N, 4.09.

Properties

The formula weight is 1008.3 g mol^{-1}. It displays the following spectroscopic properties: ^{1}H NMR (C_7D_8 -30°C) δ 2.44 (s, 3H), 2.55 (s, 6H), 4.31 (m, 4H), 7.2–8.1 (m, 40H); ^{31}P{^{1}H} NMR (C_7D_8, 25°C) 18.21(s); IR (toluene ν_{CN} 2075, 1717 cm^{-1}.

References

1. J. K. Gong and C. P. Kubiak, *Inorg. Chim. Acta* **162**, 19 (1989).
2. D. L. Delaet, R. del Rosario, P. Fanwick, and C. P. Kubiak, *J. Am. Chem. Soc.* **109**, 754 (1987).
3. J. K. Gong, J. Huang, P. E. Fanwick, and C. P. Kubiak, *Angew. Chem., Int. Ed., Engl.* **29**, 396 (1990).

4. K. S. Ratiff, G. K. Broeker, P. E. Fanwick, and C. P. Kubiak, *Angew. Chem. Int. Ed. Engl.* **102**, 405 (1990).
5. D. A. Morgenstern, R. E. Wittring, P. E. Fanwick, and C. P. Kubiak, *J. Am. Chem. Soc.* **115**, 6470 (1993).
6. K. S. Ratliff, P. E. Fanwick, and C. P. Kubiak, *Polyhedron* **9**, 1487 (1990).
7. J. A. Osborn, F. F. Stanley, and P. H. Bird, *J. Am. Chem. Soc.* **110**, 2117 (1988).
8. R. A. Schunn, S. D. Ittel, and M. A. Cushing, *Inorg. Synth.* **28**, 94 (1990).
9. K. W. Barnett, *J. Chem. Educ.* **51**, 422 (1997).

36. PREPARATION OF A SYNTHETIC RIBONUCLEASE: THE EUROPIUM(III) COMPLEX OF 1-(4-NITROBENZYL)- 4,7,10-TRIS(CARBAMOYLMETHYL)-1,4,7,10- TETRAAZACYCLODODECANE

Submitted by SHAHID AMIN, LARA L. CHAPPELL, and JANET R. MORROW[*]
**Checked by T. ANDREW TSENG, BRETT R. BODSGARD,
and JUDITH N. BURSTYN**[†]

Lanthanide(III) complexes of tetraazamacrocycles are efficient catalysts for RNA cleavage by phosphate ester transesterification, provided they supply a sufficient number of coordination sites for binding and catalysis.[1] For lanthanide(III) complexes of tetraazamacrocycles with amide pendant groups, a minimum of two coordination sites are necessary for the preparation of an active catalyst.[1,2] The europium (III) complex of the heptadentate ligand 1-(4-nitrobenzyl), 4,7,10-tris(carbamoylmethyl-1,4,7,10-tetraazacyclododecane binds two water molecules at pH 6.39 and is an active catalyst for phosphate diester transesterification.[3] The nitro group can be converted into an isothiocyanate group for conjugation to a deoxyribonucleotide or other RNA recognition agent.[4]

Macrocyclic ligands with pendant amide groups have received little attention in comparison to macrocyclic ligands bearing carboxylate groups. Early examples of macrocyclic ligands containing multiple pendant amide groups were prepared by treating a tetraazamacrocycle with acrylamide or alternately with acrylonitrile followed by metal-ion-promoted hydrolysis of the nitrile group.[5-7] These macrocycles had pendant amide groups that coordinated to the metal ion to form six-membered chelate rings. Examples of ligands with pendant amide groups that form five-membered chelate rings include the series of macrocycles with *N,N*-dimethyl- or *N*-methylcarbamoylmethyl pendant groups[8-10] and the tetrakis (carbamoylmethyl)[1,11] derivatives of 1,4,7,10-tetraazacyclododecane (cyclen).

[*] Department of Chemistry, State University of New York at Buffalo, Buffalo, NY 14260.
[†] Department of Chemistry, University of Wisconsin—Madison, Madison, WI 53706-1396.

We present the preparation of a new heptadentate ligand containing pendant amide groups in addition to a pendant 4-nitrobenzyl. The selective monoalkylation of cyclen is accomplished by a method similar to that published previously.[12]

Materials and General Procedures

The free-base form of cyclen(1,4,7,10-tetraazacyclododecane) was generated by dissolving the tetrahydrochloride salt (Parish Chemicals) in Milli-Q water (1.5 g per 8 mL) and adjusting the pH to 12.5 with NaOH pellets. Extraction of the aqueous solution with CHCl$_3$ (6 × 100 mL) and evaporation of the solvent gave the free-base form of cyclen in 93% yield. Amylene stabilized CHCl$_3$ was purchased from Aldrich and used as received for the synthesis of the NBC macrocycle. Merck grade 9385, 230–400-mesh silica gel, 60 Å (Aldrich) was used for preparative chromatography. Thin-layer chromatography (TLC) plates (Aldrich) were of silica gel (60F$_{254}$, 0.2 mm) thickness. Reagent-grade absolute ethanol was used. Ethanol and methylene chloride used in the preparation of [Eu(NBAC)](CF$_3$SO$_3$)$_3$ were dried over 3-Å molecular sieves. Eu(CF$_3$SO$_3$)$_3$ was obtained by treating Eu$_2$O$_3$ with concentrated trifluoromethane sulfonic acid as reported previously.[13] NMR spectra were recorded by use of a Varian Gemini-300 spectrometer (^{13}C NMR) and a Varian 400 XL spectrometer (^{1}H NMR). Chemical shifts are reported as parts per million downfield of SiMe$_4$. NMR sample temperatures were measured by use of a thermocouple located near the probe. A VG 70-SE mass spectrometer with fast-atom bombardment (FAB) was utilized.

A. 1-(4-NITROBENZYL)-4,7,10-TETRAAZACYCLODODECANE (NBC)

Procedure

A 25-mL, two-necked, round-bottom flask is equipped with a Teflon-coated magnetic stirring bar and a gas inlet tube. To a solution of cyclen (0.500 g of free base, 2.90 mmol) in CHCl$_3$ (7 mL) is added 4-nitrobenzyl bromide (0.419 g, 1.94 mmol). The reaction mixture is stirred for 3.5 h, during which time a precipitate forms. The solution volume is reduced under vacuum to approximately 1 mL, and the resulting slurry is applied to a silica gel chromatography column (1×8 in.). Elution with CHCl$_3$/MeOH/conc. NH$_4$OH

12 : 4 : 1 gives 7 (1 mL) fractions that contain solely the monoalkylated product. Following this, 50 mL of the solvent system is passed down the column to ensure that all the monoalkylated product has been collected. All fractions are analyzed by TLC (solvent system identical to that of the column, $R_f = 0.83$ trialkylated, $R_f = 0.73$ dialkylated, and $R_f = 0.67$ monoalkylated. The unreacted cyclen remains adhered to the column. Fractions containing the NBC ligand are combined and the solvent removed under vacuum to give a yellow powder. The product is further dried under vacuum overnight at 50°C. Yield: 63% based on 4-nitrobenzyl bromide. Melting point: 125–127°C (decomposition). FABMS m/e: 308.2 (ligand + H).

Anal. Calcd. for $C_{15}H_{25}N_5O_2$: C, 58.61; H, 8.19; N, 22.78. Found: C, 58.39; H, 8.29; N, 23.06.

Properties

A significantly shorter reaction time was used compared to that reported previously by Kruper and co-workers.[12] ^1H NMR (CD_3CN): 2.21 (broad, 3H, NH), 2.45-2.51 (m, 8H, NCH_2), 2.59 (t, 4H, NCH_2), 2.69 (t, 4H, NCH_2) 3.66 (s, 2H, $ArCH_2$), 7.56 (d, 2H, Ar). 8.16 (d, 2H, Ar). ^{13}C ($CDCl_3$): 44.93, 46.14, 46.97, 51.57 (NCH_2 ring) 58.74 ($ArCH_2$), 123.83, 129.64, 147.44 (Ar).

B. 1-(4-NITROBENZYL)-4,7,10-TRIS(CARBAMOYLMETHYL)-1,4,7,10-TETRAAZACYCLODODECANE (NBAC)

The structures for cyclen, NBC, $[Eu(NBAC)]^{3+}$, and NBAC are shown in Scheme 1.

Procedure

A 25-mL, single-necked, round-bottom flask is equipped with a Teflon-coated stirring bar, a reflux condenser, and a gas inlet tube. NBC (0.156 g, 0.508 mmol), absolute ethanol (10 mL), triethylamine (0.50 mL, 3.6 mmol), and 2-bromoacetamide (0.239 g, 1.73 mmol) are added to the flask, and the reaction mixture is refluxed under a N_2 atmosphere for 3 h. The product precipitates during the course of the reaction and after cooling is filtered and washed

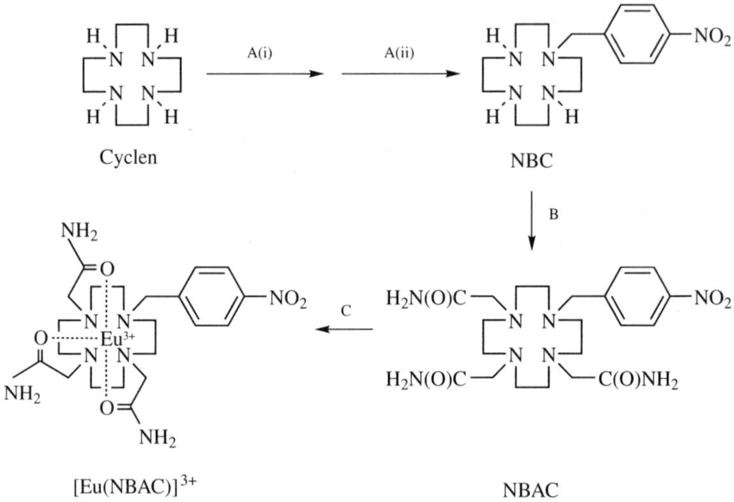

Cyclen

NBC

B

[Eu(NBAC)]$^{3+}$

NBAC

A(i). $BrCH_2C_6H_4NO_2$ / $CHCl_3$
A(ii).   Flash Silica Chromatography, solvent system: $CHCl_3$ / CH_3OH/conc.NH_4OH
B. $BrCH_2C(O)NH_2$ / $N(C_2H_5)_3$ / C_2H_5OH
C. $Eu(CF_3SO_3)_3$ / C_2H_5OH

Scheme 1

with water (5 mL) and finally with ethanol (5 mL). The NBAC ligand is dried in vacuum at 50°C overnight. Yield: 75%. Melting point: 205–207°C (decomposition). FABMS m/e: 479.3 (ligand+H).

Anal. Calcd. for $C_{21}H_{34}N_8O_5$: C, 52.72; H, 7.11; N, 23.43. Found C, 52.50; H, 7.29; N, 23.26.

Properties

The NBAC ligand is sparingly soluble in most organic solvents and water and is most soluble in DMSO. ^1H NMR δ, ppm (DMSO-d_6) (45°C); 2.54 (m, 4H, NCH$_2$ ring), 2.60 (s, 12H, NCH$_2$ ring), 2.83 (s, 4H, CH$_2$C(O)NH$_2$), 2.95 (s, 2H, CH$_2$C(O)NH$_2$), 3.62 (s, 2H, CH$_2$Ar), 6.85 and 7.45 (two m, 6 H, C(O)NH$_2$), 7.59 (d, 2H Ar), and 8.15 (d, 2H, Ar). ^{13}C NMR (DMSO-d_6): δ, ppm 52.60, 53.13, 53.50 (NCH$_2$ ring), 57.61, 58.13, (CH$_2$C(O)NH$_2$), 58.60 (CH$_2$Ar), 123.56, 129.67, 146.73, 148.01 (Ar), 173.08, 173.20 [C(O)].

C. 1-(4-NITROBENZYL)-4,7,10-TRIS(CARBAMOYLMETHYL)-
1,4,7,10-TETRAAZACYCLODODECANE[EUROPIUM(III)]
[[Eu(NBAC)](CF$_3$SO$_3$)$_3$]

Procedure

A 100-mL, three-necked, round-bottom flask is equipped with a Teflon-coated stirring bar, a reflux condenser, and a gas inlet tube. NBAC (0.120 g, 0.251 mmol) and Eu(CF$_3$SO$_3$)$_3$ (0.150 g, 0.250 mmol) are added to the flask. The remaining open necks are sealed with rubber septa and the apparatus evacuated and then flushed with N$_2$. Ethanol (70 mL) is added by syringe and the mixture refluxed for one hour. The solution is concentrated in vacuum to <1 mL, and methylene chloride is added until an oil forms. The supernatant is decanted and the oil dried in vaccuo to give a powder. Yield: 88%. Melting point: 212–215°C. FABMS *m/e*: 926.9 and 928.9 (−CF$_3$SO$_3$).

Anal. Calcd. for C$_{24}$H$_{34}$N$_8$O$_{14}$F$_9$S$_3$Eu: C, 26.74; H, 3.16; N 10.40. Found: C, 26.85; H, 3.21; N, 10.17.

Properties

In some instances after reflux and cooling to room temperature, a precipitate forms. Removal of the precipitate by filtration (identified as excess NBAC) followed by the workup described above yields analytically pure [Eu(NBAC)] (CF$_3$SO$_3$)$_3$. [The [Eu(NBAC)] (CF$_3$SO$_3$)$_3$ complex is extremely soluble in ethanol, which facilitates its separation from the free ligand.] The precipitate appears to form when the Eu(CF$_3$SO$_3$)$_3$ salt is not thoroughly dried. For work with RNA, it is advisable to have a slight excess of NBAC over Eu(CF$_3$SO$_3$)$_3$ to ensure that the sample is not contaminated with traces of Eu(CF$_3$SO$_3$)$_3$.

References

1. S. Amin, J. R. Morrow, C. H. Lake, and M. R. Churchill, *Angew. Chem., Int. Ed. Engl.* **33**, 773 (1994).
2. S. Amin, D. A. Voss, Jr., W. DeW. Horrocks, Jr., C. H. Lake, M. R. Churchill, and J. R. Morrow, *Inorg. Chem.* **34**, 3294 (1995).

3. S. Amin, D. A. Voss, Jr., W. DeW. Horrocks, Jr., and J. R. Morrow, *Inorg. Chem.* **35**, 7466 (1996).
4. L. Huang, L. L. Chappell, B. F. Baker, and J. R. Morrow, *J. Biol. Inorg. Chem.* **5**(1), 85 (2000).
5. K. P. Wainright, *J. Chem. Soc., Dalton Trans.* 2117 (1980).
6. E. K. Barefield, K. A. Foster, G. M. Freeman, and K. D. Hodges, *Inorg. Chem.* **25**, 4663 (1986).
7. J. R. Morrow, S. Amin, C. H. Lake, and M. R. Churchill, *Inorg. Chem.* **32**, 4566 (1993).
8. R. Kataky, D. Parker, A. Teasdale, J. P. Huthchinson, and H.-J Buschmann, *J. Chem. Soc., Perkin Trans.* **2**, 1347 (1992).
9. R. Kataky, K. E. Mattes, P. E. Nicholson, D. Parker, and H.-J. Buschmann, *J. Chem. Soc., Perkin Trans.* **2**, 1425 (1990).
10. H. Tsukube, K. Adachi, and S. Morosawa, *J. Chem. Soc., Perkin Trans.* **1**, 1537 (1989).
11. L. Carlton, R. D. Hancock, Maumela, and K. P. Wainright, *J. Chem. Soc., Chem. Commun.*, 1007 (1994).
12. W. J. Kruper, Jr., P. R. Rudolf, and C. A. Langhoff, *J. Org. Chem.* **58**, 3869 (1993).
13. P. H. Smith and K. N. Raymond, *Inorg. Chem.* **24**, 3469 (1985).

37. CARBONYL HALIDE TUNGSTEN(II) COMPLEXES OF TRIS(3,5-DIMETHYLPYRAZOLYL)HYDROBORATE

Submitted by SIMON THOMAS and CHARLES G. YOUNG[*]
Checked by BRADFORD BROCE and JOSEPH L. TEMPLETON[†]

In 1969, Trofimenko reported the first organometallic compounds to contain the tris(3,5-dimethylpyrazolyl)hydroborate ligand (Tp*); these included salts of the group 6 tricarbonyl anions, $[Tp^*M(CO)_3]^-$ (M = Cr, Mo, W).[1] Since then the coordination and organometallic chemistry of scorpionate ligands has developed into an important subdiscipline of inorganic chemistry.[2–4] In contrast to the very considerable synthetic utility of the tricarbonyl anions, carbonyl halide complexes have played only a limited role in the development of the group 6 chemistry of Tp*.

However, syntheses for carbonyl halide complexes of tungsten, e.g., $Tp^*WX(CO)_3$ and $Tp^*WX(CO)_2$, are now available[5–9] and a comprehensive account of the properties of these compounds has appeared.[9] As expected, compounds of this type are valuable starting materials for low-valence organometallic and high-valence coordination compounds (upon oxidative decarbonylation). The acetonitrile-κ^2N,C complexes, $Tp^*WX(MeCN)(CO)$, reported by Thomas et al., are also valuable precursors for a variety of organometallic complexes; the iodide is the most readily accessible derivative of this type.[10–12] We report here syntheses for $NEt_4[Tp^*W(CO)_3]$, the chloro, bromo, and iodo members

[*] School of Chemistry, University of Melbourne, Victoria 3010, Australia.
[†] Department of Chemistry, University of North Carolina, Chapel Hill, NC 27599-3290.

of the series Tp*WX(CO)$_3$ and Tp*WX(CO)$_2$, and Tp*WI(MeCN)(CO). Details of the syntheses of 3,5-dimethylpyrazole and KTp* employed in our laboratory are also presented.

Materials and General Procedures

Starting materials (AR grade) and anhydrous *N,N*-dimethylformamide were purchased from Aldrich Chemical Co. and were used without further purification. Solvents for inert-atmosphere syntheses were dried and distilled under dinitrogen. The following drying agents are suitable: acetonitrile (CaH$_2$), dichloromethane (P$_2$O$_5$), and methanol (Mg). The syntheses detailed for KTp*[13] and NEt$_4$[Tp*W(CO)$_3$][1] are modifications of those reported by Trofimenko.

A. POTASSIUM TRIS(3,5-DIMETHYLPYRAZOLYL)-HYDROBORATE (KTp*)

■ **Caution.** *The reaction of 2,4-pentanedione and hydrazine is highly exothermic and potentially explosive. Ensure continuous cooling and efficient stirring as the reaction proceeds. Explosive hydrogen gas is evolved in the synthesis of KTp*, and precautions to prevent ignition should be taken. The hot melt obtained in the KTp* synthesis can cause serious burns and should be handled only with tongs or thick oven gloves.*

I. 3,5-Dimethylpyrazole

$$[(CH_3)CO]_2CH_2 + N_2H_4 \rightarrow C_5H_8N_2 + 2H_2O$$

Procedure

This starting material is readily prepared by condensation of 2,4-pentanedione and hydrazine. However, the reaction is very exothermic and should be carefully controlled. A 1.5-L Erlenmeyer flask is charged with 2,4-pentanedione (425 mL, 4.25 mol), ethanol (80 mL), and a large magnetic stirring bar. The flask is placed in a large ice bath on top of a magnetic stirring unit. The mixture is stirred and allowed to cool to ~0°C. A 500-mL dropping funnel containing hydrazine hydrate (55%, 244 mL, 4.20 mol) is clamped above the Erlenmeyer flask, and the hydrazine solution is added dropwise over a 1-h period to the cold, stirred 2,4-pentanedione solution. Regular replenishment of the ice is necessary because of the exothermic nature of the reaction—do not allow unstirred or uncooled addition of the hydrazine solution to occur while replenishing the ice. On completion of the addition, the mixture is allowed to stir in ice and after one hour the solid isolated by vacuum filtration using a large Büchner funnel. The solid is not washed but is pressed onto the filter paper to remove the mother liquor. The product is dried by continuing pumping. If the filtrate is chilled at $-4°C$, a second crop of crystals can be obtained, and these can be isolated as described above. The compound is recrystallized from hot cyclohexane. The crystals are placed in a 1-L Erlenmeyer flask and then dissolved in a minimum amount of boiling cyclohexane. The solution is then allowed to cool, with stirring, then chilled in ice. The crystalline product is isolated by filtration, washed with ice-cold cyclohexane (20 mL), and dried at the pump. The yield of white crystals is about 380 g (95%).

II. Potassium Tris(3,5-dimethylpyrazolyl)hydroborate, KTp*

$$3C_5H_8N_2 + KBH_4 \rightarrow K[(C_5H_7N_2)_3BH] + 3H_2$$

Procedure

Separate samples of dry 3,5-dimethylpyrazole (100 g, 1.04 mol) and potassium borohydride, KBH_4, (10.81 g, 0.2 mol) are ground to powders in a mortar and pestle. These powders are intimately mixed and reground. Ensure minimal exposure of the KBH_4 to the atmosphere (work quickly or prepare the mixture in an inert atmosphere). The solid mixture is transferred to a 250-mL, three-necked, round-bottomed flask fitted with a (well-greased) glass stopper, a 250°C thermometer, and a wide-bore, glass gas outlet tube connected via transparent tubing to an oil bubbler. Since hydrogen gas is evolved, the oil bubbler should be positioned well away from the reaction vessel in a well-ventilated fume hood. The flask is placed in a heating mantle, and the temperature of the mixture is steadily raised to 220–230°C. The temperature should not exceed 240°C. Because of vigorous evolution of hydrogen, the maximum temperature should be attained in about three stages of approximately 20 min each; this reduces uncontrolled gas evolution and blocking of the gas exit tube by sublimed dimethylpyrazole (the exit tube should be inspected regularly, and any blockage should be rectified). The mixture is heated with occasional swirling for approximately 4 h. When the reaction is complete and hydrogen evolution has ceased, quickly pour the melt into stirred, ice-cold ethanol (100 mL), contained in a 250-mL beaker. Ethanol can be used to wash any residual product from the reaction vessel into the beaker. Transfer the cooled mixture to a 500-mL pear-shaped flask and remove the solvent by rotary evaporation ($T \sim 35°C$) and finally, after fitting an adapter to the flask, dry the residue under high vacuum. It is advisable to break up the solid just prior to complete drying to prevent the formation of a solid mass, which can prove difficult to extricate.

The dry solid is then finely ground and loaded into a large sublimation apparatus. The excess dimethylpyrazole is removed by sublimation under high vacuum at 110°C. The sublimation apparatus and its contents are allowed to cool to room temperature before opening to the atmosphere. The white, nonvolatile residue is then dissolved in acetone ($\sim$600–800 mL; not all of the solid will dissolve) and filtered to yield a clear solution. The acetone is removed on a rotary evaporator (do not allow the water-bath temperature to exceed 30°C, or decomposition will take place), then diethyl ether (100 mL) is added to assist homogenization and filtration of the solid product. The solid is washed with diethyl ether and air dried. The yield is typically 52 g (77%, cf. *lit.*[13] 34%).

Properties

The white, air-stable crystalline solid exhibits the physical and spectroscopic properties reported by Trofimenko.[13] Infrared: KBr, ν_{BH} 2438(s); ν_{CN} 1540(s) cm^{-1}.

B. TETRAETHYLAMMONIUM TRICARBONYL[TRIS-(3,5-DIMETHYLPYRAZOLYL)HYDROBORATO]-TUNGSTATE(0) {NEt$_4$[Tp*W(CO)$_3$]}

$$W(CO)_6 + KTp^* + NEt_4Cl \rightarrow NEt_4[Tp^*W(CO)_3] + KCl + 3CO$$
$$[Tp^* = (3,5\text{-dimethylpyrazolyl)hydroborate}]$$

General Procedures

In this synthesis, and those to follow, it is possible that local conditions, particularly atmospheric pressure, or changes to the scale of the syntheses may affect the reaction times. The course of the reactions may be conveniently assayed by solution infrared spectroscopy.

■ **Caution.** *Metal carbonyls are volatile and toxic and should be handled in a fume hood. This reaction shoud be performed in a fume hood due to the evolution of toxic carbon monoxide. The reaction may be performed in the air.*

A single-necked, 500-mL round-bottomed flask is charged with W(CO)$_6$ (20 g, 57 mmol), KTp* (19.11 g, 57 mmol), and a large magnetic stirring bar. Anhydrous *N,N*-dimethylformamide (200 mL) is added to the mixture and the flask is fitted with an air condenser. The flask is then placed in a (preheated) oil bath at 110°C for 6 h or until solution infrared spectoscopy reveals complete conversion of $W(CO)_6$ to $[Tp^*W(CO)_3]^-$. The $W(CO)_6$ that sublimes out of the reaction mixture is periodically returned by vigorously swirling the flask. On completion of the reaction, the yellow-brown solution is reduced in volume to 100 mL on a rotary evaporator (bath temperature of 80°C), then the solution is poured into a 500-mL beaker containing a stirred solution of tetraethylammonium chloride (28 g, excess) in 300 mL of distilled water. The yellow suspension is stirred for 1 h (but can be left stirring overnight). The precipitate is filtered and washed with distilled water (80 mL). A hot recrystallization is then carried out. While slightly wet, the filtered yellow solid is transferred to a 2-L Erlenmeyer flask, then dissolved in ~500–600 mL of boiling acetonitrile. On almost complete dissolution of the solid, the mixture is hot-vacuum-filtered (use a glass filtration frit and 2-L Büchner flask preheated to ~80°C) to give a homogeneous yellow filtrate. The filtrate is quickly brought back to the boil, and distilled water (500 mL) is added in three equal portions; homogeneity is maintained by stirring and/or heating. The suspension is stirred and allowed to cool at room temperature for 30 min and then in ice for a further 20 min [coprecipitation of $Tp^*WO(CO)$ $(\mu\text{-}O)WO_2Tp^{*14}$ occurs if stirring is continued for more than an hour]. The yellow crystalline product is isolated by filtration and washed with 1 : 1 acetonitrile/water (25 mL), methanol (25 mL), then diethyl ether (25 mL). If the product is discolored (orange or pink-yellow), it should be washed with chloroform to remove pink $Tp^*WO(CO)(\mu\text{-}O)WO_2Tp^*$. The yield of the bright yellow product is 28 g (70%).

Properties

Yellow, diamagnetic $NEt_4[Tp^*W(CO)_3]$ is air stable and can be stored without precautions. An alternative synthesis of the compound is available;[15] it offers improved yields but is longer and more involved. The compound is slightly soluble in acetonitrile but insoluble in water, chlorinated solvents, alcohols and diethyl ether. It is soluble in hot acetonitrile. Infrared: MeCN, ν_{CO} 1877(s), 1742(s) cm^{-1}. KBr, ν_{BH} 2528(w); ν_{CO} 1874(s), 1736(s); ν_{CN} 1540(m) cm^{-1}.

C. TRICARBONYLHALO[TRIS(3,5-DIMETHYLPYRAZOLYL)-HYDROBORATO]TUNGSTEN(II) [Tp*WX(CO)₃] (X = Cl, Br)

$$NEt_4[Tp^*W(CO)_3] + S\text{-}X \xrightarrow{CH_2Cl_2} Tp^*WX(CO)_3 + NEt_4[S]$$

[S = Succinimide]

I. Tp*WCl(CO)₃

Procedure

A 50-mL Schlenk flask is charged with $NEt_4[Tp^*W(CO)_3]$ (4.0 g, 5.75 mmol) and a magnetic stirring bar. The flask is fitted with a rubber septum cap and then evacuated and backfilled with dinitrogen. Dry, deoxygenated dichloromethane (30 mL) is then added to the flask and the mixture is stirred. Under a flow of dinitrogen, four portions of *N*-chlorosuccinimide (total <0.85 g, 6.33 mmol) are added at ~5-min intervals and stirring is continued for 1 h following the final addition. (If the *N*-chlorosuccinimide is added too quickly or in one portion, then >0.85 g is required to complete the reaction; this results in contamination of the product on precipitation with methanol in the step that follows.) The volume of the resulting red solution is reduced to ~10 mL, and then deoxygenated methanol (40 mL) is added to precipitate orange crystals. These are filtered, washed with methanol, and dried in vacuo. The yield of orange Tp*WCl(CO)₃ is 2.95 g (85%). The compound can be recrystallized from dicloromethane/methanol or dichloromethane/hexane.

An alternative synthesis from $NEt_4[Tp^*W(CO)_3]$ and PhICl₂ is available but is less convenient.[9]

Properties

Orange, diamagnetic Tp*WCl(CO)₃ is air-stable for short periods but should be stored under an inert atmosphere. Solution infrared samples prepared in air can show an additional peak at ~1980 cm^{-1}, due to the formation of the carbonyl-oxo complex, Tp*WOCl(CO), during sample preparation.[6] Infrared: CH₂Cl₂, ν_{CO} 2027(s), 1936(sh)* 1912(s) cm^{-1}. KBr, ν_{BH} 2550(m); ν_{CO} 2016(s), 1929(s), 1899(s); ν_{CN} 1542(m) cm^{-1}.

All the tricarbonyl complexes are soluble in chlorinated solvents, acetonitrile (partially), and tetrahydrofuran. They are insoluble in water, alcohols, diethyl ether, and hydrocarbons.

II. Tp*WBr(CO)₃

Procedure

A 50-mL Schlenk flask is charged with $NEt_4[Tp^*W(CO)_3]$ (4.0 g, 5.75 mmol) and a magnetic stirring bar. The flask is fitted with a rubber septum cap and

---

* Shoulder.

then evacuated and backfilled with dinitrogen. Dry, deoxygenated dichloro-methane (30 mL) is then added to the flask and the mixture is stirred. Under a flow of dinitrogen, *N*-bromosucccinimide (1.14 g, 6.40 mmol) is added and the mixture is stirred for 1 h. The volume of the resulting red solution is reduced to ~ 10 mL, then deoxygenated methanol (40 mL) is added to precipitate orange crystals. These are filtered, washed with methanol, and dried in vacuo. The yield of orange Tp*WBr(CO)$_3$ is 3.10 g (83%). The compound can be recrystallized from dichloromethane/methanol or dichloromethane/hexane.

Attempts to prepare the compound from NEt$_4$[Tp*W(CO)$_3$] and elemental bromine result in the formation of mixtures of Tp*WBr(CO)$_3$ and Tp^{Br}WBr(CO)$_3$ [Tp^{Br} = tris(3,5-dimethyl-4-bromopyrazolyl)hydroborate].[9]

Properties

Orange, diamagnetic Tp*WBr(CO)$_3$ is air-stable for short periods but should be stored under an inert atmosphere. Infrared: CH$_2$Cl$_2$, ν_{CO} 2025(s), 1933(sh), 1915(s) cm^{-1}. KBr, ν_{BH} 2560(m), ν_{CO} 2022(s), 1913(s), 1897(s); ν_{CN} 1546(m) cm^{-1}.

D. TRICARBONYLIODO[TRIS(3,5-DIMETHYLPYRAZOLYL)-HYDROBORATO]TUNGSTEN(II) [Tp*WI(CO)$_3$]

$$NEt_4[Tp^*W(CO)_3] + I_2 \xrightarrow{MeCN} Tp^*WI(CO)_3 + NEt_4I$$

Procedure

The method of Feng et al.[5] was modified to allow the synthesis to be performed in air using reagent-grade solvents, albeit in a slightly lower yield.

A 50-mL Erlenmeyer flask is charged with NEt$_4$[Tp*W(CO)$_3$] (4.0 g, 5.75 mmol), a magnetic stirring bar, and acetonitrile (30 mL). Solid iodine (1.47 g, 5.79 mmol) is added to the stirred suspension of NEt$_4$[Tp*W(CO)$_3$], and the reaction mixture is allowed to stir for 1 h. Methanol (10 mL) is added and the reaction allowed to stir for a further 10 minutes in ice. The dark brown crystalline solid is filtered off and washed with ice-cold methanol (5 mL) and dried in vacuo. The yield is 3.00 g (75%).

Properties

Brown, diamagnetic Tp*WI(CO)$_3$ is air-stable and can be stored for weeks in air without precautions. Infrared: CH$_2$Cl$_2$, ν_{CO} 2020(s), 1920(s) cm^{-1}. KBr, ν_{BH} 2550(m); ν_{CO} 2012(s), 1915(s), 1872(s); ν_{CN} 1545(m) cm^{-1}.

E. DICARBONYLHALO[TRIS(3,5-DIMETHYLPYRAZOLYL)-HYDROBORATO]TUNGSTEN(II) [Tp*WX(CO)₂] (X = Cl, Br, I)

$$\text{Tp}^*\text{WX(CO)}_3 \xrightarrow{\text{MeCN}, 80°C} \text{Tp}^*\text{WX(CO)}_2 + \text{CO}^\uparrow$$

General Procedures

The following general method is applicable. If very different scales are employed, the completion of the reactions should be assessed using solution infrared spectroscopy.

Procedure

A suitably sized Schlenk flask containing Tp*WX(CO)₃ (typically 2.0 g) and a magnetic stirring bar is fitted with a water condenser and an oil bubbler (isolable from the reaction by a stopcock). The apparatus is charged with dinitrogen and dry, deoxygenated acetonitrile (10–15 mL per gram of tricarbonyl) is added via a gas-tight syringe. The stirred reaction mixture is then heated at 80°C for 30 min under a slow flow of dinitrogen gas. The mixture is allowed to cool briefly, and the resulting pale brown crystalline solid is quickly filtered in air, washed with ice-cold acetonitrile (2 mL), and dried in vacuo. The yields are ~80% in all three cases.

Properties

The pale brown, paramagnetic compounds are moderately air-sensitive and should be stored under a dinitrogen atmosphere. In air, solid samples can be kept without apparent decomposition, but solutions rapidly decompose. All the dicarbonyl complexes are soluble in chlorinated solvents and tetrahydrofuran but only sparingly soluble in acetonitrile, alcohols, diethyl ether, and hydrocarbons. Solution infrared samples prepared in air show an additional peak at ~1980 cm⁻¹ due to the formation of Tp*WOX(CO) during sample preparation.

Infrared spectral data are as follows: Tp*WCl(CO)₂: CH₂Cl₂, ν_{CO} 1925(s), 1834(s) cm⁻¹. KBr, ν_{BH} 2550(m); ν_{CO} 1919(s), 1822(s); ν_{CN} 1540(m) cm⁻¹. Tp*WBr(CO)₂: CH₂Cl₂, ν_{CO} 1937(s) 1841(s) cm⁻¹. KBr, ν_{BH} 2550(m); ν_{CO} 1933(s), 1826(s); ν_{CN} 1540(m) cm⁻¹. Tp*WI(CO)₂: CH₂Cl₂, ν_{CO} 1946(s), 1845(s) cm⁻¹. KBr, ν_{BH} 2550(m); ν_{CO} 1927(s), 1817(s); ν_{CN} 1540(m) cm⁻¹.

F. (ACETONITRILE-$\kappa^2 N,C$)CARBONYL[TRIS-(3,5-DIMETHYLPYRAZOLYL)HYDROBORATO]-TUNGSTEN(II) [Tp*WI(MeCN-$\kappa^2 N,C$)(CO)]

$$\text{Tp*WI(CO)}_3 \xrightarrow{\text{MeCN, reflux}} \text{Tp*WI(MeCN)(CO)} + 2\text{CO}^\uparrow$$

Procedure

A 50-mL Schlenk flask is charged with Tp*WI(CO)$_3$ (2.24 g, 3.24 mmol) and a magnetic stirring bar. A water condenser with an oil bubbler/stopcock attachment is fitted to the flask. The apparatus is charged with a dinitrogen atmosphere and dry, deoxygenated acetonitrile (30 mL) is then added. The mixture is stirred and heated under reflux for 4 h under a slow flow of nitrogen gas. After this time, the reaction mixture is allowed to cool to room temperature, then stored overnight at $-4°C$. The gold-colored, crystalline solid is filtered off, washed with cold acetonitrile (5 mL), and then dried in vacuo. The yield is 1.36 g (62%).

The compound prepared as described above is suitable for synthetic use. Analytically pure samples can be obtained following column chromatography (silica support, 2 : 1 CH$_2$Cl$_2$/hexane eluent) and recrystallization from dichloromethane/hexane.

Properties

Tp*WI(MeCN)(CO) is air-stable for long periods but should be stored under an inert atmosphere. It is soluble in chlorinated solvents, tetrahydrofuran, acetonitrile (partially), and aromatic hydrocarbons but insoluble in aliphatic hydrocarbons and alcohols (it is decomposed by the latter solvents). Infrared: CH$_2$Cl$_2$, ν_{CO} 1907(s); $\nu_{C=N}$ 1688(w) cm^{-1}. KBr, ν_{BH} 2553(m); ν_{CO} 1896(s); $\nu_{C=N}$ 1687(w); ν_{CN} 1545(s) cm^{-1}.

References

1. S. Trofimenko, *J. Am. Chem. Soc.* **91**, 588 (1969).
2. S. Trofimenko, *Chem. Rev.* **72**, 497 (1972).
3. S. Trofimenko, *Prog. Inorg. Chem.* **34**, 115 (1986).
4. S. Trofimenko, *Chem. Rev.* **93**, 943 (1993).
5. S. G. Feng, C. C. Philipp, A. S. Gamble, P. S. White, and J. L. Templeton, *Organometallics* **10**, 3504 (1991).
6. S. G. Feng, L. Luan, P. White, M. S. Brookhart, J. L. Templeton, and C. G. Young, *Inorg. Chem.* **30**, 2582 (1991).
7. C. C. Philipp, C. G. Young, P. S. White, and J. L. Templeton, *Inorg. Chem.* **32**, 5437 (1993).

8. S. Thomas, E. R. T. Tiekink, and C. G. Young, *Inorg. Chem.* **33**, 1416 (1994).
9. C. G. Young, S. Thomas, and R. W. Gable, *Inorg. Chem.* **37**, 1299 (1998).
10. S. Thomas, E. R. T. Tiekink, and C. G. Young, *Organometallics* **15**, 2428 (1996).
11. S. Thomas, C. G. Young, and E. R. T. Tiekink, *Organometallics* **17**, 182 (1998).
12. S. Thomas, P. J. Lim, R. W. Gable, and C. G. Young, *Inorg. Chem.* **37**, 590 (1998).
13. S. Trofimenko, *J. Am. Chem. Soc.* **89**, 6288 (1967).
14. C. G. Young, R. W. Gable, and M. F Mackay, *Inorg. Chem.* **29**, 1777 (1990).
15. A. J. M. Caffyn, S. G. Feng, A. Dierdorf, A. S. Gamble, P. A. Eldredge, M. R. Vossen, P. S. White, and J. L. Templeton, *Organometallics* **10**, 2842 (1991).

38. A LUMINESCENT COMPLEX OF Re(I): *fac*-[Re(CO)₃ (BPY)(PY)](CF₃SO₃) (BPY = 2.2′-BIPYRIDINE; PY = PYRIDINE)

Submitted by ERICK J. SCHUTTE and B. PATRICK SULLIVAN*
Checked by CHRIS CHANG and DAN G. NOCERA†

Re(I)tricarbonyl complexes containing α-diimine ligands have played a central role in understanding the photoreactivity of charge transfer excited states of metal comlexes.[1] Because of their intense charge transfer luminescence, they have found application as molecular probes and as photosensitizers for solar energy conversion,[2] in addition to being the subject of a large number of fundamental photophysical studies.[1] In general, complexes of the type *fac*-[Re(CO)₃(chelate)](L)⁺ (chelate is a polypyridine, L is a neutral donor, and the counterion is generally $CF_3SO_3^-$ or PF_6^-) are extremely air/moisture-stable, which lends a great deal of flexibility to the study of both their ground- and excited-state chemistry. The procedure here describes the preparation of *fac*-[Re(CO)₃(bpy)](py)](CF₃SO₃)[3] from the precursor *fac*-[Re(CO)₃ (py)₃](CF₃-SO₃)[4] via the very labile complex Re(CO)₅(CF₃SO₃) (py is pyridine).[5] The example provides a general preparation of many structurally similar α- diimine Re(I)tricarbonyl complexes containing ligands such as 1,10-phenanthroline and 2,2′-biquinoline, as well as tetraazines such as 2,2′-bipyrazine, 2,2′-bipyrimidine, and 1,10,4,7′-tetraazaphenanthrene. Previous methods for preparation of *fac*-[Re(CO)₃(chelate)(py)]⁺ complexes can be found in the literature.[6–12] The uncomplicated, high-yield, synthesis of *fac*-[Re(CO)₃(py)₃](CF₃SO₃), and the complex of study, *fac*-[Re(CO)₃(bpy)(py)](CF₃SO₃), both follow from the trans-labilizing ability of the facial carbonyl groups.

*University of Wyoming, Laramie, WY 82071-3838.
†Massachusetts Institute of Technology Cambridge, Massachusetts, 02139-4307.

A. *fac*-TRICARBONYL(TRISPYRIDINE)RHENIUM(I) TRIFLUOROMETHANESULFONATE {*fac*-[Re(CO)₃(py)₃](CF₃SO₃)}

$$Re(CO)_5O_3SCF_3 + 3py \rightarrow fac\text{-}[Re(CO)_3(py)_3](CF_3SO_3) + 2CO$$

Procedure

To a 100-mL round-bottom flask containing 15 mL of neat pyridine (185 mmol), 200 mg of $Re(CO)_5O_3SCF_3$ (0.42 mmol) is added. The solution is heated at reflux (under a N_2 blanket) with stirring for 5 h. A white precipitate is evident after reflux. After cooling to room temperature, 50 mL of Et_2O is added to precipitate the product contained in the pyridine. The complex is recovered by vacuum filtration and washed with two 25-mL aliquots of Et_2O. After drying the solid overnight under vacuum, 0.22 g (0.336 mmol, 80%) of product is obtained.

Properties

fac-[Re(CO)₃(py)₃](CF₃SO₃) is an off-white powder, which luminesces faint green when exposed to a "black light" source. It is air-, moisture-, and light-stable, and has excellent solubility in medium to high-polarity solvents such as acetonitrile, acetone, and 1,2-dichlorobenzene. It is insoluble in low-polarity solvents such as diethyl ether. The infrared spectrum in CH_2Cl_2 shows a sharp single stretch at 2040 cm^{-1} and a broad stretch at 1934 cm^{-1}, which supports the assignment of a facial structure. The first-order ^1H NMR spectrum in CD_3CN shows two triplets (8.10 and 7.55 ppm) and a doublet (8.53 ppm) associated with the coordinated pyridine. The electronic absorption spectrum shows a band at 262 nm ($\varepsilon = 18,300$ M^{-1} cm^{-1}) and at 196 nm ($\varepsilon = 30,900$ M^{-1} cm^{-1}), both of which can be assigned to internal pyridine $\pi{-}\pi^*$ transitions based on oscillator strength arguments.

B. *fac*-TRICARBONYL(2,2'BIPYRIDINE)(PYRIDINE)RHENIUM(I) TRIFLUOROMETHANESULFONATE {*fac*-[Re(CO)₃(bpy)(py)](CF₃SO₃)}

$$fac\text{-}[Re(CO)_3(py)_3](CF_3SO_3) + 2,2'\text{-bpy} \rightarrow$$

$$fac\text{-}[Re(CO)_3(bpy)(py)](CF_3SO_3)$$

Procedure

A mixture containing 100 mg of *fac*-[Re(CO)$_3$(py)$_3$](CF$_3$SO$_3$) (0.15 mmol), 50 mg of 2.2′-bipyridine (0.32 mmol), and 5 mL of 1,2-dichlorobenzene is placed in a 50-mL round-bottomed flask. The solution is then heated at reflux with stirring for 15 min. The orange-red solution is allowed to cool to room temperature, after which time 40 mL of Et$_2$O is added to precipitate the yellow product. The product is then filtered by suction and washed with two 25 mL aliquots of Et$_2$O. After drying overnight under vacuum, 92 mg (0.14 mmol; 92%) of product is obtained. Recrystallization of *fac*-[Re(CO)$_3$(bpy)(py)](CF$_3$SO$_3$) is as follows. In a 100-mL round-bottomed flask, 50 mg of the complex is brought into solution with a minimum amount of a 2 : 1 toluene/CH$_3$CN mixture. The resultant solution is slowly heated until slight boiling is noticed. The solution is then removed from the heat source, and Et$_2$O is slowly added until streaking is observed. The flask is then capped and placed in the freezer overnight to produce light yellow-green crystals of the complex. The crystals are carefully filtered by suction filtration, washed with generous amounts of Et$_2$O, and dried under vacuum, yielding 46 mg (0.070 mmol) of product. Elemental analysis, calculated (with $\frac{2}{5}$ toluene molecule to every complex, which is found in the ^{1}H NMR spectrum): C, 38.12%; H, 2.32%; N, 6.06%. Found: C, 38.24%; H, 2.41%; N, 6.24%.

Properties

fac-[Re(CO)$_3$(bpy)(py)](CF$_3$SO$_3$) is light yellow in powder form and light yellow-green in crystalline form. It is air-, moisture-, and light-stable, soluble in high-polarity solvents such as acetonitrile and acetone, but only slightly soluble in water. It is also soluble in dichloromethane at low concentrations or high temperatures. It is insoluble in low-polarity solvents such as diethylether, toluene, and hexanes. The infrared spectrum shows a single stretch at 2036 cm^{-1}, and broad stretch at 1931 cm^{-1}, which is consistent with the facial symmetry. In water, the electronic absorption spectrum shows a band at 318 nm ($\varepsilon = 11,600$ M^{-1} cm^{-1}) and a shoulder at 342 nm ($\varepsilon = 3590$ M^{-1} cm^{-1}). In addition, bands at 306 nm ($\varepsilon = 11,000$ M^{-1} cm^{-1}), and 248 nm ($\varepsilon = 16,900$ M^{-1} cm^{-1}) are evident, as well as a shoulder at 264 nm ($\varepsilon - 15,800$ M^{-1} cm^{-1}). The first-order ^{1}H NMR spectrum in CD$_3$CN shows the two triplets (8.25 and 7.79 ppm) and two doublets (9.22 and 8.35 ppm) associated with the 2,2-bipyridine, as well as the two triplets (7.86 and 7.30 ppm) and doublet (8.38 ppm) associated with the pyridine. Toluene of crystallization appears in the region 7.1–7.30 ppm.

The luminescent properties evident in rhenium(I)tricarbonyl complexes arise from the metal-to-ligand charge transfer (MLCT) between the $d\pi$ (Re) orbitals and the π^* orbitals on the corresponding chelate (342 nm). The emitting states, which are primarily triplet in character, are typified by a broad emission that occurs at 584 nm in water and 565 nm in CH_2Cl_2 (20°C, corrected for detector response). The emission and the corresponding absorption bands are sensitive to changes in solvent and temperature. The lifetime and quantum yield of emission in deoxygenated CH_2Cl_2 is 646 ns and 0.077, respectively. Substitution of bpy for other substituted polypyridines produces large changes in lifetimes, quantum yields, and luminescence energies, making this class of molecule useful as excited-state energy or electron transfer probes.[6–12]

References

1. K. Kalyanasundaram, *Photochemistry of Polypyridine and Porphyrin Complex*, Academic Press, New York, 1992.
2. K. D. Karlin, *Progress in Inorganic Chemistry*, Vol. 44, *Molecular Level Artificial Photosynthetic Materials*, J Wiley, New York, 1997.
3. S. M. Fredricks, J. C. Luong, and M. S. Wrighton, *J. Am. Chem. Soc.* **101**, 7415 (1979).
4. D. A. Edwards and J. Marshalea, *J. Organomet. Chem.* **131**, 73 (1977).
5. S. P. Schmidt, J. Nitschke, and W. C. Trogler, *Inorg. Synth.* **26**, 113 (1989).
6. J. V. Caspar and T. J. Meyer, *J. Phys. Chem.* **87**, 952 (1983).
7. L. Salsteder, A. P. Zipp, E. A. Brown, J. Streich, J. N. Demas, and B. A. Degraff, *Inorg. Chem.* **29**, 4335 (1990).
8. J. M. Lang, Z. A. Dregler, and H. G. Drickamer, *Chem. Phys. Lett.* **192**, 299 (1992).
9. J. K. Hino, L. D. Ciana, W. J. Ciana, W. J. Dressick, and B. P. Sullivan, *Inorg. Chem.* **31**, 1072 (1992).
10. R. J. Shaver and D. P. Rillema, *Inorg. Chem.* **31**, 4101 (1992).
11. L. Wallace and D. P. Rillema, *Inorg. Chem.* **32**, 3836 (1993).
12. R. J. Shaver, M. W. Perkovic, D. P. Rilema, and C. Woods, *Inorg. Chem.* **34**, 5446 (1995).

39. $RSi(OH)_3$ AND $RSi(NH_2)_3$ [R = 2,6-$iPr_2C_6H_3N(SiMe_3)$]: SYNTHESIS OF STABLE ORGANOSILANES WITH THREE FUNCTIONAL GROUPS

Submitted by **HENDRIK DORN** and **HERBERT W. ROESKY***
Checked by **PHILIP P. POWER**†

The chemistry of silanetriols and triaminosilanes is relatively new compared to other types of silanols or aminosilanes. The isolation of organosilanes with three hydroxy or amino groups has been achieved by the use of sterically demanding substituents that offer a high degree of kinetic stability. The bulky amido

*Institut für Anorganische Chemie der Universität Göttingen, Tammannstr. 4, D-37077 Göttingen, Germany.
†Department of Chemistry, University of California, Davis, CA 95616.

ligand 2,6-*i*Pr$_2$C$_6$H$_3$N(SiMe$_3$) affords the desired steric protection to avoid self-condensation reactions.

In this contribution we describe a facile, high-yield synthesis of RSi(OH)$_3$ **(1)**[1] and RSi(NH$_2$)$_3$ **(2)**[2] R = 2,6-*i*Pr$_2$C$_6$H$_3$N(SiMe$_3$) [throughout this text].

The silanetriol **(1)** shows an interesting reactivity. It has been successfully used as starting material for the preparation of soluble zeolite precursors or three-dimensional metallasiloxanes. A review of the chemistry of silanetriols has appeared in the literature.[4]

The valence-isoelectronic triaminosilane **2** was prepared according to a method first reported by P. P. Power et al. for the synthesis of the compound [2,4,6-Ph$_3$C$_6$H$_2$]Si(NH$_2$)$_3$.[4]

Selective substitution of the amino groups can be achieved.[2] Furthermore, triaminosilanes are possible precursors for dendrimers and ceramics.

Both of the procedures described here start from the trichlorosilane RSiCl$_3$, which can be obtained from the reaction of RLi and SiCl$_4$ in excellent yield.[1]

Materials and General Procedures

All experiments were performed under a nitrogen atmosphere by Standard Schlenk techniques. Solvents (including NMR solvents) were distilled from an appropriate drying agent.

The ^{1}H and ^{29}Si NMR data presented here were obtained in (D$_6$)DMSO **(1)** or C$_6$D$_6$ **(2)** on a Bruker AM 250 spectrometer. Infrared spectra were recorded on a Bio-Rad FTS-7 spectrometer (Nujol mulls between KBr plates), mass spectra on Finnigan MAT 8230, and Varian MAT CH5 instruments (EI, 70 eV). Melting points were obtained using a Büchi 510 or HWS-SG 300 apparatus. Elemental analyses were performed by the Analytical Laboratory of the Inorganic Institute at the University of Göttingen.

A. [2,6-*i*Pr$_2$C$_6$H$_3$N(SiMe$_3$)]Si(OH)$_3$ (1)

$$RSiCl_3 + 3H_2O + 3PhNH_2 \rightarrow RSi(OH)_3 + 3PhNH_3Cl$$

Procedure

A 2-L two-necked reaction flask, equipped with a dropping funnel, a reflux condenser, and a strong magnetic stirrer, is charged with water (8.7 g, 0.49 mol) and aniline (45.1 g, 0.49 mol) in diethyl ether (~1.3 L). The reaction flask is cooled to 0 °C. To the vigorously stirred emulsion a solution of [2,6-*i*Pr$_2$C$_6$H$_3$N

(SiMe₃)]SiCl₃[1] (61.9 g, 0.16 mol) in diethyl ether (400 mL) is added dropwise over a period of 3 h. After warming to room temperature the reaction mixture is stirred for a further 12 h. The solution is concentrated under vacuum to a third of its volume, the precipitated aniline hydrochloride is filtered off, and the remaining solvent is removed under vacuum. The pale yellow crude product is washed with two 100-mL portions of cold pentane to obtain a pure product as a white solid with a yield of 45.8 g (88%).

Characterization Data. Melting point: 131°C. ^{1}H NMR δ 0.05 [s, 9H, Si(CH₃)₃], 1.10 [d, $^3J_{HH} = 6.8$ Hz, 6H, CH(CH₃)(CH₃)], 1.15 [d, $^3J_{HH} = 6.8$ Hz, 6H, CH(CH₃)(CH₃)], 3.65 [sept. $^3J_{HH} = 6.8$ Hz, 2H, CH(CH₃)₂], 5.74 (s, 3H, OH), 6.92 (m, 3H, H$_{arom}$) ppm. ^{29}Si NMR δ −67.3 [Si(OH)₃], 3.8 [Si(CH₃)₃] ppm. IR: ν 3344(s), 1584(w), 1315(w), 1250(s), 1178(m), 987(s), 920(s), 836(m), 814(s) cm^{-1}. Mass spectrum (*m/z*): 327 (M, 54%), 312 (M-Me, 100%).

Elemental Analysis Data. Calcd. For C₁₅H₂₉NO₃Si₂ (327.57): C, 55.00%; H, 8.92%; N, 4.27%. Found: C, 56.3%; H, 8.8%; N, 4.1%.

Properties

The silanetriol (**1**) is air-stable as a solid and can be manipulated without special care. It is very soluble in most organic solvents. The crystal structure of the *N*-bonded silanetriol {[2,4,6-Me₃C₆H₂N(SiMe₃)]Si(OH)₃} has been determined by X-ray diffraction and shows an interesting formation of silanetriol tubes in the crystal that contain a hydrophilic interior and a hydrophobic exterior.[5]

B. [2,6-*i*Pr₂C₆H₃N(SiMe₃)]Si(NH₂)₃ (2)

$$RSiCl_3 + 6NH_3 \rightarrow RSi(NH_2)_3 + 3NH_4Cl$$

Procedure

■ **Caution.** *Ammonia gas is toxic and corrosive. All manipulations should be performed in a well-ventilated fume hood, and adequate precautions should be taken to prevent contact with skin and eyes.*

A 100-mL Schlenk flask is charged with [2,6-*i*Pr₂C₆H₃N(SiMe₃)]SiCl₃[1] (1.2 g, 3.1 mmol) and equipped with a magnetic stirring bar and a Dewar condenser filled with powdered dry ice/ethanol. The flask is then cooled in a dry-ice/ethanol bath ($-78\,°C$) and ~25 mL of ammonia is condensed into the flask. Prior to use the ammonia is kept for 24 h in a monel cylinder (300 mL) over metallic sodium (1 g).[6] After stirring the reaction mixture for 4 h at $-78\,°C$, the cooling bath is removed and the ammonia is allowed to evaporate off. The residue is taken up in diethyl ether (20 mL), and the ammonium chloride formed is removed by filtration through a glass frit. Removal of the solvent in vacuum leaves a colorless solid, which is recrystallized from hexane (5 mL) at $0\,°C$ to give 0.8 g (79%) of **2**.

Characterization Data. Melting point: 73°C. ^{1}H NMR δ 0.2 [s, 9H, Si(CH₃)₃], 0.5 [s, 6H, Si(NH₂)₃], 1.22 [d, $^3J_{HH} = 6.9$ Hz, 6H, CH(CH₃)(CH₃)], 1.25 [d, $^3J_{HH} = 6.9$ Hz, 6H, CH(CH₃)(CH₃)], 3.65 [sept. $^3J_{HH} = 6.9$ Hz, 2H, CH(CH₃)₂], 7.05 (m, 3H, H$_{arom}$) ppm. ^{29}Si NMR: δ -41.0 [Si(NH₂)₃], 3.6 [Si(CH₃)₃] ppm. IR: ν 3472(m), 3392(m), 3345(m), 1363(m), 835(s), 751(w) cm^{-1}. Mass spectrum (m/z): 324 (M, 12%), 307 (M-NH₂, 100%).

Elemental Analysis Data. Calcd. for C₁₅H₃₂N₄Si₂ (324.62): C, 55.50%; H, 9.94%; N, 17.26%; Si, 17.30%. Found: C, 54.9%; H, 9.3%; N, 14.7%; Si, 17.5%.

Properties

The triaminosilane **2** is somewhat air- and moisture-sensitive but can be stored indefinitely in an inert atmosphere. Minor deviations in the elemental analysis data are due to the formation of silicon carbide and silicon nitride. It is highly soluble in common organic solvents. The solid-state structure of **2** has been determined by single-crystal X-ray analysis and shows nitrogen–hydrogen bridges.[2]

References

1. N. Winkhofer, A. Voigt, H. Dorn, H. W. Roesky, A. Steiner, D. Stalke, and A. Reller, *Angew. Chem.* **106**, 1414 (1994); *Angew. Chem., Int. Ed. Engl.* **33**, 1352 (1994).
2. K. Wraage, A. Künzel, M. Noltemeyer, H.-G. Schmidt, and H. W. Roesky, *Angew. Chem.* **107**, 2954 (1995); *Angew. Chem., Int. Ed. Engl.* **34**, 2645 (1995).
3. R. Murugavel, A. Voigt, M. G. Walawalkar, and H. W. Roesky, *Chem. Rev.* **96**, 2205 (1996).
4. K. Ruhlandt-Senge, R. A. Bartlett, M. M. Olmstead, and P. P. Power, *Angew. Chem.* **105**, 459 (1993); *Angew. Chem., Int. Ed. Engl.* **32**, 425 (1993).
5. R. Murugavel, V. Chandrasekhar, A. Voigt, H. W. Roesky, H.-G. Schmidt, and M. Noltemeyer, *Organometallics* **14**, 5298 (1995).
6. D. Nichols, *Inorganic Chemistry in Liquid Ammonia*, Elsevier, Amsterdam, 1979.

40. FLUORINATION WITH Me₃SnF: SYNTHESIS OF CYCLOPENTADIENYL-SUBSTITUTED GROUP 4 METAL FLUORIDE COMPLEXES

Submitted by HENDRIK DORN and HERBERT W. ROESKY*
Checked by ROBERT J. MORRIS†

Trimethyltin fluoride (Me_3SnF) is a versatile fluorinating reagent for the preparation of cyclopentadienyl-substituted fluoride complexes of titanium, zirconium, and hafnium.[1-8] Starting from the corresponding chlorides and stoichiometric amounts of Me_3SnF, group 4 organometallic fluorides can be synthesized in high yields. The resulting Me_3SnCl in the reaction is easily removable in vacuo and reconverted with aqueous KF or NaF to the starting material.[9]

This is a striking improvement over other fluorinating reagents. In addition, this elegant method of fluorination can be applied to complexes exhibiting an oxo or imido function and to titanium(III) compounds.

Herein we report the syntheses and characterization of selected organometallic group 4 fluorides as examples for the versatility of the fluorinating reagent.

Materials and General Procedures

All experiments were performed under a nitrogen atmosphere by standard Schlenk techniques. Toluene was dried prior to use over sodium by refluxing the solvent for 10 h. Me_3SnF was synthesized by published methods[9] and sublimed at $100\,°C/10^{-4}$ mbar prior to use.

The ^1H and ^{19}F NMR data presented here were obtained in C_6D_6 solutions on a Bruker AM 250 spectrometer. Infrared spectra were recorded on a Bio-Rad FTS-7 spectrometer (Nujol mulls between KBr or CsI plates), mass spectra on Finnigan MAT 8230, and Varian MAT CH5 instruments (EI, 70 eV). Melting points were obtained using a Büchi 510 or HWS-SG 3000 apparatus. Elemental analyses were performed by the Beller Laboratory (Göttingen) or by the Analytical Laboratory of the Inorganic Institute at the University of Göttingen.

A. PENTAMETHYLCYCLOPENTADIENYL(TITANIUM TRIFLUORIDE) [(η^5-C₅Me₅)TiF₃]

$$3Me_3SnF + (\eta^5\text{-}C_5Me_5)TiCl_3 \rightarrow (\eta^5\text{-}C_5Me_5)TiF_3 + 3Me_3SnCl$$

* Institut für Anorganische Chemie der Universität Göttingen, Tammannstr. 4, D-37077 Göttingen, Germany.
† Department of Chemistry, Ball State University, Muncie, IN 47306.

Procedure

■ **Caution.** *Because of its toxicity and volatility, care should be taken to avoid inhalation of trimethyltin chloride or contact of its solutions with the skin. All reactions should be carried out in a well-ventilated hood.*

The reaction vessel consists of a two-necked 250-mL flask fitted with a septum, a T-shaped N_2 inlet, a dropping funnel, and a magnetic stirrer. In a drybox the flask is charged with freshly sublimed Me_3SnF.

To a suspension of the freshly sublimed Me_3SnF (5.49 g, 30.0 mmol) in toluene (30 mL) is added dropwise a solution of $(\eta^5\text{-}C_5Me_5)TiCl_3$[10] (2.89 g, 10.0 mmol) in toluene (50 mL). The resulting mixture is stirred at room temperature for 5 h. The solvent is removed under vacuum and the orange residue sublimed at $110\,°C/10^{-2}$ mbar to yield 2.2 g (92%) of orange $(\eta^5\text{-}C_5Me_5)TiF_3$.

Characterization Data. Melting point: $180\,°C$. ^1H NMR: δ 1.93 (s, C_5Me_5) ppm. ^{19}F NMR: δ 124.0(s) ppm. IR (CsI): ν 1072(s), 1023(s), 704(vs), 648(vs), 595(vs), 590(vs), 486(vs), 341(vs) cm^{-1}.

Mass spectrum (*m/z*): 240 (M, 38%), 135 (C_5Me_5, 100%).

Elemental Analysis Data. Calcd. for $C_{10}H_{15}F_3Ti$ (240.11): C, 50.02%; H, 6.29%; F, 23.74%. Found: C, 49.8%; H, 6.1%; F, 23.6%.

B. PENTAMETHYLCYCLOPENTADIENYL(ZIRCONIUM TRIFLUORIDE) {[(η^5-C₅Me₅)ZrF₃]₄}

$$4(\eta^5\text{-}C_5Me_5)ZrCl_3 + 12Me_3SnF \rightarrow [(\eta^5\text{-}C_5Me_5)ZrF_3]_4 + 12Me_3SnCl$$

Procedure

An oven-dried two-necked 250 mL round-bottomed flask equipped with a magnetic stirrer, a septum, a reflux condenser, and a T-shaped N_2 inlet is charged in a drybox with $(\eta^5\text{-}C_5Me_5)ZrCl_3$ (5.00 g, 15.0 mmol) and Me_3SnF (8.23 g, 45.0 mmol). Toluene (80 mL) is added via a syringe. The suspension is heated for 0.5 h under reflux. After the mixture is cooled to room temperature, a small quantity of solid is retained by filtration. The filtrate is evaporated to dryness in vacuo and the solid residue washed with hexane (25 mL) to yield 4.0 g (94%) of a white solid of $(\eta^5\text{-}C_5Me_5)ZrF_3$.

Characterization Data. Melting point: $>350\,°C$. ^1H NMR: δ 2.02 (s, C_5Me_5) ppm. ^{19}F NMR: δ 97.5 (m, 4F), -26.3 (m, 4F), -50.7 (m, 2F),

-72.7 (m, 2F) ppm. IR (CsI): ν 1094(s), 1070(s), 1030(s), 574(vs), 456(vs), 425(vs), 367(vs) cm^{-1}. Mass spectrum (m/z): 999 (M$_4$-C$_5$Me$_5$, 4%), 831 (M$_3$-F, 100%), 545 (M$_2$-F, 40%).

Elemental Analysis Data. Calcd. for C$_{10}$H$_{15}$F$_3$Zr (283.44): C, 42.38%; H, 5.33%; F, 20.11%. Found: C, 42.2%; H, 5.5%; F, 19.5%.

It should be noted that (η^5-C$_5$Me$_5$)ZrF$_3$ and (η^5-C$_5$Me$_5$)HfF$_3$ form tetrameric units in the solid state as well as in solution. This was shown by single-crystal X-ray diffraction, ^{19}F NMR spectroscopy, and mass spectrometry.[1]

C. PENTAMETHYLCYCLOPENTADIENYL(HAFNIUM TRIFLUORIDE) {[(η^5-C$_5$Me$_5$)HfF$_3$]$_4$}

$$4(\eta^5\text{-C}_5\text{Me}_5)\text{HfCl}_3 + 12\text{Me}_3\text{SnF} \rightarrow [(\eta^5\text{-C}_5\text{Me}_5)\text{HfF}_3]_4 + 12\text{Me}_3\text{SnCl}$$

Procedure

The reaction vessel consists of a two-necked 250-mL flask fitted with a reflux condenser, a septum, a T-shaped N$_2$ inlet, and a magnetic stirrer. The flask is charged in a drybox with a mixture of (η^5-C$_5$Me$_5$)HfCl$_3$[12] (4.60 g, 11.0 mmol) and Me$_3$SnF (6.03 g, 33.0 mmol). Toluene (3.0 mL) is added via a syringe.

Reaction conditions and workup are the same as described for (η^5-C$_5$Me$_5$) ZrF$_3$ to yield 3.7 g (93%) of a white solid of (η^5-C$_5$Me$_5$)HfF$_3$.

Characterization Data. Melting point: $>350\,°$C. ^{1}H NMR: δ 2.07 (s, C$_5$Me$_5$) ppm. ^{19}F NMR: δ 41.8 (m, 4F), -54.0 (m, 4F), -77.3 (m, 2F), -97.8 (m, 2F) ppm. IR (CsI): ν 1097(s), 1075(s), 1029(s), 804(s), 572(vs), 563(vs), 454(vs), 384(vs) cm^{-1}. Mass spectrum (m/z): 1348 (M$_4$-C$_5$Me$_5$, 3%), 1093 (M$_3$-F, 100%), 977 (M$_3$-C$_5$Me$_5$, 30%), 723 (M$_2$-F, 20%).

Elemental Analysis Data. Calcd. for C$_{10}$H$_{15}$F$_3$Hf (370.72): C, 32.40%; H, 4.08%; F, 15.37%. Found: C, 31.1%; H, 4.3%; F, 15.2%.

D. PENTAMETHYLCYCLOPENTADIENYL(TITANIUM OXOFLUORIDE) {[(η^5-C$_5$Me$_5$)TiFO]$_4$}

In this metathesis reaction the cyclic chloride complex [(η^5-C$_5$Me$_5$)TiClO]$_3$ is converted to give the eight-membered cyclic compound [(η^5-C$_5$Me$_5$)TiFO]$_4$.[13] Identification of the tetramer was confirmed by mass spectrometry [m/z 872 (M)].

Procedure

An oven-dried, two-necked, 100-mL round-bottomed flask equipped with a reflux condenser, a dropping funnel, a magnetic stirrer, and a T-shaped gas inlet is charged in a drybox with freshly sublimed Me_3SnF. To the suspension of Me_3SnF (0.37 g, 2.1 mmol) in toluene (20 mL) is added dropwise a solution of $[(\eta^5\text{-}C_5Me_5)TiClO]_3{}^{14}$ (0.48 g, 0.7 mmol) in toluene (40 mL). The mixture is stirred for 2 h at room temperature and then for 0.5 h at 50 °C. The solution is filtered, and the solvent is removed under vacuum from the filtrate. The resulting light yellow solid is recrystallized from hexane (30 mL) to give 0.39 g (87%) of $[(\eta^5\text{-}C_5Me_5)TiFO]_4$.

Characterization Data. Melting point: 325 °C. ^1H NMR: δ 2.09 (s, C_5Me_5) ppm. ^{19}F NMR: δ 83.17(s) ppm. IR (KBr): ν 1026(m), 922(m), 863(s), 813(s), 605(s), 472(m), 383(s) cm^{-1}. Mass spectrum (m/z): 872 (M, 100%).

Elemental Analysis Data. Calcd. for $C_{40}H_{60}F_4O_4Ti_4$ (872.42): C, 55.07%; H, 6.93%; F, 9.36%. Found: C, 55.1%; H, 6.9%; F, 9.2%.

E. PENTAMETHYLCYCLOPENTADIENYL(TITANIUM IMIDOPHENYL FLUORIDE) $\{[(\eta^5\text{-}C_5H_4Me)TiF(NPh)]_2\}$

The reaction must be carried out in the presence of pyridine (py); Otherwise, no metathesis reaction takes place. It seems to be necessary to cleave first the imido-bridged dimer $[(\eta^5\text{-}C_5H_4Me)TiCl(NPh)]_2$ in order to generate in situ the mononuclear pyridine adduct $(\eta^5\text{-}C_5H_4Me)TiCl(NPh)py$. This compound can be easily fluorinated with Me_3SnF, and the coordinated pyridine is displaced.[4]

Procedure

An oven-dried, two-necked, 100-mL round-bottomed flask equipped with a magnetic stirring bar, a reflux condenser, a T-shaped N_2 inlet, and a septum in a glovebox is charged with a mixture of $[(\eta^5\text{-}C_5H_4Me)TiCl(NPh)]_2{}^{15}$ (1.50 g, 1.96 mmol) and Me_3SnF (1.08 g, 5.92 mmol). Toluene (50 mL) and pyridine (0.5 mL) are added via a syringe. The mixture is stirred for 8 h at 60 °C. All volatile contents are removed carefully under vacuum, yielding an orange solid. The solid is washed once with cold hexane (20 mL). Recrystallization from toluene (20 mL) at -24 °C affords $[(\eta^5\text{-}C_5H_4Me)TiF(NPh)]_2$ as yellow-orange crystals that are filtered off and dried in vacuo, yield 1.26 g (90%).

Characterization Data. Melting point: 153 °C. ^1H NMR data: δ 7.10–6.79 (m, 5H, NPh), 6.14 (t, $J_{HH} = 2.7$ Hz, 2H, C_5H_4Me), 5.65 (t, $J_{HH} = 2.7$ Hz, 2H,

C_5H_4Me), 1.96 (s, 3H, C_5H_4Me) ppm. ^{19}F NMR: δ 95.9(s) ppm. Mass spectrum (m/z): 474 (M, 60%), 395 (M-C_5H_4Me, 100%).

Elemental Analysis Data. Calcd. for $C_{24}H_{24}F_2N_2Ti_2$ (474.22): C, 60.79%; H, 5.10%; F, 8.01%; N, 5.91%. Found: C, 60.7%; H, 5.0%; F, 8.0%; N, 5.9%.

F. BISPENTAMETHYLCYCLOPENTADIENYL(TITANIUM FLUORIDE) (η^5-C_5Me_5)$_2$TiF

Procedure

A 100-mL, two-necked, round-bottomed flask is fitted with a magnetical stirrer, a nitrogen gas inlet tube, and a dropping funnel. The flask is charged with Me_3SnF (0.55 g, 3.0 mmol) in a drybox. Toluene (30 mL) is added via a syringe. A solution of (η^5-C_5Me_5)$_2$TiCl (0.06 g, 3.0 mmol) in toluene (30 mL) is added dropwise under stirring. The solid disappears after stirring for 3 h at room temperature. The solvent is removed in vacuo and the residue sublimed at 94 °C/10^{-2} mbar to give 0.88 g (88%) of (η^5-C_5Me_5)$_2$TiF as a green solid.

Characterization Data. Melting point: 184 °C. IR (KBr): ν 1261(m), 1022(s), 562(s), 445(s) cm^{-1}. Mass spectrum (m/z): 337 (M, 65%), 135 (C_5Me_5, 100%).

Elemental Analysis Data. Calcd. for $C_{20}H_{30}FTi$ (337.34): C, 71.21%; H, 8.96%; F, 5.62%. Found: C, 71.0%; H, 8.9%; F, 5.8%.

References and Notes

1. A. Herzog, F.-Q. Liu, H. W. Roesky, A. Demşar, K. Keller, M. Noltemeyer, and F. Pauer, *Organometallics* **13**, 1251 (1994).
2. H. W. Roesky, A. Herzog, and F.-Q. Liu, *J. Fluorine Chem.* **71**, 161 (1995).
3. H. W. Roesky, A. Herzog, and F.-Q. Liu, *J. Fluorine Chem.* **72**, 183 (1995).
4. F.-Q. Liu, A. Herzog, H. W. Roesky, and I. Usón, *Inorg. Chem.* **35**, 741 (1996).
5. E. F. Murphy, T. Lübben, A. Herzog, H. W. Roesky, A. Demşar, M. Noltemeyer, and H.-G. Schmidt, *Inorg. Chem.* **35**, 23 (1996).
6. E. F. Murphy, P. Yu, S. Dietrich, H. W. Roesky, E. Parisini, and M. Noltemeyer, *J. Chem. Soc., Dalton Trans.*, 1983 (1996).
7. S. A. A. Shah, H. Dorn, A. Voigt, H. W. Roesky, E. Parisini, H.-G. Schmidt, and M. Noltemeyer, *Organometallics* **15**, 3176 (1996).
8. M. G. Walawalkar, R. Murugavel, and H. W. Roesky, *Eur. J. Solid State Inorg. Chem.* **33**, 943 (1996).
9. (a) E. Krause, *Ber. Dtsch. Chem. Ges.* **51**, 1447 (1918); (b) W. K. Johnson, *J. Org. Chem.* **25**, 2253 (1960); (c) L. E. Levchuk, J. R. Sams, and F. Aubke, *Inorg. Chem.* **11**, 43 (1972).
10. A. M. Cardoso, R. J. H. Clark, and S. Moorhouse, *J. Chem. Soc., Dalton Trans.*, 1156 (1980).

11. P. T. Wolczanski and J. E. Bercaw, *Organometallics* **1**, 793 (1982).

12. D. M. Roddick, M. D. Fryzuk, P. F. Seidler, G. L. Hillhouse, and J. E. Bercaw, *Organometallics* **4**, 97 (1985).

13. H. W. Roesky, I. Leichtweis, and M. Noltemeyer, *Inorg. Chem.* **32**, 5102 (1993).

14. F. Palacios, P. Royo, R. Serrano, J. L. Balcázar, I. Fonseca, and F. Florencino, *J. Organomet. Chem.* **375**, 51 (1989).

15. [(η^5-C₅H₄Me)TiCl(NPh)]₂ is prepared according to the methods in Y. Bai, H. W. Roesky, H.-G. Schmidt, and M. Noltemeyer, *Z. Naturforsch.* **47B**, 603 (1992).

16. J. W. Pattiasina, H. J. Heeres, F. van Bolhuis, A. Meetsma, J. H. Teuben, and A. L. Spek, *Organometallics* **6**, 1004 (1987).

41. SEVEN-COORDINATE [MI₂(CO)₃(NCMe)₂], [MI₂(CO)₃ (NCMe)(PPh₃)], AND ALKYNE [MI₂(CO)(NCMe)(η^2-RC₂R)₂] COMPLEXES OF MOLYBDENUM(II) AND TUNGSTEN(II)

Submitted by PAUL K. BAKER* and MARGARET M. MEEHAN
Checked by HAIDOO KWEN, AMY ABBOTT, and ERIC A. MAATTA†

The importance of seven-coordinate halocarbonyl complexes has been highlighted by their applications as homogeneous catalysts. For example, Bencze and Kraut-Vass[1] have shown that the complexes [MX₂(CO)₃L₂] (M = Mo,W; X = Cl,Br; L = PPh₃,AsPh₃) are active single-component catalysts for the ring-opening polymerization of norbornene and norbornadiene. The polymerization mechanism involves initial displacement of L by the alkene, followed by a 1,2-hydrogen shift to give a carbene intermediate.[2] In 1966, Colton and Tomkins[3] described the synthesis of the highly reactive and relatively unstable, chloro-bridged dimer {[Mo(μ-Cl)Cl(CO)₄]₂}. They followed this up by investigating the reactions of [M(CO)₆] (M = Mo,W) with X₂ (X = Br,I) to give the dimeric complexes {[M(μ-X)X(CO)₄]₂}.[4–6] The diiodo complexes {[M(μ-I)I (CO)₄]₂} were difficult to isolate, and only relatively low yields of the complexes were obtained from the photochemical reaction of [M(CO)₆] with I₂ in Et₂O at room temperature.[6] However, Calderazzo et al.[7] have described an improved synthesis of {[Mo(μ-I)I(CO)₄]₂} by reaction of [MoI(CO)₃(η^6-arene)] [Mo₂I₅ (CO)₆] (arene = C₆H₅Me, 1,4-C₆H₄Me₂, 1,3,5-C₆H₃Me₃) with CO at atmospheric pressure and room temperature to afford {[Mo(μ-I)I(CO)₄]₂} in 88% yield.

In 1986,[8] we reported the highly versatile seven-coordinate dihalocarbonyl complexes [MX₂(CO)₃(NCMe)₂] (M = Mo,W; X = Br,I). These important starting materials and related derivatives have been shown to react with an extremely

* Department of Chemistry, University of Wales, Bangor, Gwynedd LL57 2UW, Wales, UK.
† Department of Chemistry, Kansas State University, Manhattan, KS 66506.

wide range of reagents, and we have prepared and characterised over 2000 new organomolybdenum and tungsten complexes derived from the complexes $[MXY(CO)_3(NCMe)_2]$ ($M = Mo,W$; $X = Y = Br,I$;[8,9] $M = Mo$, $X = I$, $Y = Cl$,[10] Br;[11] $M = Mo$, $X = Cl$, $Y = GeCl_3$,[12] $SnCl_3$[13]). Two review articles have been published describing this and related work.[14,15] Other research groups[16,17] have used the complexes $[MI_2(CO)_3(NCMe)_2]$ as starting materials in their research work. Hence, it is important to describe a detailed synthesis of $[MX_2(CO)_3(NCMe)_2]$ and related complexes. A BBC Open University video[18] for the Open University 3rd-level CHEM777, S343 Inorganic Chemistry course also shows in detail how the seven-coordinate complexes $[MI_2(CO)_3(NCMe)_2]$ and several complexes derived therefrom are made, including the complexes $[WI_2(CO)(NCMe)(\eta^2\text{-}PhC_2Ph)_2]$[19] and $[WI_2(CO)_3(NCMe)(PPh_3)]$,[20] which are described as part of this series of detailed syntheses.

The synthesis of $fac\text{-}[M(CO)_3(NCMe)_3]$ ($M = Mo,W$), first described by Tate et al. in 1962,[21] has been modified by us. It is important to note that in our preparation of $fac\text{-}[M(CO)_3(NCMe)_3]$, it is not essential that all the $[M(CO)_6]$ be converted to $fac\text{-}[M(CO)_3(NCMe)_3]$, since the zero-valence bis (acetonitrile) complexes $cis\text{-}[M(CO)_4(NCMe)_2]$ will also be oxidised by X_2 to yield $[MI_2(CO)_3(NCMe)_2]$ with elimination of CO rather than NCMe. Analytically pure, quantitative yields of $[MI_2(CO)_3(NCMe)_2]$ can be obtained as described in the detailed procedure presented below.

A. DIIODO(TRICARBONYL)BIS(ACETONITRILE) MOLYBDENUM(II)/TUNGSTEN(II)[8]

$$M(CO)_6 + CH_3CN + I_2 \rightarrow [MI_2(CO)_3(NCMe)_2](M = Mo \text{ or } W)$$

Procedure

■ **Caution.** *$M(CO)_6$ and CO are toxic by inhalation. An efficient fume hood must be used. As acetonitrile is highly toxic, gloves (not latex) must be worn). Iodine is toxic and gloves (not latex) must be used. As dichloromethane is flammable and a suspected carcinogen gloves (not latex) should be worn and a well-ventilated fume hood should be used.*

Amounts of $Mo(CO)_6$ (5 g, 0.02 mol) and 200 mL of acetonitrile are placed into a 250-mL round-bottomed Schlenk flask containing a magnetic stirring bar. A silicone oil bath is put in place, and the flask is attached to the nitrogen line. A reflux condenser is put in place and also connected to the nitrogen line via a Teflon tap, making sure that both outlets to the nitrogen line are closed. The suspension is stirred during the degassing of the system by evacuating and refilling with nitrogen. This procedure is repeated 6 times. The silicone oil bath mantle is set at 50°C for 3 h, then at 60°C 2 h. Mild reflux allows the hexacarbonyl to dissolve and prevents it from subliming in the condenser. Upon reaction a pale yellow solution, forms and heating is continued at 85°C for 24 h. After 24 h, the reaction of $Mo(CO)_6$ to *fac*-[Mo(CO)₃(NCMe)₃] in acetonitrile is complete. Then the nitrogen supply to the condenser is increased, and heating is discontinued. When the deep yellow solution has cooled down, the nitrogen supply at the flask is opened and the condenser removed and the flask stoppered. The temperature of the solution is lowered to 0°C in an ice bath and 1 equiv of iodine [4.8 g (0.02 mol)] is added (no effervescence should be observed) and the deep red reaction mixture stirred for 15 min, brought to room temperature, and stirred for a further 30 min. The solvent is removed in vacuo, and after drying the product overnight in vacuo, yields an analytically pure red powder in 92% yield of [MoI₂(CO)₃(NCMe)₂].

Using 5.0 g (0.014 mol) of $W(CO)_6$ in an analogous preparation that was maintained at 85°C for 72 h, the checkers obtained 8.2 g (0.0136 mol, 97%) of [WI₂(CO)₃(NCMe)₂].

Anal. Calcd. for $C_7H_6I_2N_2O_3Mo$: C, 16.3; H, 1.2; N, 5.4. Found: C, 16.3; H, 1.2; N, 5.2.

Properties

[MoI₂(CO)₃(NCMe)₂] is a red-brown powder stable in air for short periods (several minutes) but should be stored under nitrogen. The IR shows carbonyl bands at 2038(s), 1968(s), and 1940(m) cm^{-1} and nitrile bands at 2355(w) and 2316(w) cm^{-1} in chloroform solution.[8] The uses of the complexes [MI₂(CO)₃(NCMe)₂] (M = Mo or W) in syntheses are described in two review articles.[14,15]

B. DIIODO(CARBONYL)(ACETONITRILE) BIS(DIPHENYLACETYLENE)[TUNGSTEN(II)][19]

Procedure

■ **Caution.** *WI₂(CO)₃(NCMe)₂ and CO are toxic by inhalation. An efficient fume hood must be used.*

A 250-mL round-bottomed Schlenk flask containing a magnetic stirring bar is connected to a nitrogen Schlenk line and degassed by pumping in vacuo and refilling with nitrogen. Then 150 mL of dry dichloromethane is degassed in a

separate flask by blowing nitrogen through a coarse sinter (porosity 0), while 4 g
(6.62 mmol) of $WI_2(CO)_3(NCMe)_2$ is placed in the flask. The dichloromethane is
added to the flask and the solution stirred. To the flask is added 2 equiv [2.36 g
(13.24 mmol)] of diphenylacetylene, and the reaction mixture is stirred. After
24 h the resultant yellow-orange solution is filtered through Celite and the solvent volume reduced in vacuo to 40 mL. A layer of dried, degassed diethyl ether
(15 mL) is added gently down the side of the flask via syringe and the flask is
placed in the freezer ($-17\,^\circ$C) for 24 h to yield orange crystals of [$WI_2(CO)$
$(NCMe)(\eta^2\text{-}PhC_2Ph)_2$]. Yield $= 4.55$ g, 79%.

The analogous 2-butyne complex [$WI_2(CO)(NCMe)(\eta^2\text{-}MeC_2Me)_2$][19] is prepared and recrystallized in a similar manner. The molybdenum complexes are
also prepared in a similar manner but they are not light-stable, and the flask
must be covered in foil. In addition, the molybdenum bis(diphenylacetylene)
complex [$MoI_2(CO)(NCMe)(\eta^2\text{-}PhC_2Ph)_2$][22] forms after stirring for one hour
and after recrystallization from a concentrated solution of the complex in
CH_2Cl_2, [$MoI_2(CO)(NCMe)(\eta^2\text{-}PhC_2Ph)_2$] was obtained as maroon-colored
crystals in 74% yield. Both the molybdenum bis(alkyne) complexes are recrystallized from concentrated solutions in dichloromethane.

Anal. Calcd. For $C_{31}H_{23}NOI_2W$: C, 43.1; H, 2.7; N, 1.6. Found: C, 42.9; H, 2.7;
N, 1.5

Properties

$WI_2(CO)(NCMe)(\eta^2\text{-}PhC_2Ph)_2$ is an orange crystalline solid stable in air for prolonged periods, but should be stored under nitrogen. The IR shows a carbonyl
band at 2090(s) cm^{-1} and nitrile bands at 2320(w) and 2300(w) cm^{-1} in chloroform solution[19]. The uses of the complexes [$WI_2(CO)(NCMe)(\eta^2\text{-}RC_2R)_2$]
($R = Me, Ph$) in synthesis are described in two review articles.[14,15]

C. DIIODOTRICARBONYL(ACETONITRILE)
TRIPHENYLPHOSPHINE[TUNGSTEN(II)][20]

Procedure

■ **Caution.** *The synthesis described herein should be carried out in a well-
ventilated fume hood because of the toxicity of metal carbonyl complexes and
carbon monoxide. Triphenylphosphine is toxic, and gloves (not latex) should
be worn and use a well-ventilated fume hood.*

A 250-mL round-bottomed Schlenk flask containing a magnetic stirring bar is
connected to a nitrogen Schlenk line and degassed by pumping in vacuo and

refilling with nitrogen. Then 100-mL of dry dichloromethane is degassed by blowing nitrogen through a coarse sinter (porosity 0), while 0.5 g (0.83 mmol) of WI$_2$(CO)$_3$(NCMe)$_2$ is placed in the flask. The dichloromethane is added to the flask and the solution stirred. To the flask is added 1 equiv [0.22 g (0.83 mmol)] of triphenylphosphine, and the reaction mixture is stirred for 30 s. The resulting yellow solution is filtered through Celite and reduced to minimum volume, and a layer of degassed diethyl ether (10 mL) is added down the side of the flask and the solution placed in the freezer for 24 h to give orange crystals of WI$_2$(CO)$_3$(NCMe)(PPh$_3$). Yield: 0.51 g, 75%. (However, this reagent is often used in situ, and its uses are described in two review articles.[14,15])

Anal. Calcd. for C$_{23}$H$_{18}$NO$_3$PI$_2$W: C, 33.5; H, 2.2; N, 1.7. Found: C, 33.7; H, 2.1; N, 1.5.

Properties

The complex WI$_2$(CO)$_3$(NCMe)(PPh$_3$) is an orange powder, that is stable in air for prolonged periods of time. The IR shows carbonyl bands at 2040(s), 1950(s), and 1918(s), cm^{-1} and a nitrile band at 2325(w) cm^{-1} in chloroform solution.[20] WI$_2$(CO)$_3$(NCMe)(PPh$_3$), and the closely related complexes, [MI$_2$(CO)$_3$(NCMe)L] (M = Mo, W; L = PPh$_3$, AsPh$_3$, SbPh$_3$), which are prepared in a similar manner, have been used extensively to prepare a wide range of mixed-ligand complexes MI$_2$(CO)$_3$LL'. For example, reaction of [MI$_2$(CO)$_3$(NCMe)L] (prepared in situ[20]) with L' [L' = SC(NH$_2$)$_2$, SC(NMe$_2$)$_2$, SC(NH$_2$)Me] affords the mixed ligand complexes, [MI$_2$(CO)$_3$LL'].[23]

References

1. L. Bencze and A. Kraut-Vass, *J. Mol. Catal.* **28**, 369 (1985).
2. L. Bencze, A. Kraut-Vass, and L. Prókai, *J. Chem. Soc., Chem. Commun.*, 911 (1985).
3. R. Colton and I. B. Tomkins, *Aust. J. Chem.* **19**, 1143 (1966).
4. R. Colton and I. B. Tomkins, *Aust. J. Chem.* **19**, 1519 (1966).
5. M. W. Anker, R. Colton, and I. B. Tomkins, *Aust. J. Chem.* **20**, 9 (1967).
6. R. Colton and C. J. Rix, *Aust. J. Chem.* **22**, 305 (1969).
7. F. Calderazzo, R. Poli, and P. F. Zanazzi, *Gazz. Chim. Ital.* **118**, 583 (1988).
8. P. K. Baker, S. G. Fraser, and E. M. Keys, *J. Organomet. Chem.* **309**, 319 (1986).
9. P. K. Baker, M. B. Hursthouse, A. I. Karaulov, A. J. Lavery, K. M. A. Malik, D. J. Muldoon, and A. Shawcross, *J. Chem. Soc., Dalton Trans.*, 3493 (1994).
10. P. K. Baker, T. Birkbeck, S. Bräse, A. Bury, and H. M. Naylor, *Trans. Met. Chem.* **17**, 401 (1992).
11. P. K. Baker, K. R. Flower, H. M. Naylor, and K. Voigt, *Polyhedron* **12**, 357 (1993).
12. P. K. Baker and D. Ap. Kendrick, *J. Organoment. Chem.* **466**, 139 (1994).
13. P. K. Baker and A. Bury, *J. Organomet. Chem.* **359**, 189 (1989).
14. P. K. Baker, *Adv. Organomet. Chem.* **40**, 45 (1996), and references cited therein.
15. P. K. Baker, *Chem. Soc. Rev.* **27**, 125 (1998), and references cited therein.
16. M. S. Balakrishna, S. S. Krishnamurthy, and H. Manohar, *Organometallics* **10**, 2522 (1991).

17. M. Cano, J. A. Campo, J. V. Heras, E. Pinilla, and A. Monge, *Polyhedron* **15**, 1705 (1996).
18. *Open University S343 Laboratory Techniques of Inorganic Chemistry*, 1990, Vol. 4 (a BBC video).
19. E. M. Armstrong, P. K. Baker, and M. G. B. Drew, *Organometallics* **7**, 319 (1988).
20. P. K. Baker and S. G. Fraser, *Trans. Met. Chem.* **12**, 560 (1987).
21. D. P. Tate, W. R. Knipple, and J. M. Augl, *Inorg. Chem.* **1**, 433 (1962).
22. N. G. Aimeloglou, P. K. Baker, M. M. Meehan, and M. G. B. Drew, *Polyhedron* **17**, 3455 (1998).
23. P. K. Baker, K. R. Flower, and S. M. L. Thompson, *Trans. Met. Chem.* **12**, 349 (1987).

42. CHLOROTHIOCARBONYL-BIS(TRIPHENYLPHOSPHINE) IRIDIUM(I) [IrCl(CS)(PPh₃)₂]

Submitted by ANTHONY F. HILL[*] and JAMES D. E. T. WILTON-ELY
Checked by BRIAN K. BREEDLOVE and CLIFFORD P. KUBIAK[†]

Previous routes to the thiocarbonyl[1] analog of Vaska's compound $[IrCl(CS)(PPh_3)_2]^2$ have involved the use of either organic azides and carbon disulfide under anaerobic conditions[3] or methyltriflate.[4] A newer route to this useful precursor has been developed,[5] which can be performed in air with reagents requiring comparatively minimal health and safety precautions. An essentially single-pot synthesis of the title complex using this method is described here on a 4-g scale. All intermediates have been previously isolated and characterized[5]; however, this is not necessary.

Procedure

■ **Caution.** *Dichloromethane is harmful if inhaled or absorbed through the skin. p-Tolylthionoformate is corrosive, flammable, and an irritant. Concentrated hydrochloric acid and DBU (1,8-diazobicyclo[5.4.0]undec-7-ene) are both corrosive as is sodium borohydride, which also liberates hydrogen gas on contact with water. All the reagents mentioned above should be used in a well-ventilated fume hood; protective gloves and goggles are recommended.*

All manipulations are carried out aerobically using solvents and reagents as obtained from commercial sources (Aldrich). Chlorocarbonylbis(triphenylphosphine)iridium(I) is prepared by the published method.[6] All intermediates are stable in solution for a reasonable period of time (hours) and can be stored indefinitely as solids. Light petroleum ether refers to that of the fraction 40–60°.

[*] Department of Chemistry, Imperial College of Science, Technology and Medicine, London SW7 2AY, UK, email: *a.hill@ic.ac.uk*.
[†] Department of Chemistry, University of California, San Diego, CA 92093-0358.

A suspension of [IrCl(CO)(PPh$_3$)$_2$][6] (4.00 g, 5.13 mmol) in tetrahydrofuran (100 mL) is treated with ClC(=S)OC$_6$H$_4$Me-4[*] (1.20 mL, 1.43 g, 7.67 mmol) and the mixture stirred for 1 h. Ethanol (100 mL) is then added, followed by a filtered solution of sodium borohydride (0.50 g, 13.0 mmol, excess) in ethanol (50 mL), and the mixture stirred for 1 h. The solvent is then removed using a rotary evaporator. The residue is extracted with dichloromethane, and the combined extracts are filtered through diatomaceous earth into a 250-mL round-bottomed flask. To this is added concentrated hydrochloric acid (0.5 mL) and the solution stirred for 20 min. or until the solution is colorless. Ethanol (100 mL) is then added, and the total solvent volume reduced to ~80 mL (by rotary evaporator) and the supernatant is decanted from the resultant colorless precipitate. The precipitate is washed by decantation with ethanol (3×20 mL) and then suspended in dichloromethane (100 mL) and treated with DBU (4.00 mL, 3.93 g, 25.8 mmol) and stirred until all the white solid has dissolved to give a bright orange solution (~1 h). This solution is then diluted with ethanol (100 mL) and concentrated to ~60 mL. The bright orange crystals of the title complex are isolated by filtration and washed with ethanol (4×20 mL) and light petroleum ether (20 mL) and dried in vacuo. Yield: 3.00 g (75% based on [IrCl(-CO)(PPh$_3$)$_2$]). A further small crop of product can be isolated from the filtrate after being cooled overnight at −20°C. The product can be recrystallized aerobically from dichloromethane and ethanol to give bright orange crystals.

[*]*p*-tolylthionoformate (Aldrich, 42,061).

Properties

The complex [IrCl(CS)(PPh$_3$)$_2$] is κκ indefinitely stable under air as a solid and shows little sign of decomposition in solution even over a period of days. The complex shows a characteristically intense thiocarbonyl absorption in the infrared spectrum (Nujol) at 1328 cm^{-1}. The ^{31}P-{^{1}H} NMR spectrum (CDCl$_3$, 25°C) is a singlet at 8.0 ppm. A molecular ion is observed in the FAB mass spectrum (NBA matrix) at *m/z* 796 with the complex showing little assignable fragmentation.

References

1. For a review of the chemistry of the title complex and thiocarbonyl complexes in general, see P. V. Broadhurst, *Polyhedron* **4**, 1801 (1985).
2. M. P. Yagupsky and G. Wilkinson, *J. Chem. Soc. A*, 2813 (1985).
3. M. Kubota, *Inorg. Synth.* **19**, 206 (1979).
4. T. J. Collins, W. R. Roper, and K. G. Town, *J. Organomet. Chem.* **121**, C41 (1976).
5. A. F. Hill and J. D. E. T. Wilton-Ely, *Organometallics* **15**, 3791 (1996).
6. K. Vrieze, J. P. Collman, C. T. Sears, Jr., and M. Kubota, *Inorg. Synth.* **11**, 101 (1968).

CONTRIBUTOR INDEX
Volume 33

SUBJECT INDEX

Names used in this Subject Index are based upon IUPAC *Nomenclature of Inorganic Chemistry,* Second Edition (1970), Butterworths, London.

Inverted forms of the chemical names (parent index headings) are used for most entries in the alphabetically ordered index.

FORMULA INDEX
Volume 33

The formulas for compounds described in volume 33 are entered in alphabetical order. They represent the total composition of the compounds, e.g., $BF_{24}KC_{38}H_{21}$ for potassium tetra-3,5-bis(trifluoromethyl)phenylborate. The elements in the formulas are arranged in alphabetical order, with carbon and hydrogen listed last. All formulas are permuted on the symbols other than carbon and hydrogen representing organic groups in coordination compounds. Thus potassium tetra-3,5-bis(trifluoromethyl)phenylborate can be found under B, F, and K in this index.

Water of hydration and other solvents found in crystal lattice are not added into formulas of the compounds listed, e.g., $C_{36}H_{46}N_2O_4 \cdot (C_2H_5)_2O$.

255